From Cave Man to Cave Martian

Living in Caves on the Earth, Moon and Mars

The *Springer-Praxis Space Exploration* program covers all aspects of human and robotic exploration, in earth orbit and on the Moon and planets. Books tell behind the scenes stories of the early missions, both manned and unmanned, covering the human and engineering aspects of the space programs of all the leading spacefaring nations. Accounts of planetary exploration encompass the very early missions through to the very latest results received from space probes to the planets and their moons.

The books are well illustrated with figures and photographs, with targeted use of color throughout. They feature recommended further reading and glossaries and appendices where appropriate.

The books are written in a style that space enthusiasts and historians, readers of popular magazines such as *Spaceflight* and readers of *Popular Mechanics* and *New Scientist* will find accessible.

More information about this series at http://www.springer.com/series/4138

Other Springer-Praxis books by Manfred "Dutch" von Ehrenfried

Stratonauts: Pioneers Venturing into the Stratosphere, 2014
ISBN: 978-3-319-02900-9

The Birth of NASA: The Work of the Space Task Group,
America's First True Space Pioneers, 2016
ISBN: 978-3-319-28426-2

Exploring the Martian Moons: A Human Mission to Deimos and
Phobos, 2017
ISBN: 978-3-319-52699-7

Apollo Mission Control: The Making of a National Historic Landmark
ISBN: 978-3-319-76683-6

Manfred "Dutch" von Ehrenfried

From Cave Man to Cave Martian

Living in Caves on the Earth, Moon and Mars

 Springer

Published in association with
Praxis Publishing
Chichester, UK

Manfred "Dutch" von Ehrenfried
Leander, TX, USA

SPRINGER-PRAXIS BOOKS IN SPACE EXPLORATION

Springer Praxis Books
ISBN 978-3-030-05407-6   ISBN 978-3-030-05408-3   (eBook)
https://doi.org/10.1007/978-3-030-05408-3

Library of Congress Control Number: 2019930634

Cover design: Jim Wilkie. Cover Images used under license from Shutterstock.com.
Project Editor: David M. Harland

This Springer imprint is published by the registered company Springer Nature Switzerland AG
The registered company address is: Gewerbestrasse 11, 6330 Cham, Switzerland

Hole in Mars. Art by Ron Miller, 2014.

Dedication

This book is dedicated to all those scientists and students who have gotten "down and dirty" crawling around caves and lava tubes seeking life and understanding of the subterranean world. Call them cavers, spelunkers, geologists, biologist, or scientists, they are the ones looking for knowledge and understanding. Hopefully, some of that knowledge will help future astronauts crawling around the Moon and Mars. Just as the Apollo astronauts walked around sites ranging from Hawaii to Iceland studying geology half a century ago, future crews will do the same thing one day. In fact, this book describes what some of the ESA astronauts are doing now; getting down and dirty in caves and lava tubes conducting analog studies to benefit future astronauts on the Moon and Mars.

This book is also dedicated to the hundreds of scientists, engineers, and mission planners defining the next steps back to the Moon and the first steps to Mars and its small moonlets. Hopefully, they will conduct the initial studies of "precursor" missions as well as the "grand" missions that seem to drive the imagination. My previous book, *Exploring the Martian Moons: A Human Mission to Deimos and Phobos*, defined a precursor mission to the satellites of Mars. Likewise, this book describes a precursor mission to the Moon and to Mars; but one that utilizes the natural environment for protection rather than solely relying upon extensive and costly "Made on Earth" resources.

This book is also dedicated to those people in a position to guide NASA in its planning; be they members of the National Space Council, legislators, politicians, administrators, or advisory councils. As the details are worked out by the various NASA working groups, there are others that guide the space policy.

If indeed the next humans to go back to the Moon will be approximately in the year 2023 and to Mars in 2033, then those lunar astronauts are currently aged about 30 and the Martian astronauts are about 25; and neither group has been selected yet. If the missions slip, they could be even younger now. They actually could be reading this book in the future. If so, then my thoughts will have taken the ride with them. That would make me very happy, because I knew I would get there one way or another!

Acknowledgements

It takes a lot of input from people all over the world to write a book about a topic that is barely on NASA's radar. In some corners, the mission concept proposed is hardly given much thought, let alone serious consideration. Nevertheless, it will clearly make sense to use caves and lava tubes as protection from the hazards of space radiation, micrometeoroids, thermal extremes and, in the case of Mars, dust storms. This will be particularly attractive for the early missions to the Moon and to Mars. The concept seems to solve a lot of problems, can be implemented years ahead of the current schedule, and certainly saves many billions of dollars in total mission costs.

This book provides some thoughts for consideration when NASA is updating their Design Reference Mission document; that is, to conduct the requisite studies that identify "Short-Stay" precursor missions to the Moon and Mars which utilize caves and lava tubes for the initial protection shelters. I find encouraging support from some in the scientific community to do just that. Because such a mission is many years out, the timing is right to formalize the detailed mission planning and modify the likely flight schedule. Moreover, the economic timing is also good, in that money which has already been spent for the major space exploration elements and systems is directly applicable to a precursor mission. The size of the national debt could also be a driver that curtails NASA's budget and concepts and designs for deep space missions. In fact, the economic and political "stars" might also be aligned in support of this idea.

Firstly, let me acknowledge those that reviewed my proposal to the publisher, Springer-Praxis. They are: Dr. David M. Harland, Glasgow, Scotland; Dr. Pascal Lee, Director, Mars Institute; Dr. J. Judson Wynne, Northern Arizona University; and one other reviewer who remained anonymous. Also, many thanks to Dr. Lee for his Foreword and thoughts about the use of caves in exploring the Moon and Mars.

There were many others who provided me input, or leads to others who are, or where active in cave and lava tube research. To my surprise, there are hundreds of people in this field all across the world. Not all are thinking about the Moon and Mars but many are engaged in space analogs, not only here in the U.S.A. but also in Europe. Many more people are working in the field of inflatables and habitats for in-space and surface applications without realizing that these structures might also be perfect for subsurface use. The same is true for the hundreds of people all over the country who are working to develop space related technologies, many of them state-of-the-art, some of which hold out the prospect of making space travel safer and possibly cheaper.

I would also like to acknowledge all those that have contributed to hundreds of research papers related to cave and lava tube research; many of these are listed in the Reference section. While most are focused on geology and biology, they have relevance to the exploration of the Moon and Mars because they investigate space analogs and add elements to the knowledge base for future expeditions.

List of Contributors
(In alphabetical order by group)

NASA/JPL
Dr. Dean Eppler, Johnson Space Center, Houston, Texas
Marc A. Gibson, Glenn Research Center, Cleveland, Ohio
Dr. Robert L. Howard, Johnson Space Center
Dr. Diana Northrup, Jet Propulsion Laboratory, Pasadena, California
Dr. Aaron Parness, Jet Propulsion Laboratory
Dr. Noah E. Petro, Goddard Space Flight Center, Greenbelt, Maryland
Larry Toups, Johnson Space Center
Rob Wyman, Langley Research Center, Langley, Virginia

USGS
Dr. Glen E. Cushing, Astrogeology Science Center, Flagstaff, Arizona
Dr. Laslo Kestay, Astrogeology Science Center, Flagstaff, Arizona

Institutes
Sandra Cherry, Lunar and Planetary Institute, Houston, Texas
Dr. Sarah A. Fagents, Hawaii Institute of Geophysics and Planetology, Honolulu, Hawaii
Dr. Pascal Lee, Mars Institute/SETI Institute, Moffett Field, California
Dr. Peter J. Mouginis-Mark, Hawaii Institute of Geophysics and Planetology, University of Hawaii, Manoa
Dr. George Veni, National Cave and Karst Research Institute, Carlsbad, New Mexico

Universities
Dr. Philip Christensen, Arizona State University, Tempe, Arizona
Dr. Saugata Datta, Kansas State University, Manhattan, Kansas
Dr. Nels Forsman and Dr. Jaakko Putkonen, University of North Dakota, Grand Forks, North Dakota
Dr. Yongli Gao, University of Texas, San Antonio, Texas
Dr. Pablo de León, University of North Dakota, Grand Forks, North Dakota
Dr. Evelynn J. Mitchell, St. Mary's University, San Antonio, Texas
Dr. Thomas Turner and Dr. Diana Northup, University of New Mexico, Albuquerque, New Mexico
Dr. William "Red" Whittaker, Carnegie Mellon University, Pittsburgh, Pennsylvania

Dr. David A. Williams, Arizona State University, Tempe, Arizona

Dr. J. Judson Wynne, Northern Arizona University, Flagstaff, Arizona

Corporations

Bret G. Drake and Jeff Hanley, Aerospace Corporation, Houston, Texas

Mike Dunn, 4th Planet Logistics, Illinois

Ron Miller, South Boston, Virginia

Pat Rawlings, Visioneering, LLC, Kimberly, Texas

William Studebaker, Richard Lightbound, Erin Sheperd and Frank Tobe, ROBO Global, Dallas, Texas

Dr. R. Roy Whitney, BNNT, LLC, Newport News, Virginia

Keith Splawn, ILC Dover, Houston, Texas

Foreign

Monica Fischer, Austrian Space Forum, Innsbruck, Austria

Bernhard Kaliauer Design Studio, Linz, Austria

Many other scientists are mentioned in the Reference section, along with their reports. In addition to contributions from many individuals, I acknowledge the help of Wikipedia and Google, which allowed me to fill in the pieces of the puzzle on just about any subject. Their inputs are woven into many sections. I also thank NASA and ESA for their websites and helpful inputs.

Finally, many thanks go to the people who helped me in turning my ideas into this book, particularly Maury Solomon and Hannah Kaufman of Springer in New York, Clive Horwood of Praxis in Chichester, England, and cover designer Jim Wilkie in Guildford, England. A special thanks to David M. Harland in Glasgow, Scotland, who edited this, my fifth Springer-Praxis book. After over five years of communications solely by email, I hope finally to meet him in Glasgow in 2019.

Foreword

We were at Lofthellir in Iceland. The entrance to the lava tube was a circular hole in the ground, approximately 20 meters across. It was in the middle of a young lava flow, with volcanoes in the distance but none immediately nearby. Using a ladder, we climbed down into the pit to its sandy floor. Below the mounds of windblown sand were large blocks of rock which had fallen when the roof of the lava tube collapsed and created the "skylight" that now served as its entrance.

Beneath the skylight, the lava tube extended in two directions. The darkness beyond the twilight zone in either direction looked ominous. As we made our way into the cave in its upstream direction, the passage quickly narrowed from 20 meters to about 1 meter. This lava tube was no subway tunnel. Instead, over its accessible length of 200 meters it would prove to be a 3D maze of crawl spaces, narrows, attics, cellars, ledges, ramps, and cavernous chambers.

But the challenge of squeezing through was amply rewarded. Soon after entering the darkness, the beams of our helmet lamps caught intensely bright reflections. We had encountered ice. *Massive Ice!* Not just veneers on the walls, and icicles dangling from the ceiling, or ice stalagmites rising from the floor, but entire underground mini-glaciers, with accumulation, ablation and melt zones, lateral moraines, and terminal aprons. This was the stuff of Jules Verne.

It is difficult to describe the awe, thrill, and wonder of caving, particularly inside lava tubes. But lava tubes filled with massive ice add an entirely new dimension to the whole experience. More importantly, had we been on the Moon or Mars, such ice would have been a holy grail. Finding massive ice in a cave could potentially mean having readily accessible water for hydration, fuel production (by breaking down H_2O into hydrogen and oxygen), cleaning, diluting, irrigation, heat exchange, and more. In the case of Mars, finding ice might also mean the possibility of finding extant *alien* Life.

We know for certain today, thanks to remote-sensing imaging from orbital spacecraft, that there are caves on the Moon, and also on Mars. Many if not most of these caves are lava tubes formed in volcanic lava fields or in impact-melt lava sheets.

Earlier this year, after examining hundreds of images of the Moon's polar regions, both north and south, taken by NASA's Lunar Reconnaissance Orbiter (LRO), I reported finding candidate skylights and associated lava tubes in the impact-melt deposits within Philolaus crater, a 70-km-wide impact structure located just 500 km from the north pole of the Moon. If confirmed, these features would be the highest latitude caves known on the Moon. They would be at such a high latitude that the Sun's grazing rays would never enter the caves and warm up the rocks on their floor. Instead, the caves would remain in perpetual, complete darkness, the underground equivalent of the permanently shadowed regions at the actual lunar poles. The Philolaus caves would be so cold that, if water were available, it could be cold-trapped as ice in these caves and remain stable for eons.

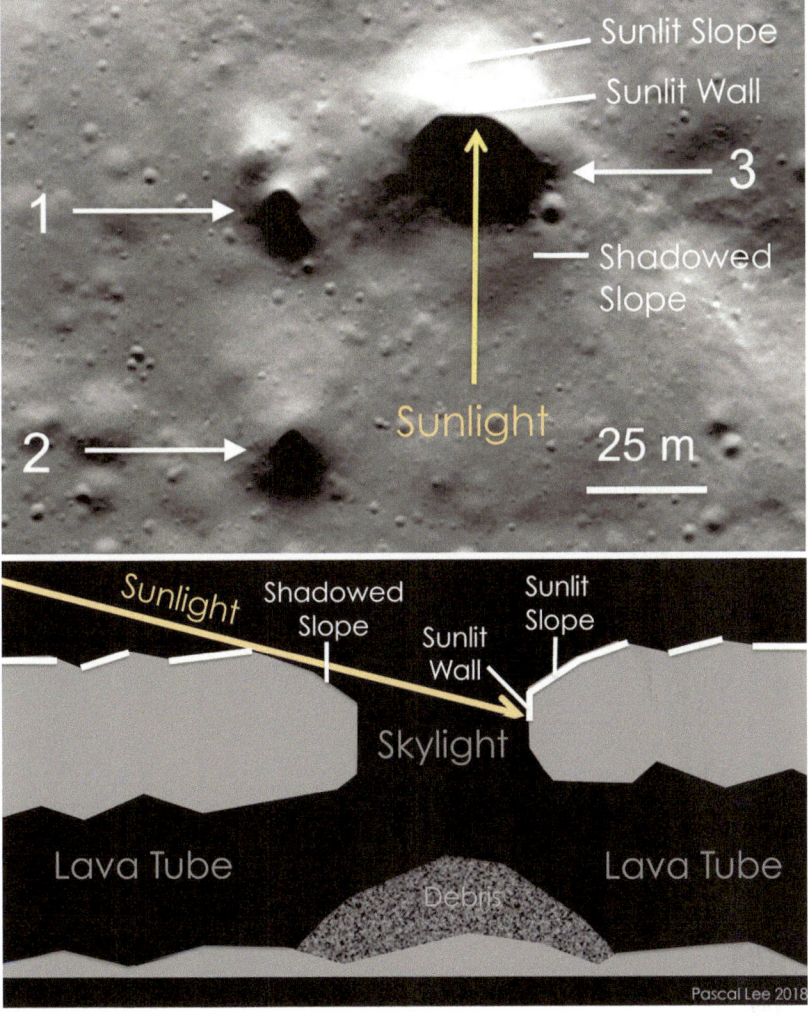

Possible skylights into lava tubes on the floor of the crater Philolaus, located near the north pole of the Moon. Image courtesy of Pascal Lee.

But caves, including lava tubes, even with ice, are not all good news. Many are prone to collapse, possess terribly jumbled floors, abrasive ceilings and walls, and downright awkward geometries. While caves on the Moon or Mars are often touted as obvious natural shelters, as they would protect humans and their assets against ionizing radiation, wide diurnal temperature swings, micrometeorite bombardment, and even rocket exhaust sandblasting, not all are ready for us to just move into. In fact, based on our terrestrial experience (but with the caveat that caves on the Moon and Mars might be different from those on Earth) most caves might not be occupation friendly. Finding the right caves on the Moon and Mars could take some time and a good measure of resources. And we must recognize there is no guarantee of finding a suitable location. Of course, this is not meant to discourage the search, rather it is to instill a dose of realism in our expectations and thus our planning.

In this exceptional book, the first one dedicated to the topic of exploring and settling caves on other worlds, "Dutch" von Ehrenfried has a plan, and takes us on an exciting journey through space and time to show us how it could unfold. It is a well-researched account of how caves provided natural shelters to early humans before they developed abodes of their own, and proposes that caves on the Moon and Mars be considered as analogous first dwellings for future space explorers. How caves will be scouted out, then explored and prepared in advance of human occupation is examined in unprecedented detail. Throughout the book, Dutch's brilliant mind shines.

What I find compelling in Dutch's ideas, insights, and proposals are the pragmatic thinking that he puts behind them and the depth of his firsthand real-life experience with space exploration. Dutch was directly involved in Mercury, Gemini, Apollo, Skylab, and the International Space Station programs, including being an Apollo Pressure Suit Test Subject and Mission Control flight controller. He was there when it happened. He helped make it happen. He knows how to make it happen again! Dutch's book is a must-read for all forward thinkers, as it offers truly new perspectives, prospects, and priorities for our human future in space. Amazingly, while it is well known that history repeats itself, Dutch shows us that prehistory might too.

Pascal Lee
Director, Mars Institute
Planetary Scientist, SETI Institute
NASA Ames Research Center, Moffett
Field, CA, USA
December 2018

Preface

Isn't it ironic that in planning to leave Earth and explore our solar system we find ourselves reconsidering the practicality (if not the necessity) of returning to living in caves. While most people cannot comprehend living in caves, currently over 30 million people do so all across the world! Indeed, some caves have been occupied for generations.

We regard living in a beautiful home as the highest form of human habitation evolution. Some people have large and beautiful houses equipped with all of the latest amenities, sometimes including several robots such as a smart refrigerator, smart washer/dryer, and a smart air conditioner to relieve them of undue work or stress. We might also have a beautifully landscaped yard, a pool, and clubhouse for our personal use. We have arrived! We don't need these things; we just desire them. Most people do not have this luxury. Some people just want to have a roof over their heads, even if it is a cave.

But no so fast! Now we want to go to the Moon and Mars, and maybe beyond. What do we need now? Well, there are a few little problems such as there being no air, no water, little or no pressure, large temperature extremes, a lot of deadly radiation, and a flux of micrometeoroids. On Mars, there are blowing winds and dust storms that last for very long times. Forget the amenities; we need survival gear! But we shall still require the robots and all of the advanced technology. We are going to need all the help we can get if we are to live on these hostile worlds; even for short periods of time.

Prehistoric man didn't have the luxury of the very best of caves to shelter his entire family from the elements and predators, they needed to make do with what they could find.

Spacefaring humans will find that in order to provide the absolute basic life-support elements for extended operations, even the simplest of shelters will be complex and expensive. Although we don't think we will need protection from predators, we will need protection from the potentially deadly environment. And, in reverse, any indigenous life on the planets we settle will certainly require to be protected from us.

It is evident that habitats of some form or another will be needed on the Moon and Mars. Efforts are now well underway in the USA, Europe, and Asia to define them. Unlike Earth, astronauts on the surfaces of the Moon or Mars will not be protected from space radiation by global magnetic fields. While astronauts will necessarily work on the surface, they will be obliged to retreat into a protected environment for some activities, particularly to eat and sleep. The choices are to make use of the natural geology to find shelter, to manufacture and send habitats from Earth, or to build them in-situ once there.

Actually, working underground may provide us the best chance of finding life, because the radiation on the surface may well have either killed it off or driven it deep into the soil eons ago. Similarly, water ice is more likely to be found below the surface. Ice that is accessible to a habitat is key to obtaining drinking water, as well as fuel for vehicles. Having a readily available water source means less mass needs to be transported from Earth, and so be less costly. Exploration of caves on Earth has found water and life in abundance and in very strange forms, as well as yielding insight to geological structures not damaged by the surface environment. Why not first go to where you can get shelter and find water ice and perhaps even indigenous life; underground in caves or lava tubes, not on the surface! Ideally, NASA will take note, and consider using caves in mission planning for the return to the Moon and the early trips to Mars.

Finding a suitable underground place for habitation will require a considerable effort. The low gravities of the Moon and Mars may enable some of the lava tubes and caves to be huge. On the Moon, they could be measured in kilometers (miles) and on Mars up to 250 m (820 ft); as large as two World Cup football fields. One report estimates lunar caves could be large enough to hold a city! We do not need that much, though. For those early missions, we would want the "optimum" cave. The cave research effort will go on for a very long time before actual exploration gets underway on the Moon and Mars. Satellite imagery has already shown many potential caves on both bodies, as well as elsewhere. We see collapsed sections of lava tubes that open up "skylights" into black pits. When the illumination is right, we can see details of the interior. The satellites which discovered these structures were not designed to seek caves. There are proposals for satellites that will carry special radar instruments just for that purpose.

And there are efforts underway to develop specific and unique technologies to work in these caves. Earthly analogs are already being explored to determine how caves can be formed from volcanic lava tubes or by water dissolving the geology. These can easily be seen in Hawaii, Iceland, the Canary Islands, Australia, Spain, Germany, Sicily, and the Galápagos Islands. Of course, there are caves elsewhere on Earth; probably thousands of them. As it turns out, some of these caves can be very hostile and cannot be entered without wearing protective gear, and some can be downright deadly and have proven to be so.

The optimum caves on the Moon and Mars must be deep enough to provide a level of radiation protection that is equivalent to that here on Earth. There is also extensive research into developing radiation shielding materials that can be used in transit, as well as on and in the Moon and Mars. When the crew enters a cave, how can they assure themselves that it is safe from collapse and be safely cleared of rubble? There are other concerns. Special tools and equipment may be needed. In most concepts, a suitable cave will be outfitted with portable inflatable habitats that are pressurized with airlocks and then interconnected for various functions such as sleep quarters, personal hygiene, research, galleries, and food production. Over time, this type of habitat system can be expanded by later missions. There is even one study group that is looking at special coatings that can be applied to the cave in order to seal and pressurize it.

There are many lessons being learned from current cave exploration that may be useful on the Moon and Mars. Terrestrial cave research, called "speleology," has been underway all across the world for decades. It is primarily for studies of biology and geology, but there is an emphasis on advance technology to enable the research. NASA, the Mars Institute, the SETI Institute, the Planetary Society, the US Geological Survey, ESA, Japanese institutions, and many universities and organizations are actively involved. Examples of their research projects are also included in this book. An extensive Reference section is provided which also includes links and videos. Not only have new life-forms been discovered but new methods for exploration. Although most of these people are scientists, some are recreational cavers. There is even an International Speleological Society and an International Union of Speleology. Their members include explorers as well as entire groups of scientists that are exploring caves with the latest technologies at their command. It is most likely that the first crews to return to the Moon and go on to Mars will include a geologist/biologist who has spent considerable time in caves as part of their education and training.

This book starts with a short history on the use of caves by humans. It spans a great period of time. We have been using caves much longer than was originally believed; perhaps 300,000 years. It will also briefly describe the types of human beings that anthropologist have found in caves. It will describe various types of caves and in order to give the reader a sense of scale, it will highlight those which set the records for being the deepest, the longest, and the largest. Other caves of interest are also mentioned. Even caves occupied by present-day "Troglodytes" are included.

This book will describe current research into caves, pits, tunnels, lava tubes, and skylights, and the associated technologies that pertain to potential lunar and Mars exploration and habitation. The work of noted scientists and technologists will be described with emphasis on extraterrestrial applications. This continuing work is more extensive than one would think, and is directly applicable to longer-term habitation and exploration of the Moon and Mars. While directly related to the search for life, emphasis is also given to the operational aspects of working and living in caves and lava tubes on the Moon and Mars. Just how will the first crew find the right cave and prepare it for habitation? What are the dangers for EVA astronauts poking around holes, seeking an entrance to a potential place in which to work and live? What kind of robots will they need to explore unknown areas? What kind of tools will they need? How will they get inside safely? Once inside, how are they to clear away rubble and prepare the surface for equipment and supplies? How are they to prepare a volume for a pressurized environment? Will they use an inflatable structure fitted with an airlock? Will they be able to get support from Mission Control? These and other questions will require to be answered before we can realistically plan to go there.

So let us look for shelter in the caves and lava tubes of the Moon and Mars, and then propose a precursor mission that will make use of them!

Manfred "Dutch" von Ehrenfried
Leander, TX, USA
Nearing the 50th Anniversary of the first lunar explorers!

Contents

1

Introduction

Isn't it ironic that as mankind settles upon new worlds, he may once again seek to live in caves? Our ancient ancestors lived in caves for tens of thousands of years prior to creating something more habitable and comfortable. Hopefully, mankind will not have to spend that long in caves on the Moon or Mars before he achieves more comfortable surroundings on a permanent basis. We modern Homo sapiens, a term coined by Carl Linnaeus (1707–1758), have certainly evolved technology-wise since that term was introduced. Looking at the youth of the world today in comparison to early man it is evident that we have evolved physically, but are we smarter? Did you know that modern humans have a 3 pound brain about 1300 cc in size, but that Neanderthals had a larger brain of about 1500 cc? Was this just because they needed to see and smell better in order to hunt and survive? Or was their brain adapted to a harsh environment and life style, while ours is just more evolved and efficient in our modern environment? It's interesting to note that our imagined impressions of advanced aliens have huge heads, presumably because they have huge brains, but this might not be the case at all for the next species of humans.

To put this subject into perspective, I found it fascinating to learn how early humans lived and where. This led to an understanding of the size and scope of caves and how they were used for shelters, burial grounds, and even rituals. Did you know that millions (yes, millions) of people still live in caves? Chapter 2 is a review of how humans have lived in caves, and why it will be feasible to occupy them on the Moon and Mars in the future.

I propose that the first person born on Mars be called homospatium-viatorem (space travelling man). Perhaps I ought to leave that to the future anthropologist to define. I wonder how far into the future that will be? For now, let's just look at what mankind will be up against when occupying the Moon and Mars in the near future, as opposed to hundreds, if not thousands of years from now, as some like to dream.

To date, humans have spent only 75 hours at a time on the surface of another celestial body! That is no longer than you might have spent at the beach during a long weekend. Only a few astronauts have spent more than a year orbiting Earth on a single flight. That puts our space travelling experience into perspective, as great an achievement as it is. But in our mind's eye, we can see not only space colonies on Mars but terraforming that entire

© Springer Nature Switzerland AG 2019
M. von Ehrenfried, *From Cave Man to Cave Martian*,
Springer Praxis Books, https://doi.org/10.1007/978-3-030-05408-3_1

planet. That's called imagination! Although there is some truth to the expression, "What man can perceive, man can achieve," this does not say anything about how long it will take or what it will costs in resources and lives to achieve the dream. Mission planners have shown some imagination, but they are limited by the state of the art in key technologies and the physical laws which constrain them. Flight operations people will have to limit their scope to achieving the assigned mission at hand, and getting the crew home safely. This book discusses the evolving technologies which will eventually get us to the state of being able to achieve the first missions and return their crews home safely.

This book will review the ongoing research, programs, projects, engineering, technology, and mission planning in relation to the first "baby steps" required to allow humans to spend extended periods of time on the Moon and Mars. Perhaps they will live in caves once more and for a very long time. Let's just focus on this century, for it will be many years until we go back to the Moon and decades until we place the first human foot prints on Mars or even one of its moonlets, Deimos and Phobos.

The amount of research related to the initial attempts to live on the Moon and Mars is fairly extensive and addresses many areas, disciplines, and technologies. It has been going on for decades; a couple of generations. I have seen references to exploring lava tubes on the Moon and Mars that date back to the 1950's. It is currently being carried out by geologists, astrobiologists, engineers, technologists, roboticists, mission planners and even astronauts. Rest assured that, just as there are thousands of caves on Earth, there are plenty on the Moon and Mars available as potentially acceptable sites for a safe habitat. It will take a lot of searching by better satellites and rovers than we currently have, before we will be able to select a suitable landing site with a high probability that it will contain suitable caves or lava tubes. It was only in the last decade that scientists became certain that there are caves on the Moon and Mars. Furthermore, they now suspect that conditions in caves are conducive for the accumulation of water ice, and there is the enticing prospect that where there is water there could be indigenous life.

Although humans and robots have been to the Moon, only robots have been to Mars. We have not really explored the Moon, however; nor have we, as humans, pioneered it. In the 2014 report, "Pioneering Space: NASA's Next Steps on the Path to Mars" NASA announced a new philosophy for humans on Mars named "pioneering." This philosophy advocates building toward a permanent presence of humanity off-Earth. The document distinguished exploration from pioneering, saying, "Explorers go with the intent of returning to tell their story and point the way for future forays. Pioneers go with the intent to establish a permanent presence. Pioneering space requires we progress from Earth-dependent to Earth-independent." Of course, independence will not be feasible for a very long time, probably several generations, so the initial missions will continue to require the support of Mission Control on Earth.

The human spaceflight outlined in this book will take place during that interim period, between dependence and independence. This is the type of mission which will seek shelter, water, and resources in support of a relatively brief "stay" in the range of 30 to 60 days. Doesn't that seem like the type of mission that prehistoric man was on too? The first exploration missions will utilize the technologies that were introduced decades prior to the astronauts starting their journeys. Robotic missions will have carried out the preliminary exploration to identify the site for the first real camp; albeit it a rather primitive temporary encampment more akin to a bivouac than a base. Perhaps, as the book suggests, it will be a suitable cave inside which inflatable habitats can be pressurized, equipped, and expanded over periods of decades. This won't be a mission of the "independence" type but it will establish an interim shelter for early exploration prior to creating a long-term base of operations and exploration.

One group of scientists and engineers are focused on a more immediate use of "in-situ resources" to create the initial habitats for protection against the radiation and environment. In their view, a large number of habitat structure configurations can be developed from a relatively small set of in-situ resource-based construction products such as blocks, raw regolith, and reinforced concrete and glass products. This would need more resources to construct than a shelter that exploits a cave or lava tube. I believe this is too ambitious an approach for those early missions and might not even be required if a suitable cave or lava tube can be converted into a shelter.

Having reviewed what the NASA Centers and contractors are doing, I see that there is a lack of focus on those first missions; instead, the emphasis is placed on the "grand" missions. The current design references envisage "gateways" to the Moon and beyond, to eventually establish a long-term presence on the Moon and Mars. There appears to be little focus on just the next few missions. You can see this with all the studies and research on "farming" the regoliths of the Moon and Mars, and on building habitats and bases. Certainly, the first couple of missions back to the Moon, or the first to Mars, will not rely upon extensive processing of regolith and undertake the construction of habitats and bases. It seems logical to initially use what is there in the form of caves and lava tubes to establish a simple temporary shelter. Bear in mind that these first missions will most likely not be back-to-back as in the case of the Apollo era; many years could elapse between missions, particularly those to Mars. What we learn on those first missions may radically change how we undertake later missions.

Another group is focused on importing materials from Earth, including thin films, liners, and foldable or erectable metal structures. A great deal of thought has been given to this fundamental protection issue. Many of the tradeoffs are likely to be dollar-driven, since the cost to send a pound of material to either the Moon or Mars is huge.

One NASA study classified the class of habitat construction as follows:

Type 1 Pre-Integrated	Completely built and integrated on Earth. Lands on the surface and stays in one piece.
Type 2 Deployable	Built on Earth but may be integrated, assembled, deployed, erected, inflated, moved or reconfigured on the Moon/Mars.
Type 3 In-Situ Resources	May be built on Earth but incorporates in-situ materials on the surface, or the primary structure may employ in-situ construction.

It seems clear that there will be great potential for savings of energy, mass and dollars if mission planners can avoid, as much as possible, Type 1 construction of the habitation elements by locating the shelter inside a lava tube or cave. In turn, a greater fraction of the payload mass which is landed on a planetary surface will be available to life support and science mission support. It also seems clear that total dependence on Type 3 would require more expenditure in resources in order to fly construction equipment and robots to the surface. That appears to imply a Type 2 mission. Alas, the devil is in the details. This book will cover many aspects of all three types of mission and consider combinations of types.

Irrespective of the path chosen, there will be operational considerations. How do the astronauts select the cave site, enter it, and create the conditions for even a short term stay? What technologies will they require to assist them in that effort? How will they go about their tasks, and do they need mission operations support? Given the amount of ground support that astronauts receive when spacewalking from the International Space Station, it is difficult to imagine how astronauts on Mars could safely operate without the assistance of Mission Control, despite the long communications time delays of 4 to 24 minutes, one way. (As I write this, Mars is in opposition and only 35.8 million miles away with a communications delay of about 4 minutes.) Will robots be sufficiently advanced by then to assist the astronauts without help from Earth? How autonomous will they be? Or will there be a human/robot cooperative team? How many different types of robots will be required? Perhaps that is why some people want to restart missions to the Moon to develop the capabilities that will be required for a mission to Mars. This book will describe current robotics research which, although it was designed for servicing satellites, should be able to be applied to surface operations in support of a crew. These robots range from the very small to human-sized, and all of them could assist a crew in exploring caves and lava tubes and setting up an inflatable habitat to enable a Short-Stay mission.

One of my favorite ideas mentioned at the First International Caves Workshop in 2011 was the use of "Moon Bats" and "Mars Bats," fueled robots that fly from the crew's rover into the candidate caves equipped with state-of-the-art sensors to characterize the dimensions of a cave and transmit the data back to the crew/rover for forwarding to Earth for analysis. This concept keeps the astronauts safe until a cave is cleared for them to proceed with further evaluation and exploration. Keep in mind that from a crew safety point of view caves can range from treacherous to benign.

This book will describe many other technologies which will be able to support the first missions, including lasers, inflatables, crew tools, nuclear power systems, and shielding. Most promising are those efforts to create materials that are better at protecting against radiation and micrometeoroids. Others include the design of new and more effective nuclear power systems for human exploration. New kinds of "smart" robot will ease the work load on astronauts while exploring caves and lava tubes and, later, preparing them for habitation. While these technologies are part of NASA's Technology Roadmap, many could be adapted to support a cave-oriented mission. Much of the technology work on board the International Space Station will also be applicable to such missions.

I will describe what the NASA Centers and other government agencies, ESA, and affiliated organizations are doing in related research as well as what various universities are doing. This includes government sponsored analog projects that have been underway for many years. Many universities supporting the unmanned programs at JPL are also working on sensors and sciences related to future Moon and Mars missions. Non-profit organizations such as the Mars Institute, the SETI Institute, the Planetary Society, and 4th Planet Logistics are also covered.

Space travel is a dangerous business. There are hazards everywhere you look, and also where you can't look! This book will cover those in summary and point out how current analog projects face real hazards even though they conduct their activities on Earth, or should I say "in" the Earth. Several projects involve people living and working in caves for long periods to identify related the hazards. Even now, ESA astronauts are actually being trained in caves. Obviously, not all of the hazards foreseen on the Moon or Mars can be addressed in analog situations. For example, radiation, low pressures, dust storms, micrometeoroids, and extremes of temperature cannot be simulated in caves, but these concerns are being addressed in different ways. Caves do provide the astronaut test subjects with stressors such as sensory deprivation, the lack of a day-night cycle, restricted hygiene, and even some behavioral problems. Certainly, in Earthly cave exploration there have been situations of panic and death. Multinational teams have been trained to cope with these issues during simulated space missions that focused on geology and biology activities.

The sensors on existing orbiters are unable to fully answer all of the questions that will face mission planners or scientists. Because they primarily look straight down (nadir) they can see pits and holes but cannot really look for cave entrances in the way that an astronaut or robot would on the surface. Another generation of orbiters are needed with sensors to provide low altitude, oblique views to reveal cave entrances. Also, the shadows that are created lend to measuring techniques of the caves. It is already clear that some of the skylights observed on the Moon and Mars are really giant caverns that may well be too dangerous for humans to enter, and therefore are not suitable for habitation. Some of them could swallow even the largest football stadium.

This book will discuss how future astronauts will seek the first desirable cave and perform the selection process. What will be their criteria, and how will they determine whether a cave meets the criteria? What robotic capabilities will they require, and what other tools will they need in order to obtain access to the cave, verify its safety, and set up an initial habitat? To supplement robotic assistance, there are scientists devising ways to give astronauts enhanced muscular strength with more advanced pressure suits and, perhaps, exoskeletons to enable them to accomplish more work without getting so tired. This might come in handy when exploring potential entrances to caves. The mission planners and scientists who select the first landing site must consider these new and enhanced capabilities in addition to balancing all of the objectives and the many constraints and hazards.

I will discuss those constraints and hazards. They cut across many disciplines, but to give the crew their best chance to achieve the objectives on that first flight the mission design must be driven by crew safety and operational factors. So it is envisaged that the first missions will target landing sites that are known to offer a high probability of finding a satisfactory cave or lava tube. Most likely, they will utilize the first cave that meets their criteria and then enhance its habitability and utility over the course of several missions. We would be fortunate if the first cave selected were able to be used for an extended period with repeated upgrades. The durations between missions could be many years, so the cave and habitat must be able to be left in an inactive state, perhaps being monitored by Mission Control, awaiting the next crew.

One constraint that has been difficult to address is contamination, considered both forward and backward. While we don't want to contaminate Mars with our Earthly microbes, we also don't want to bring back any that may cause us harm. This complicates our desire to determine whether life exists on other planets and, if so, how we can study it? How do we protect ourselves and also assure that we don't populate the planet with our own, potentially harmful bacteria. Perhaps the cave habitat is the best place to contain and confine planetary contamination and analyze the biological interactions in-situ. I will address these issues and provide references to more detailed studies.

It is evident that the first cave must meet the safety considerations in terms of roof stability, radiation shielding, micrometeoroid protection, relative temperature stability, and, in the case of Mars, dust storm protection. Ideally, it would also be located close to, or contain water ice. It must be large enough to accommodate a specially prepared inflatable habitat equipped with airlocks and, in time, several inflatables will be added to serve different purposes. How they get deployed and installed would seem to be the role of special robotic rovers. This book discusses these key technologies. It is very clear that astronauts wearing pressure suits and back packs will not be able to undertake any strenuous work such as the clearing out of caves or anything even close to "spelunking." The Apollo astronauts were exhausted and bruised after hours of walking and working on the Moon. Robots will do the heavy and dangerous work. Astronauts will explore, supervise robots, and carry out science. It is dangerous enough working in caves here on Earth, let alone on the Moon and Mars.

This book makes a strong case that NASA should place greater emphasis on a class of Short-Stay missions that they briefly looked at but never fully embraced in their Design Reference Architecture. This type of mission is highly compatible with the idea of a precursor mission which makes use of caves and lava tubes for the first series of missions to the surface of Mars. It is believed this offers a much safer and far less costly type of mission for mankind's first steps into deep space. Furthermore, such a mission architecture can be accomplished far earlier than the currently proposed "grand" mission.

As NASA Ames astrobiologist, Dr. Penny J. Boston, explains, "The dawn of the age of 'astrospeleology' is upon us. The ramifications of this extend into all of the scientific, engineering, and exploration activities that have been conducted in Earth caves before. Now we contemplate the demanding task of applying what we have learned here on Earth to the extraterrestrial realm."

This book includes many Appendices which describe analog projects, sensors, international and national cave research, and university research. In addition, it has pictures and points to YouTube and other videos and documentaries available on the Internet. The Reference section at the end of the book gives links to more detailed reports and studies.

2

Cave Uses on Earth From Prehistory to the Present

2.1 TIME

First, let's put things into perspective. Prehistoric man is commonly related to the Pleistocene epoch of geological time, which is colloquially referred to as the "Ice Age." Lasting from about 2.6 million to 11,700 years ago, it spanned the world's most recent period of repeated glaciations. Overall, the climate was much colder and drier than it is now. Since most of the water on Earth's surface was ice, there was little precipitation, and rainfall was about half of what it is now. During peak periods, with most of the water frozen, global average temperatures were 5–10°C (9–18°F) below our temperature norms. Winters in the northern hemisphere were probably brutal, and cave shelters would have been essential for survival. So, yes there have been climate changes for millions, if not billions of years, even before modern man.

The end of the Pleistocene corresponds with the end of the last glacial period, and also with the end of the Paleolithic age, the term used in archaeology for the "Old Stone Age." It also marks the end to the woolly mammoth, the sabre tooth tiger, and some other mammals that are now extinct. The Holocene is the current geological epoch. It began after the Pleistocene, and includes the present. So we are dealing with people who lived in those prehistoric times, some of whom were cave dwellers.

Such a cave was found in Jebel Irhoud, an archaeological site 50 km (30 mi) southeast of the city of Safi in Morocco. It is noted for the hominin fossils which have been found there ever since the site's discovery in 1960. Originally thought to have been Neanderthals, the specimens have been reassigned to Homo sapiens (Latin for 'wise man') and dated at 315,000 years old. If correct, this would make them the oldest known fossil remains of an anatomically modern human. The site has since been explored by both French and American scientists.

This discovery and subsequent studies suggests that, rather than arising in East Africa around 200,000 years ago, modern humans may have already been present throughout Africa 100,000 years earlier. According to French paleoanthropologist Jean-Jacques

© Springer Nature Switzerland AG 2019
M. von Ehrenfried, *From Cave Man to Cave Martian*,
Springer Praxis Books, https://doi.org/10.1007/978-3-030-05408-3_2

Hublin, early humans may have comprised a large and interbreeding population whose dispersal across Africa was facilitated by a wetter climate that created a "green Sahara" some 330,000 to 300,000 years ago. The rise of modern humans may therefore have taken place on a continental scale rather than being confined to a particular corner of Africa. Hublin and his team made an attempt to collect a DNA sample from these fossils, but without success. Such a genomic analysis would have provided the necessary evidence to support the conclusion that these fossils are representative of the main lineage that leads up to modern humanity and that Homo sapiens had dispersed and developed all across Africa.

2.2   MANKIND

Now, let's put mankind into perspective. When we think of "Cave Man" we think of the impressions burned into our minds from books (even comics), movies and television, depicting the more archetypal Neanderthal equipped with a giant club for hunting big game. My favorite cave man is Atouk, played by Ringo Starr. In fact there can hardly be a more perfect example. Even the common cave scene is idealized, even though it probably has little semblance to reality. Artist Charles R. Knight famously painted a color mural depicting a Neanderthal family during the Ice Age.

In Knight's depiction the idealized cave not only provides some shelter, it has a porch that provides a magnificent view of a beautiful river valley. It houses an extended family that takes care of grandpa. In reality, cave use is an activity that spans the full range of Homo sapiens, likely extending over half a million years. Fossils and artefacts hint at cave use by other members of the hominid family. It is interesting to note that while cave use is a behavioral trait of many animals, it does not appear to be one that is widely shared by

Fig. 2.1 Le Moustier Neanderthals painted by Charles R. Knight in 1920. It was once part of an exhibit on Early Man at the American Museum of Natural History in New York.

Fig. 2.2 A depiction of people in a prehistoric cave. Photo from the National Art Museum of China, Beijing.

our nearest relatives, such as other higher primates. However, social mammals such as wolves and hyenas use caves regularly, and bears hibernate in caves.

Some prehistoric humans were cave dwellers, but most were not. Despite the name, only a small portion of humanity has ever dwelt in caves. While there are caves all over the world, most in the northern latitudes are dark, cold, and damp. Nevertheless, they did offer shelter from the extreme weather of the Ice Age. In the equatorial regions, caves provided shelter from the Sun and heat.

When we are told to imagine a caveman, often the first thing which comes to mind is a man wearing an animal skin who lives in a cave and carries a club. He may have a fire. This impression is probably the origin of the term "troglodyte" (not to be confused with troglobites, small creatures that are adapted to living in dark caves). The dictionary says a troglodyte is a "primitive, brutish or reclusive person." From the beginning, man sought to take shelter for protection against the inclemency of the weather and against predatory animals. New research reveals that our ancient forefathers were much more developed and capable than we've previously assumed. A reconstructed head by John Gurche may well be a more accurate representation.[1]

There is a general belief that prior to the advent of modern medicine and other modern luxuries, humans rarely lived beyond the 30–40 years of age. This might not be the case. The estimated life expectancy for prehistoric humans of 35 years is misleading. When

---

[1]For more information on Neanderthals and humans go to the Smithsonian Natural History Museum website at:

http://humanorigins.si.edu/evidence/genetics/ancient-dna-and-neanderthals

Fig. 2.3 A Neanderthal reconstruction by John Gurche based on the Shanidar 1 skeleton found in Iraqi Kurdistan. Photo courtesy of the Smithsonian National Museum of Natural History.

factoring in the higher numbers of humans who would have died during childbirth or in early life, it is apparent that this number indicates that for every prehistoric human who died very young there may have been one who achieved the grand old age of 70 years. This is corroborated by the discovery of the remains of numerous elderly prehistoric humans. One example is the remains of a prehistoric human found in France in 1908. Known as "The Old Man of La Chapelle," this skeleton was studied by scientist Pierre Marcellin Boule. In 1911, Boule reconstructed this skeleton with a severely curved spine that was indicative of a stooped, slouching stance with bent knees, forward flexed hips, and the head jutted forward. He argued the low vaulted cranium and the large brow ridge were reminiscent of large apes such as gorillas, and interpreted this to mean that early humans lacked intelligence. This led to the popular stereotype of Neanderthals as dim-witted brutes.

However, it seems that it was Boule's own preconceptions about early humans and his rejection of the hypothesis that Neanderthals were the ancestors of modern humans that led him to reconstruct a stooped, brutish creature, thereby effectively placing Neanderthals on a separate branch of the human evolutionary tree.

2.3   CAVES

Now let's put "caves" into perspective. Various kinds of cavities occur beneath the Earth's surface, both natural and artificial (created by humans). The term is generally applied to natural openings, usually in rocks, that are large enough for human entry. This definition is clearly man-centered in that we think of the cave in terms of its human occupants. They are also classified by their measurements. Caves are formed by different processes in many rock types and unconsolidated sediments, so can be classified according to their origin and the type of host rock or the type of sediment.

However, cave development processes differ greatly, and many of them do not depend strongly on host rock composition. Numerous attempts have been made to develop generic classifications that would encompass all caves and relate them to host rocks, but none of these appear to be harmonious in grouping the processes and generic types of caves. Instead, some kind of heuristic approach, only partly generic, is required to distinguish the most significant classes of caves. The great majority of natural caves, including the largest ones are solution (or karst) caves, ones which have been created principally by the dissolution of bedrock by water circulating through initial openings such as fissures and pores. Solution caves are the most important for cave and karst scientists, and they are also most important in terms of their interference with human activities. Caves may also be produced in many types of rock by other processes, for example by volcanism.

The use to which a cave can be put depends on a number of factors, of which the most important are its size and shape. A single burial can be squeezed into a small, dark crevice but residential activity requires space and light, and the more arcane of rituals seek obscurity and concealment. In terms of their suitability for human use, it is possible to classify caves according to whether they are open (or day-lit) chambers that in some cases are indistinguishable from rock shelters, or are deep fissure caves, although both of these types are often found as part of the same system. The open or day-lit caves can be, and frequently are, used for both economic and ritual activities, whereas deep caves are rarely used at all and then only for ritual purposes. A third category consists of cavities formed within rock-falls, but these features have more in common with rock shelters than real caves. Although open caves show evidence for both economic and ritual use, this rarely appears to have been simultaneous use, and in many caves there is evidence of a change from one activity to the other. One frequently encountered change is from economic to ritual use. In some cases caves have gone through several cycles of change, especially over very long periods of time, possibly thousands of years.

2.3.1 World Record and Other Caves

Kurbera Cave
In early 2017 the deepest known cave in the world was Kurbera-Voronja, in the Western Caucasus region of Georgia. It has been explored more than 2 km (1.24 mi) down, setting a record for cave penetration. It is known to be at least 2.19 km (1.36 mi) deep.[2]

Veryoukina Cave
So is the Kurbera Cave the deepest in the world?

When the "Perovo-speleo" team was exploring the Veryoukina Cave in August 2017, also in the Western Caucasus of Georgia, they reached a depth of 2.2 km (1.36 mi),

---

[2]To watch a 5 minute video of exploring the Kurbera-Voronja cave go to:
https://www.youtube.com/watch?v=R9XBm3zFG6Q

Fig. 2.4 Kurbera Cave in the Georgia Republic. Photo Courtesy of amusingplanet.com.

setting a new world depth record. A huge system of more than 6 km (3.7 mi) of sub-horizontal passages below the 2.1 km (1.3 mi) depth was found and surveyed. In March 2018 another expedition of the same team added another kilometer of tunnels to the

Fig. 2.5 Descending into the Veryoukina Cave. Photo courtesy of Andrei Buzik and the Perovo Speleo Club.

map, taking the total cave depth to 2.212 km (1.37 mi). One day, perhaps even that record will be broken.[3]

Mammoth-Flint Ridge Cave System
Mammoth Cave National Park in central Kentucky in the U.S.A. has a long cave system of chambers and subterranean passageways. With more than 643 km (400 mi) currently explored, this is the world's longest known cave system. The early guide Stephen Bishop called it a "grand, gloomy and peculiar place," but its vast chambers and complex labyrinths have earned its name: Mammoth. Its features include the Frozen Niagara section, which is known for waterfall-like flowstone formations, and the Gothic Avenue whose ceiling is covered by the signatures of 19th-century visitors. The trails take in other features of the park, including the Green and Nolin rivers and the sinkholes of Cedar Sink.[4]

Since the 1972 unification of Mammoth Cave with the even longer system of Flint Ridge to the north, the official name of the system has been the Mammoth-Flint Ridge Cave System. The national park was established on July 1, 1941. It was made a World Heritage Site on October 27, 1981, and then an international Biosphere Reserve on September 26, 1990.

Velebita Cave System
In 2004, Croatian cave explorers from several speleological clubs discovered a pit inside Velebit Mountain that is believed to have the world's deepest subterranean free-fall vertical drop, at almost 513 m (1683 ft). It is in the Rozanski Kukovi area of the North

---

[3]To watch a 16 minute video of the "Perovo-speleo" team exploring the Veryoukina Cave go to: https://www.youtube.com/watch?v=sdViBwrqa0I

[4]Go to the Mammoth Cave website for many more photos.
https://www.doi.gov/blog/mammoth-cave-explore-worlds-longest-cave

Fig. 2.6 The Mammoth Cave entrance defies its length and vastness. Photo courtesy of the NPS.

Velebit National Park in central Croatia. At the foot of the Velebita Cave are small ponds and streams, including one of the largest known colonies of subterranean leeches.

Sarawak Camber

The largest cave chamber in terms of area and second largest by volume after the Miao Room in China, is the Sarawak Chamber. It is located in Gua Nasib Bagus (Good Luck Cave) in the Gunung Mulu National Park, in the Malaysian state of Sarawak on the island of Borneo. In terms of man-made structures, the Sarawak Chamber has a volume surpassed only by the building in which Boeing builds its largest jet airliners, in Everett, Washington State, which has a usable volume of 13.3 million m^3 (470 million ft^3). But the Everett building uses internal columns to support its roof whereas the Sarawak Chamber is a clear span. A circle of 325 m (1066 ft) diameter could be fitted into the floor plan of the Sarawak Chamber, making it not only much larger than the Dallas Cowboys Stadium, which is the largest unsupported man-made dome in the U.S.A., with a diameter

Fig. 2.7 Just one of many Mammoth caverns. Photo courtesy of the NPS.

of 275 m (902 ft), but also the Singapore National Stadium which holds the world record at 310 m (1017 ft.)

Carlsbad Caverns National Park
The Carlsbad Caverns National Park is in the Guadalupe Mountains which range from west Texas into southeastern New Mexico. Elevations within the park rise from 1095 m (3595 ft) in the lowlands to 1987 m (6520 ft) atop the escarpment. The most famous of the park's geological features are its limestone caves. There are more than 100, the most famous of which is the Carlsbad Cavern. It receives more than 300,000 visitors per year and offers a rare glimpse of the underground worlds preserved under the desert above.[5]

Lechuguilla Cave
This cave is named for the canyon that provides access, which is in turn named for Agave Lechuguilla, a species of plant that is found there. It is located in the Carlsbad Caverns

---

[5]Take a two minute tour of the Carlsbad Cavern at:
https://www.youtube.com/watch?v=ob9WFa6FGDo

Fig. 2.8 Velebita cave shaft. Photo courtesy of Dr. Darko Bakšić, Zagreb Speleological Society, University of Zagreb.

Fig. 2.9 Sarawak Chamber. Photo courtesy of Robbie Shone.

Fig. 2.10 Dolls Theatre in Carlsbad Caverns. Photo courtesy of the National Park Service and Peter Jones.

National Park in Eddy County, New Mexico. Cave access is restricted to approved scientific researchers, survey and exploration teams, and National Park Service management-related trips.

Until 1986 the Lechuguilla Cave was known as a small, insignificant historic site in the park's back country. Small amounts of bat guano were mined from its entrance passages for a year in accordance with a mining claim that was filed in 1914. The historic cave contained a 27 m (90 ft) entrance pit named the Misery Hole that led to 122 m (400 ft) of dry passages which were dead ends. After the mining ceased, it was visited only infrequently. But in the 1950s, cavers heard a wind roaring up from the rubble-choked cave floor. No route was apparent, but people concluded there must be cave passages beneath the rubble. Led by Dave Allured, a group of cavers from the Colorado Grotto gained permission from the National Park Service and started to dig in 1984. They broke through into large walking passages on May 26, 1986. Since then, explorers have mapped passages extending over 222 km (138 mi), making Lechuguilla the eighth-longest cave in the world and the fourth-longest in the U.S.A. At 489 m (1604 ft) it was also the deepest cave in the continental U.S.A. until the Tears of the Turtle Cave was explored in 2014. Drawn by the Lechuguilla Cave's pristine condition and rare beauty, cavers come from around the world to explore and map its passages and geology.

In May 2012, a team led by Derek Bristol of Colorado ascended over 120 m (395 ft) into a dome and discovered several new, unexplored passages, pits, and large rooms. Many of the features in this new section were named after items in *The Wizard of Oz*. A large room measuring 180 m (590 ft) long, up to 46 m (150 ft) wide, and up to 46 m (150 ft) high was named "Munchkinland." At more than 160 m (525 ft) from floor to ceiling, the pit named "Kansas Twister" became the deepest pit in the park. During 8 days of mapping the Oz section, the team added the largest distance to the survey since 1989, taking the total length to 216.6 km (134.6 mi). Since then, further exploration has increased the total length to 222.6 km (138.3 mi), making the Lechuguilla Cave the eighth-longest explored cave in the world and the second deepest in the continental U.S.A. It is most famous for its unusual geology, rare formations, biology and pristine condition.

The chances are that you, along with millions of other people, will have seen pictures of the Lechuguilla Cave if you have watched *Planet Earth*, narrated by David Attenborough and Sigourney Weaver for the BBC/Discovery Channel.

Fig. 2.11 Lechuguilla Cave. Photo courtesy of Dave Bunnell.

Tears of the Turtle Cave

The entrance to the Tears of the Turtle Cave in the Bob Marshall Wilderness in Montana was found in 2006. The cave passage is mostly a 2.5 ft wide fissure that follows a stream down through numerous rappels. A team of nine cavers explored a new passage in 2014 and in probing to a depth of 496 m (1629 ft) they stole the record for the deepest known limestone cave in the U.S.A. from the Lechuguilla Cave. Exploration resumed in 2016 with teams reaching the current known depth of 506 m (1659 ft). On those trips, the difficulty necessitated establishing a camp inside the cave for the first time. This, together with the technical rope work, the 4°C (39°F) temperature, and the remote site of the cave made further exploration difficult, but a return trip was planned for 2018.

Exploration is done through the Caves of Montana Project, an official project of the National Speleological Society, and by a Memorandum of Understanding with the U.S. Forest Service.

Fig. 2.12 Speleologists take a break in exploring the Tears of the Turtle Cave. Photo courtesy of an unnamed caver and the U.S. Forest Service.

Fig. 2.13 Spending the night near the bottom of the Tears of the Turtle Cave. Photo courtesy of an unnamed caver and the U.S. Forest Service.

Skocjan Caves

Due to their exceptional significance, the Skocjan Caves in Slovenia were listed by UNESCO as a natural and cultural world heritage site in 1986, acknowledging their importance as one of the Earth's natural treasures. They represent the most significant underground phenomena on the Karst Plateau which crosses the border of southwestern Slovenia into northeastern Italy. Following independence from Yugoslavia in 1991, Slovenia committed to protect the Skocjan Caves by creating the Skocjan Caves Regional Park and its managing authority, the Skocjan Caves Park Public Service Agency.

The first written mention of the Skocjan Caves was the 2nd century B.C. They are certainly on the oldest published maps of this part of the world; for example, the Lazius-Ortelius map of 1561 and Mercator's Novus Atlas of 1637. The Reka River goes underground at the Big Collapse Doline and then flows underground in the Skocjan Caves for 34 km (21 mi) before surfacing near Monfalcone. Then it contributes approximately one-third of the flow of the Timavo River which runs from the Timavo Springs into the Adriatic Sea. In the rainy season it disappears underground 160 m (525 ft) below the surface.

The exceptional volume of the underground system is what distinguishes the Skocjan Cave complex. It is one of the most famous underground structures in the world. The river flowing through it is roughly 3.5 km (2 mi) long, 10–60 m (32–196 ft) wide, and over 140 m (459 ft) high. At some points, it expands into huge underground chambers. The largest of these is Martel's Chamber, with a volume of 2.2 million m^3. It is the largest known underground chamber in Europe and one of the largest in the world. The canyon of such dimensions nevertheless ends with a relatively small siphon, one that cannot deal with the enormous volume of water that pours into the cave after heavy rainfall, when

Fig. 2.14 Skocjan Caves. Photo courtesy of Borut Lozej.

water can rise by more than 100 m (328 ft) and cause major flooding. The explored length of caves is 6200 m (3.8 mi).[6]

---

[6]Watch a four minute video of the Skocjan Caves on YouTube at:

Fig. 2.15 Skocjan Caves tourist trail. Photo courtesy of Lander at Slovenian Wikipedia.

Hang Son Doong Cave

This cave is in Phong Nha-Ke Bang National Park, Bo Trach District, Quang Bình Province, Vietnam, near the Laos-Vietnam border. It has an internal, fast-flowing subterranean river and, as of 2009, the largest cross-section of any cave in the world; twice that of the next largest passage. It also has the greatest volume of any cave. Although it formed in soluble limestone, it was probably carved by fast flowing water as much, if not more so, than by the acidity of the water. It has been estimated at 2–5 million years old.

The Hang Son Doong Cave was discovered by a local man in 1991, but the steep descent and the whistling wind and the roaring water that was heard at the entrance dissuaded the local people from entering the cave. Only in 2009 did the cave become internationally known when a group of cavers from the British Cave Research Association, led by Howard Limbert, made a survey in Phong Nha-Ke Bang National Park. Their progress was halted by a 60 m (200 ft) high flowstone-coated wall which was named the Great Wall of Vietnam. This was traversed in 2010, when the group reached the end of the cave passage. At more than 200 m (656 ft) high, 150 m (492 ft) wide, and 5 km (3.1 mi) long, the Hang Son Doong Cave in Vietnam is so enormous that it has its own river, jungle, and climate.[7,8]

Lava Beds National Monument

The Lava Beds National Monument has over 800 lava tube caves, of which more than 20 can be explored by all skill levels of the general public without the need for guides or tours. Located in northeastern California, the monument lies on the northeastern flank of

---

https://www.youtube.com/watch?v=lXX7xn-WH-M

[7]For a 6 minute drone view of the Hang Son Doong cave go to:

https://video.nationalgeographic.com/video/news/150319-son-doong-vietnam-cave-vin

[8]For a 46 minute National Geographic documentary showing exploration of the Hang Son Doong cave and the violence of the river flowing through it go to:

https://www.youtube.com/watch?v=PPVYJzq_ouE

Fig. 2.16 Hang Son Doong cave doline (skylight). Photo courtesy of Dave Bonnell.

Fig. 2.17 Hang Son Doong cave interior view. Photo courtesy of John Spies.

Medicine Lake Volcano and has the largest total area of any volcano in the Cascade Range. It was established as a U.S. National Monument on November 21, 1925, and spans more than 46,000 acres (190 km^2). The flows from Mammoth and Modoc Craters comprise about two-thirds of the lava in the monument. Over 30 separate lava flows located in the park range in age from 2 million years to just over a thousand years ago. While not a world record holder, the Lava Beds National Monument boasts the greatest concentration of lava tube caves in North America. Petroglyphs are evidence of prehistoric natives in some of the caves. It is also the site for the Kansas State University lava tube research and the use of the CaveR robot (see Chapter 4.3.5).[9]

Mojave National Preserve
The Providence Mountains State Recreation Area spans almost 6000 acres in the Providence Mountains, within the Mojave National Preserve in San Bernardino County, California. It contains the Mitchell Caverns Natural Preserve and is also home to ringtail cats, desert bighorn sheep and one of the few known colonies of Gila monsters west of the Colorado River.

The isolation of the Mitchell Caverns, the only limestone caves in California's state park system, is both a blessing and a curse. Hidden in a narrow canyon in the eastern Mojave Desert, the caverns that opened to the public over 80 years ago are a lure for geology fans and outdoor adventurers. But officials closed the caverns in 2010 after the two rangers who patrolled the area retired and were not replaced amid budget cuts. That allowed vandals to move in and rip out thousands of feet of electrical copper wire, steal diesel-powered generators, metal signs and lamps, and ransack a rock-and-mortar visitor center situated 16 (desolate) miles north of Interstate 40. After extensive renovations and repairs, the caverns and the visitor center were reopened to the public 2017.[10,11]

Atacama Desert
Although the Cordillera de la Sal, which is close to San Pedro de Atacama, Chile, is one of the most arid locations on Earth, it contains extensive cave systems that have developed in halite (rock salt). The mean annual rainfall averages just 20–50 mm (0.8–2 in) per year, and sometimes there is no rainfall at all for several years at a time.

In the past 15 years, several cave expeditions have discovered and documented a subterranean karst drainage network which has over 4 km (2.5 mi) of caves and tunnels. A study has been carried out both at the surface and in the most important caves of the Cordillera de la Sal with the object of understanding the mechanisms responsible for their formation and evolution.

Radiocarbon dating of wood and bone fragments recovered from the ceilings of caves and a type of sediment that results from dry-land erosions have indicated when the cave systems were formed and when the sediments were emplaced. In some cases, huge cave passages appear to have formed in less than 2000 years by a succession of short-lived flash floods, perhaps after a single extreme rain event. The oldest cave passage appears to be slightly over 4000 years old. This area was visited in 2008 by Dr. J. Judson Wynne from the Northern Arizona University and his colleagues.[12]

---

[9]For an 8 minute video tour of the Lava Beds National Monument go to:
https://www.youtube.com/watch?v=Q1u15ryaHmw
[10]To read more: https://www.desertusa.com/mnp/mnp_mc.html#ixzz5PVnMAmph
[11]For a short two minute video go to:
https://www.desertusa.com/video_pages/mitchell-caverns.html
[12]The ages of caves in the Cordillera de la Sal are available from:
https://www.researchgate.net/publication/290889248_Age_of_caves_in_the_Cordillera_de_la_Sal_Atacama_Chile

Texas Caves

I would be remise if I did not mention the caves in my backyard, which lies on a great limestone and dolostone layer. About 20 minutes away is the Inner Space Cavern in Georgetown, Texas. During construction of Interstate 35 in 1963, the Texas Highway Department was required to carry out regulation core testing by drilling holes. In taking 20 samples, they lost 8 expensive diamond drill bits as a result of penetrating the roof of a cave. They searched the surrounding area but could not find any access into the cave, so they drilled a 2 ft wide hole through 10.2 m (33.5 ft) of limestone bedrock. The hole became known as the Discovery Hole because it was the very first entrance into the cave (and indeed the only one for several months). Exploration has revealed 8 km (5 mi) of cave, making Inner Space Cavern the fourth largest cavern in Texas![13]

Here is a list of the top ten caves in Texas:

- Airmen's Cave Barton Creek in Austin.
- Bracken Cave in San Antonio.
- Cascade Caverns near Boerne.
- Cave Without a Name also near Boerne.
- Caverns of Sonora, 15 mi from Sonora, between San Antonio and Big Bend.
- Inner Space Cavern near Georgetown.
- Jacob's Well off Cypress Creek, not far from Wimberley.
- Kickapoo Cavern State Park about 22 mi from Brackettville.
- Longhorn Cavern State Park in Burnet Country.
- Natural Bridge Caverns near San Antonio.

2.3.2 Caves Used by Prehistoric Man

The following are just a few of the magnificent caves that are known to have been occupied by humans. While they do not necessarily represent the best caves for prehistoric man, they certainly portray the scope and majesty of caves as well as their diversity. And nor do they represent the optimum cave for future lunar and Martian explorers! But remember, when these caves were initially explored they were relative small openings until someone poked around further. People have actually made their life's work studying caves on Earth. Someday, there will be real data for caves in the Moon and Mars, and new generations of scientists will be able to tackle the new field of research which Dr. Penny Boston of the NASA Ames Astrobiology Institute has called "astrospeleology."

Lascaux

The Lascaux Cave is actually a complex of caves near the village of Montignac in southwestern France. It is widely known for its almost 6000 figures (and 600 paintings) on the interior walls and ceilings. These paintings represent primarily large animals and typical local and contemporary fauna that correspond with the fossil record of the Upper Paleolithic time. It is clear that anatomically modern humans were well established in

---

[13]For a YouTube tour of Inner Space Cavern go to:
https://www.youtube.com/watch?v=kjf1DgXYsNw

Europe during that time (i.e. 17,000 to 15,000 BCE). Following the archaeological record, they seem to have been abundant in the area between southeastern France and the Cantabrian Mountains of northern Spain that includes Lascaux. The cave itself appears to have been occupied only temporarily, perhaps only for creating the art. But it is possible that the first few meters of the vestibule of the cave (the space the daylight could still reach) may well have been inhabited. Lascaux is probably what most people will imagine as a prehistoric cave that was occupied by our ancient ancestors.

On September 12, 1940, the entrance to the Lascaux Cave was discovered by 18-year-old boy and three friends. The teenagers entered it via a long shaft. They discovered that the cave walls were covered with depictions of animals. Galleries that suggest continuity, context, or simply represent a cavern were given names such as the Hall of the Bulls, the Passageway, the Shaft, the Nave, the Apse, and the Chamber of Felines. The cave complex was opened to the public on July 14, 1948. By 1955 carbon dioxide, heat, humidity, and other contaminants from the 1200 visitors per day were producing visible damage to the paintings. Fungi and lichen were increasingly infesting the walls. So the cave was closed to the public in 1963, the paintings were restored to their original condition, and a monitoring system on a daily basis was introduced.

In 2016, the International Center of Cave Art (CIAP) was opened as a tourist and cultural facility dedicated to the enhancement and popularization of cave art from the painted and engraved representations of the Lascaux Cave. The Center has an almost complete reproduction which includes a physical facsimile of the cave, as well as digital media and mobile facsimiles which accurately portray the iconic paintings of the original cave per panel.[14]

UNESCO designated the Lascaux Cave as a World Heritage Site in 1979 as representative of the Prehistoric Sites and Decorated Caves of the Vézère Valle.

Jebel Irhoud

Jebel is an archaeological site just north of the location known as Tlet Ighoud, some 50 km (30 mi) southeast of the city of Safi in Morocco. It is noted for the hominin fossils that have been found there ever since the site's discovery in 1960. Initially thought to be Neanderthals, the specimens have since been reassigned to Homo sapiens and dated at 315,000 years old. If correct, this makes them by far the oldest known fossil remains of an anatomically modern human found to date. However, it hasn't been possible to obtain any DNA to verify this theory.

The site is the remnants of a solutional type of cave that is usually formed in soluble limestone, the most frequently occurring type. It is located on the eastern side of a karstic outcrop of limestone at an elevation of 562 m (1844 ft) and has been filled with 8 m (26 ft) of deposits from the Pleistocene era. It was found in 1960 while the area was being mined for barite, a mineral consisting of barium sulfate that is the principal source of barium. A miner discovered a skull in the wall of the cave, extracted it and handed it to an engineer, who donated it to the University of Rabat. It, in turn, organized a joint French-Moroccan expedition to the site in 1961 led by the French researcher Émile Ennouchi.

---

[14]Watch a 4 minute video about the Lascaux cave complex at
https://www.youtube.com/watch?v=UnSq0c7jM-A

Fig. 2.18 An artistic reconstruction of a tunnel pathway in the Lascaux Cave. Photo courtesy of the Lascaux Visitors Center/Snøhetta Architects/contractors.

Fig. 2.19 An artist painstakingly adds the final touches. Photo courtesy of the Lascaux Visitors Center/Snøhetta Architects/contractors.

Fig. 2.20 The Jebel Irhoud site. Photo courtesy of Shannon McPherron, Max Planck Institute for Evolutionary Anthropology in Leipzig, Germany.

Ennouchi's research identified the remains of around 30 species of mammals, some of which are associated with the Middle Pleistocene, but the stratigraphic provenance is unknown. Another excavation was carried out by Jacques Tixier and Roger de Bayle des Hermens in 1967 and 1969. They identified 22 layers in the cave, the lower 13 of which contained signs of human habitation including a tool-making industry known as Levallois Mousterian, this being the name for a distinctive type of stone "knapping" developed by precursors to modern humans during the Paleolithic period.

Zhoukoudian Cave System

Zhoukoudian is a Pleistocene hominid site in China lying about 42 km (26 mi) southwest of Beijing, at the juncture of the North China Plain and the Yanshan Mountains. Adequate supplies of water and natural limestone caves provided an optimal survival environment for early humans. Scientific work at the site is still under way. Thus far, ancient human fossils, cultural remains, and animal fossils from 23 localities within the property dating from 5 million years ago to 10,000 years ago have been identified. These include Homo erectus pekinensis (Peking Man) from the Middle Pleistocene (700,000–200,000 years ago), archaic Homo sapiens from 200,000–100,000 years ago, and modern Homo sapiens dating back 30,000 years. There are also fossils of hundreds of animal species, over 100,000 fragments of stone tools and indications (hearths, ash deposits and burnt bones) of Peking Man using fire.

By revealing an evolutionary cultural sequence, Zhoukoudian is not only the most significant site of hominid remains on the Asian continent, it is also of major importance within the worldwide context. It demonstrates the process of human evolution, and so is of significant value in the research and reconstruction of early human history.

Fig. 2.21 The site of the Zhoukoudian Peking Man's cave entrance. Photo courtesy of Wikimedia Commons.

Fig. 2.22 The entrance to the Zhoukoudian museum. Photo courtesy of the museum staff.

Discovered by the Swedish archaeologist, paleontologist and geologist Johan Gunnar Andersson in 1921, the Peking Man Site was initially excavated by Otto Zdansky in 1921 and 1923, when two human teeth were unearthed. These were later identified by Davidson Black as belonging to a previously unknown species and extensive excavations followed. Fissures in the limestone containing middle Pleistocene deposits have yielded the remains of about 45 individuals, as well as animal remains and stone flake and chopping tools. A total of 13 layers have thus far been excavated to a depth of nearly 40 m (131 ft). An extensive museum now hosts many of the finds as well as beautiful statues and exhibits that have proved vital to advancing the study of human evolution. In 1987 Zhoukoudian became a UNESCO World Heritage Site.[15]

2.4　MODERN "TROGLODYTES"

Modern humans are a clever and adaptable lot. Between being desperate or just plain frugal, we can readily adapt to living in a cave and millions have done so. During war and other times of strife, small groups of people have lived in caves on a temporary basis, seeking refuge. They also have used caves for clandestine and other special purposes while living elsewhere. The DeSoto Caverns, in what became Alabama, were a burial ground for local Indians. Then, in the 1920's the same caves became a violent speakeasy. The Caves of St. Louis were used by a local brewery and were a hiding place for slaves on the Underground Railroad. From about 1000 to about 1300 AD, Pueblo people lived in villages which they built beneath cliffs in what is now the Southwestern United States. Many caves were used by outlaws such as the notorious Sam Bass. The following are just a few examples of modern uses of caves.

2.4.1　China

It is hard to believe that tens of millions of people live in caves to this very day. In China alone, more than 30 million live in caves called "yaodong," many of them in Shaanxi Province where the Loess plateau, with its distinctive cliffs of yellow, porous soil, makes digging easy and cave dwelling a reasonable option. Some caves are simple and some are beautiful; many of them are handed down for generations.

Caves have played an important role in modern Chinese history. The famous retreat of the Communist Party in the 1930s named the Long March, ended near Yanan when Mao sought refuge in caves. In *Red Star Over China*, writer Edgar Snow described a Red Army university that may well have been the world's only seat of "higher learning" whose classrooms were bombproof caves, with chairs and desks of stone and brick, and blackboards and walls of limestone and clay.

President Xi Jinping lived for 7 years in a cave when he was exiled to Shaanxi during the Cultural Revolution.

---

[15]For a three minute video of Zhoukoudian go to: http://whc.unesco.org/en/list/449/video

Fig. 2.23 An extensive yaodong cave community. Photo courtesy of Wikimedia Commons.

2.4.2   France

Troglodyte homes are not unusual in the Loire River Valley of France. Mining of sedimentary rock in the 11th century left deep cavities. People soon transformed these grottos into affordable homes which were commonplace until the early 20th century, when many fell into disrepair after people had moved into more modern apartments and houses. Many buildings in the Loire Valley are constructed from a thick bed of 100-million-year-old white stone called tuffeau or tufa. This stone is soft, manageable, and easy to work. It offered the double benefit of being able to sell the quarried stone (for chateaus and monuments) while excavating a living or working space. By remaining at a fairly constant 12°C (54°F) the troglodyte caves were a good all year round habitat that provided heat in winter and remained cool in summer, as well as giving good protection from the elements.

2.4.3   Spain

The caves at Sacromonte, near Granada in Spain, are home to about 3000 Gitano (Romani Gypsy) people, whose dwellings range from single rooms to caves of almost 200 rooms, along with churches, schools, and vast storerooms.

2.4.4   Iran

In the remote northwestern corner of Iran is a village in which the residents live as modern-day cave dwellers. Current residents of Kandovan, a tourist village in the

Fig. 2.24 The Troglodyte Village of Rochemenier. Photo courtesy of the Cave Museum of the Village.

Fig. 2.25 Caves of Sacromonte Spain. Photo courtesy of Wikiloc.

Fig. 2.26 Kandovan village in Iran. Photo courtesy of Iran Travel Center.

province of East Azerbaijan, claim that their village is more than 700 years old. It was created, they say, when those fleeing the advancing Mongol army took to the caves to hide. The homes are known as "karan" in the local Turkic dialect, a word that roughly translates as the plural of beehive.

The Kandovan village has an unusual look which resembles a gigantic termite colony more than it does traditional caves. The structures were formed by ash and debris from Mount Sahand, when the now-dormant volcano erupted sometime in the last 11,000 years. Over the course of thousands of years, the ash hardened and was carved out by the elements. The surrounding landscape is blanketed by ash in a more traditional pattern.

2.4.5 Australia

The town of Coober Pedy in South Australia is known as the "opal capital of the world" because of the precious opals that are mined there. Its population of about 3500 are renowned for having below-ground residences known as "dugouts" that protect against the scorching summer heat that can reach 45°C (113°F). The name of the town derives from the local Aboriginal term kupa-piti, which means "boys' waterhole."

2.5 IMPLICATIONS FOR SPACE EXPLORATION

We now know that the Earth has a tremendous diversity of geology and biology, and many caves of vast dimensions. Bear in mind that these large caves are on a planet that has a gravity of 1 g. We also know there are subsurface voids on both the Moon and Mars

Fig. 2.27 Coober Pedy. A family of cave dwellers (top). Photo courtesy of travelhobbies.-glogspot.com. Entrance to your B&B (middle) and full size kitchen (bottom). Photos courtesy of Venus Hill Bed & Breakfast.

created by collapsed lava tubes, impacts, or other processes. Conditions on the Moon at 1/6 g, and on Mars at 3/8 g, may well permit caves to be much larger and possibly deeper. And of course the geology and biology may be rather different. This will present many hazards to the exploring astronauts, as well as opportunities.

We don't yet know just how many voids there are on the Moon and Mars, but we certainly know enough to want more satellites with higher resolution cameras and instruments able to view the surface obliquely in order to find candidate cave sites. The first several missions to the Moon and Mars will only require cave sizes and locations that will facilitate basic habitability for small crew sizes (say 3–5 people), initially seeking the highest levels of safety from the perceived hazards. It is conceivable that the first selected cave may be used for at least a decade of exploration; maybe much longer. Once a major effort has been made to occupy a cave, it will probably be upgraded to over time, involving considerable resources and effort. The only reasons to abandon such an extensive and expensive resource would be for safety, science, or operational issues. Even after many decades of use, the site could hold its scientific value. Perhaps, over many decades, there will be a compelling reason to move to another location. It's not likely that you would load everything into a fleet of trucks and drive somewhere else. It is difficult to realistically look farther ahead than the first 25 to 100 years of initial exploration, let alone the pioneering and later colonization phases, but even then the hazards of radiation, micrometeoroids, thermal extremes, and (on Mars) winds and dust will still be there. Perhaps human explorers will choose to stay in their caves for a long time.

We also know that the currently conceived surface habitats will be extremely costly in terms of deliverable mass, numbers of required launches, development time, mission time, and money. It also seems the development of complex in-situ resource utilization concepts that are dependent on yet untested technologies will also be costly. Yet, intuitively, we know we just need shelter, however primitive. And we know we can insert an inflatable habitat into a shelter for just about all of the other needs of the crews on those early missions.

The following chapter gives information about caves on the Moon and Mars. Bear in mind that what is primarily required is protection, and that it might be a collapsed lava tube, a deep overhang, a cave entered through a doline (skylight), or a cave that is entered into horizontally or by an incline. Later chapters discuss pertinent technologies and the desirable characteristics for the "optimum" cave.

3

Lunar and Mars Caves, Lava Tubes and Pits

3.1 LUNAR

There are three reasonably well documented pits on the Moon: the Marius Hills Pit in the Oceanus Procellarum, the Mare Ingenii Pit, and the Mare Tranquillitatis Pit. The terms pit, hole, and skylight are related. What looks to be a black hole is often called just that, but from a distance it looks like a pit. If you can see into the hole, it is sometimes called a skylight. The terms all imply an interior cave that is a collapsed lava tube.

3.1.1 Marius Hills Pit

This pit is at approximately 14.09°N, 303.31°E,[1] in the Marius Hills region that was once volcanically active and has many surface expressions, including sinuous rilles which are long meandering fractures or channel-like features (for example those labeled A and B in Fig. 3.2). Prior to the Apollo missions, some scientists thought sinuous rilles were formed by running water on the surface, but that was disputed. Today, we know they are volcanic structures that were formed either as open lava channels and/or as lava tubes that subsequently collapsed. Because the Marius Hills Pit is associated with a rille, it probably represents a collapse in the roof of a lava tube. It is possible the pit was produced when an impact punched through the roof of the lava tube.

A 1985 report by Friedrich Horz of the Lunar and Planetary Science Institute at the Johnson Space Center in Houston, Texas, pointed out that lava tubes might be useful as locations for lunar bases. The interiors of lava tubes could protect human explorers from different aspects of the lunar environment, including high-energy cosmic rays,

---

[1]Longitudes on the Moon are measured both east and west from the prime meridian. When no direction is specified, east is positive and west is negative. Coordinates are usually measured positive even though it appears as though it would have been simpler to use west as in this case.

© Springer Nature Switzerland AG 2019
M. von Ehrenfried, *From Cave Man to Cave Martian*,
Springer Praxis Books, https://doi.org/10.1007/978-3-030-05408-3_3

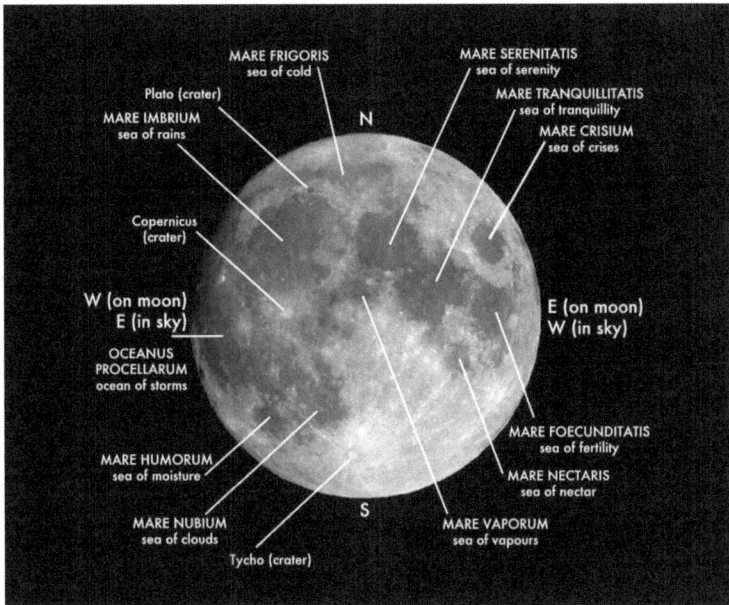

Fig. 3.1 The near side of the Moon. Note in particular Oceanus Procellarum and Mare Tranquillitatis. Mare Ingenii is on the far side, which was unknown prior to the advent of spacecraft. Photo courtesy of astronomy.swin.edu.au.

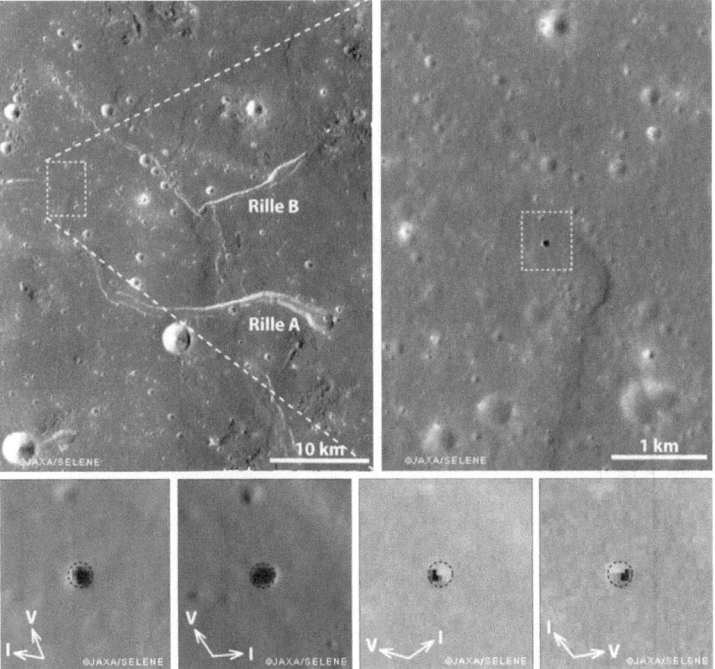

Fig. 3.2 Marius Hills long view composite. Photo courtesy of JAXA Kayuga/SELENE.

meteorite impacts, and the extreme temperature differences between the lunar day and night. Just like caves on Earth, lunar caves, including lava tubes, will maintain constant temperatures.

The Marius Hills Pit was revealed by the JAXA/SELENE Terrain Camera, and later imaged by the Lunar Reconnaissance Orbiter.[2]

The maximum and minimum pit diameters are 57 m (187 ft) and 48 m (157 ft) respectively, and the maximum depth of the floor of the pit below the surrounding terrain is 45 m (147 ft).

3.1.2   Mare Tranquillitatis (The Sea of Tranquility)

On July 20, 1969, Neil Armstrong and Buzz Aldrin left the first human footprints on the Moon near the southwestern "shore" of Mare Tranquillitatis, a dark "sea" that is approximately 873 km (543 mi) in diameter with its center at about 0.68 N, 23.43 E. The "basin" that includes the peripheral mountain ranges is believed to have been excavated by a very large impact in the Moon's early history, at least 3.9 billion years ago. Sometime later, magma welled up through the deeply fractured floor of the crater to fill the cavity with vast amounts of low-viscosity basalt that produced a relatively flat plain – as recently established by the Lunar Orbiter Laser Altimeter (LOLA). Just down the road a piece, as they say, at 8.34°N, 33.22°E, is a big pot hole called the Mare Tranquillitatis Pit. This is about 85–97 m (300 ft) across and 100 m (328 ft) deep. Equivalent to a football field across and over 30 stories deep, it is more than a cave; it is a cavern! Remember, in the low gravity of the Moon such structures might be much larger than those on Earth or Mars. The data that we obtain on the structure and material of this vast cavern might be useful in learning how to use satellites to investigate cavities with a degree of accuracy that might later be applied to smaller holes which would be better suited to human occupation.

The Lunar Reconnaissance Orbiter mission was inserted into orbit around the Moon in 2009 to perform a survey of the surface with modern sensors. The Lunar Reconnaissance Orbiter Camera (LROC), operated by Arizona State University, combined two narrow-angle cameras (NAC) and one wide-angle camera (WAC). Over a mission lasting 10 years, the spacecraft is expected to return in excess of 70 terabytes of imaging data.

When the Sun is directly overhead, the floor of the Mare Tranquillitatis Pit is illuminated. With an incidence angle of 26.5° and a shadow of 55 m (180 ft), the depth has been estimated at slightly more than 100 m (328 ft), measure from the edge of the shadow, which begins slightly downslope from the gradual margin. When measured against the surrounding plain, the depth is even greater. Compare this depth to the width, which ranges from 100–115 m (328–377 ft) spanning the sharp precipice.[3]

---

[2]JAXA prepared a video flyover of Marius Hill. Go to:
https://www.lpi.usra.edu/lunar/lunar_flyovers/marius_hills
[3]There is a YouTube video depicting a Sci-fi view of how such a cave would be used in the future. Go to YouTube and type in "Lunar Pit, Mare Tranquillitatis."

Fig. 3.3 Marius Hills Pit close up. Photo courtesy of NASA/GSFC/Arizona State University and M. S. Robinson, et al., 2012.

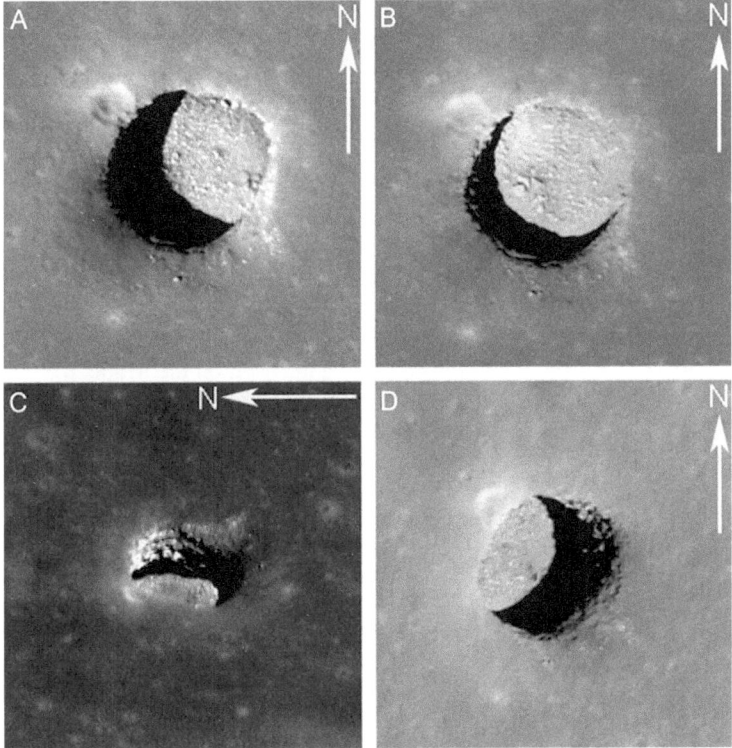

Fig. 3.4 Marius Hills Pit mosaic. Photo courtesy of NASA/GSFC/Arizona State University and M. S. Robinson, et al., 2012.

Fig. 3.5 Mare Tranquillitatis Pit long view. Photo courtesy of NASA/GSFC/Arizona State University and M. S. Robinson, et al., 2012.

3.1.3   Mare Ingenii (The Sea of Cleverness)

Located at 35.95°S, 166.06°E, Mare Ingenii is one of only a few mare features on the far side of the Moon. The mare sits in the Ingenii basin. This basin material is of the Pre-Nectarian epoch that spans from the creation of the Moon by a process of accretion some 4.5 billion years ago to about 3.9 billion years ago. The mare in Ingenii and the surrounding craters are in the Upper Imbrian epoch which dates to about 3.8 billion years ago. The dark circular feature which dominates the view is the crater Thomson, 112 km (70 mi) in diameter. The lava sheets of Mare Ingenii are thin and only partially cover the expanse. There are swirling patterns of bright material that are not associated with either topographic or volcanic features. This mare contains the second instance of a lunar pit on the Moon.

3.2   MARS

About 50 years ago, some scientists proposed that basaltic volcanism occurred on Mars in much the same way as on Earth, and hence volcanic caves were probably fairly common on Mars. Modern orbiters have confirmed that there are numerous tube-fed lava flow systems, some of which are described below. And, as it turned out, there are huge lava tubes.

A 20-year-old NASA Ames Research Center report predicted the existence of Martian caves on the basis of geology and climate history. The first approach was to consider caves as being the result of underground water activity combined with tectonic movement. They can be formed by: (1) diversion of channel courses into underground conduits,

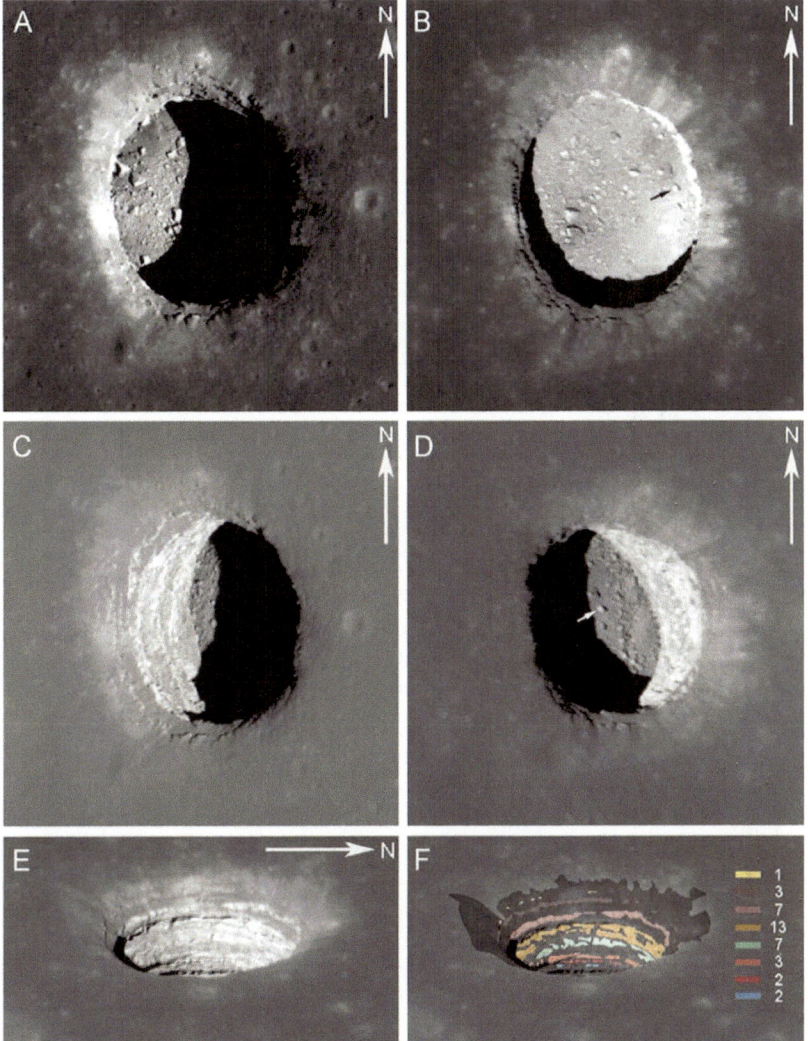

Fig. 3.6 Mare Tranquillitatis Pit mosaic. The maximum and minimum pit diameters are 100 m (328 ft) and 86 m (282 ft), and the maximum depth of the pit floor below the surrounding plain is 105 m (344 ft). A: a near-nadir image; B: an emission angle of 7°; C: an oblique view with an emission angle of 26°; D, E, & F; layering is revealed. Photo courtesy of NASA/GSFC/Arizona State University and M. S. Robinson, et al., 2012.

(2) fractures of surface drainage patterns, (3) the so-called "chaotic" terrain and collapsed areas in general, (4) seepage in valley walls and/or headwater, and (5) inactive hydrothermal vents and lava tubes.

In recent years, planetary geologists have found that extensive areas of Mars are characterized by soluble lithologies (i.e. the characteristics of rocks and rock formations)

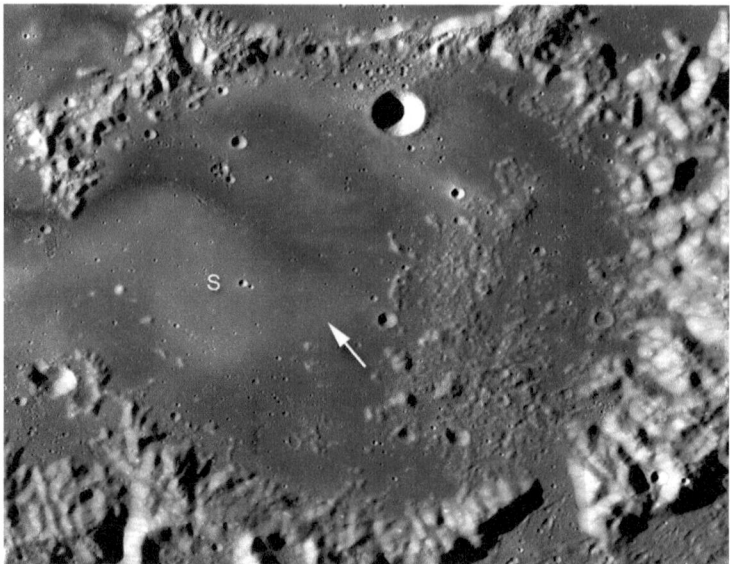

Fig. 3.7 Mare Ingenii may be best known for its prominent lunar swirls, which are high albedo surface features associated with magnetic anomalies. The arrow indicates the location of the pit. "S" indicates one of the swirls. The image was taken by the Lunar Rendezvous Orbiter Camera-Wide Angle Camera. The width of the image is 160 km (99 mi). Photo courtesy of NASA/GSFC/Arizona State University.

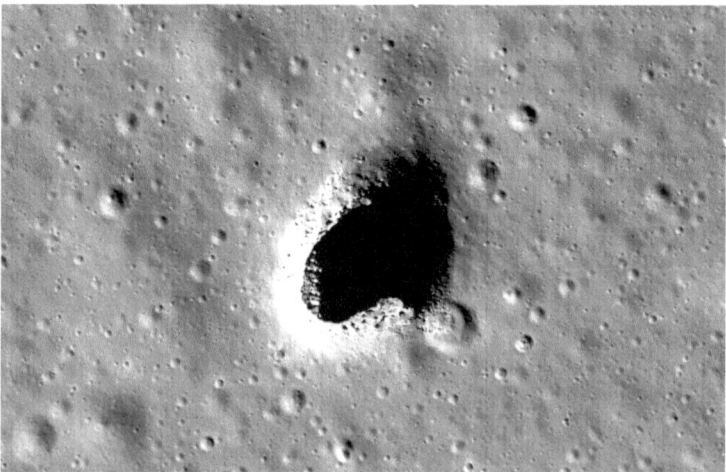

Fig. 3.8 Mare Ingenii elliptical pit. Impact craters occur everywhere on the Moon, but pits are relatively rare. This pit is about 130 m (426 ft) in diameter. The image width is 550 m (1804 ft). The illumination is from the upper right. The high-resolution cameras aboard the Japanese SELENE/Kaguya spacecraft discovered this irregularly-shaped hole, but this particular image was taken by the LROC. The boulders and debris resting on the floor of the pit are partially illuminated on the left side of the pit, and were probably originally at the surface and fell into the pit during the process of collapse that produced the opening. Photo courtesy of NASA/GSFC/Arizona State University.

such as sulfates. The surface of Mars is currently an arid and rough environment, but in the past (mostly in the Hesperian period, more than 3 billion years ago) the planet probably had oceans and underground aquifers. Back then, soluble lithologies were affected by deep solutional weathering. Therefore large cave systems could have formed. Evidence for this kind of process was found in sulfate terrains and diapirs (a domed rock formation in which a core of rock has moved upward and pierced the overlying strata) in the Arabia Terra region, with surface structures matching their equivalents on Earth.

The Tharsis region that straddles the Martian equator features a trio of large volcanoes on a line running southwest to northeast, from Arsia Mons to Pavonis Mons to Ascraeus Mons. The tallest volcano in the solar system, Olympus Mons, lies some distance away to the northwest. Tharsis has been studied by a range of sensors from orbit and found to contain many skylights and caves.

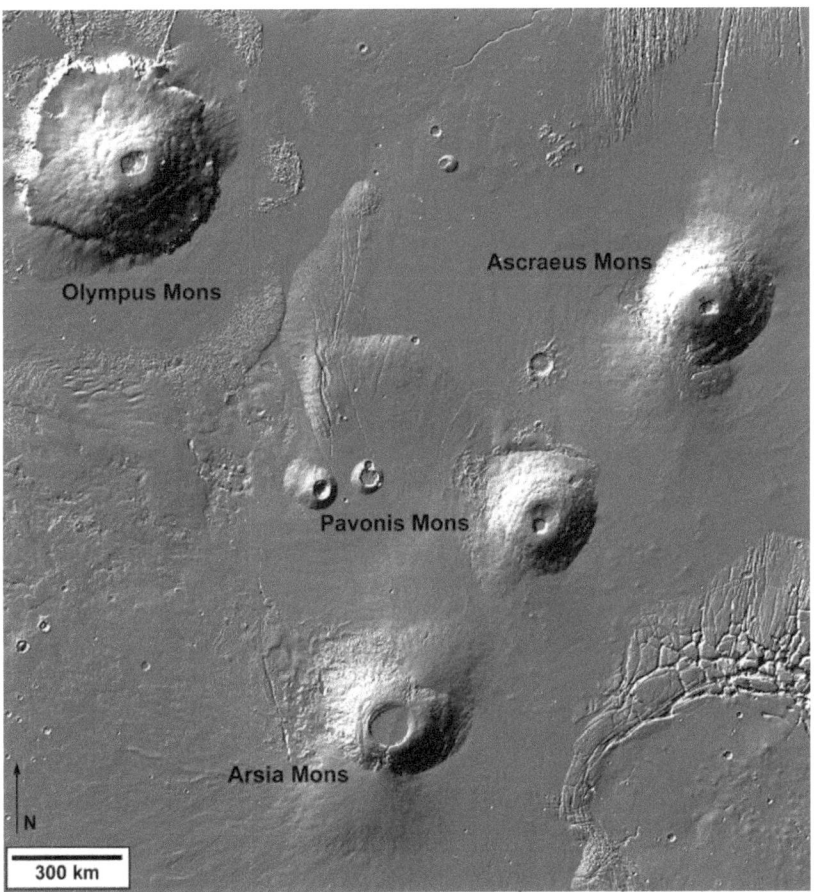

Fig. 3.9 The Tharsis region viewed by the Mars Express High Resolution Stereo Camera (HRSC). The area circled is the area of cave interest. Photo courtesy of ESA/DLR/FU Berlin (G. Neukum).

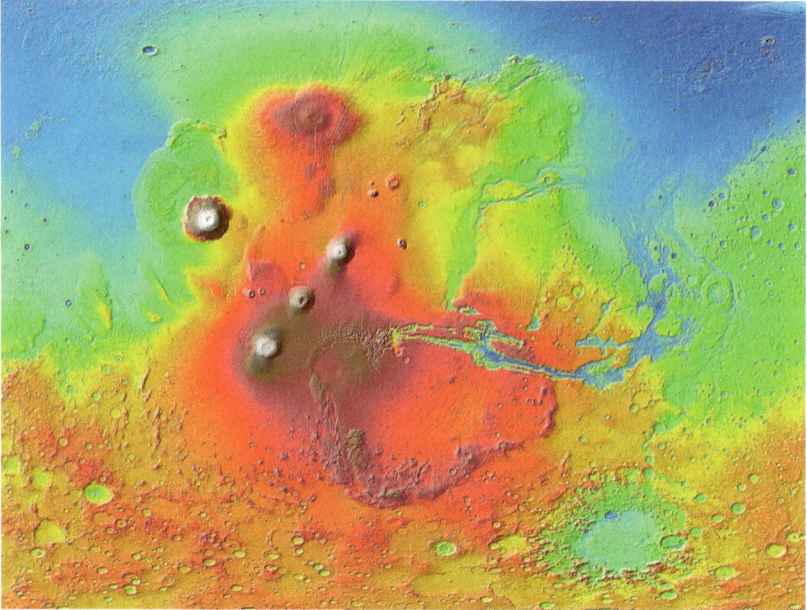

Fig. 3.10 The Mars Orbiter Laser Altimeter (MOLA) mapped the vertical profile of the Tharsis region. Photo courtesy of NASA/JPL-Caltech/Arizona State University.

Arsia Mons, located at 9°S, 239°E, is the primary area of interest for potential caves. It is a shield volcano, a type of volcano usually composed almost entirely of fluid lava flows. This reflects its low profile, similar to a warrior's shield on the ground. Such a volcano has a relatively low slope and an enormous caldera at its summit. The base is 435 km (270 mi) across, and the summit rises to an elevation of almost 20 km (12 mi); more than 9 km (5.6 mi) above the surrounding terrain, which is itself an enormous bulge in the crust. The caldera spans 110 km (72 mi). Excluding Olympus Mons, Arsia is the largest known volcano in terms of volume, having 30 times that of Mauna Loa in Hawaii, which is the largest volcano on the Earth. The summit caldera formed when the mountain collapsed in on itself after its reservoir of magma was exhausted. There are many other collapse features on its flanks. The floor of the caldera is believed to have formed around 150 million years ago.

The rifts (faults) to the southwest have been imaged in detail by Mars Express, a European Space Agency (ESA) orbiter. In 2004, a 3D map was created at high resolution, revealing cliffs, landslides, and numerous collapse features. Combined with the extensive lava flows at the terminations of the rifts, this may reveal areas that drained the caldera lavas and contributed to the collapse. The northwest flank is significantly different and rougher than the southeastern one. It is believed that many of its features were produced by glaciation. Seven very dark holes on the northern slope have been proposed as possible cave skylights, based on day-night temperature patterns that suggest they are openings into subsurface spaces.

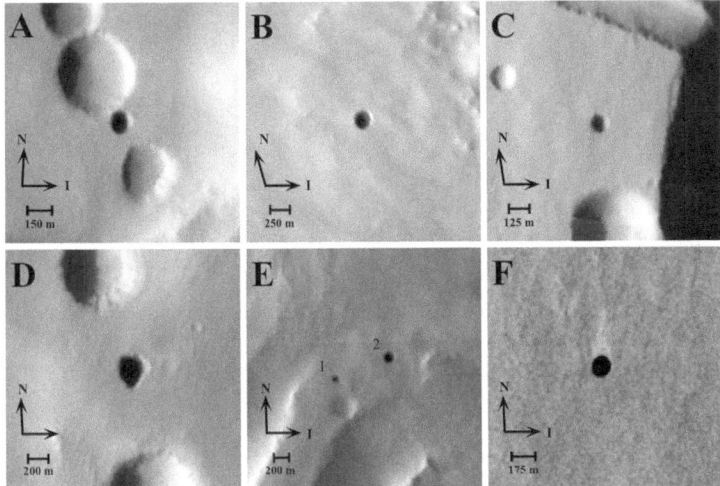

Fig. 3.11 Six images taken in visible-wavelength light by the Thermal Emission Imaging System (THEMIS) camera on NASA's Mars Odyssey orbiter showing seven possible cave skylights. The features have been assigned informal names to aid discussion. They range in diameter from about 100 m (328 ft) to about 225 m (738 ft). The candidate cave skylights are (A) "Dena," (B) "Chloe," (C) "Wendy," (D) "Annie," (E) "Abby" (left) and "Nikki," and (F) "Jeanne." Arrows indicate north and the direction of illumination. These are known as the "seven sisters." The pit "Jeanne" is essentially a vertical shaft cut through the lava flows on the flank of the volcano. Such pits form on similar volcanoes in Hawaii and are called "pit craters." They generally do not connect to long open caverns, but are the result of deep underground collapse. From the shadow of the rim cast onto the wall of the pit, Laszlo P. Keszthelyi (a.k.a. Kestay) of the USGS calculated the pit to be 150 x157 m (492x 515 ft) across and at least 178 m (584 ft) deep; almost 6 stories! Photo courtesy of NASA/JPL-Caltech/ASU.

Fig. 3.12 Arsia Mons cave "Jeanne." Photo courtesy of NASA/JPL/ASU.

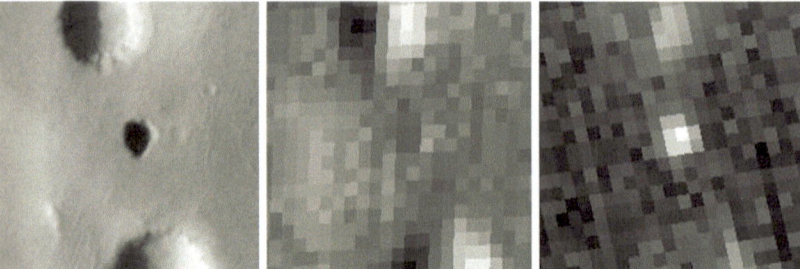

Fig. 3.13 Three images of "Annie" by NASA's Mars Odyssey orbiter showing its diurnal thermal behavior. The image on the left is in the visible spectrum and was taken in the late afternoon; the middle image, which was taken at the same time, is in the thermal infrared; and the image on the right is an infrared image at 4 a.m. Scientists can infer that "Annie" is warmer in the afternoon than the shadows of adjacent collapse pits and cooler than the sunlit portions. At night it is warmer than its surroundings. This would give astronauts living in caves some protection from the diurnal temperature extremes that occur on the surface. This behavior is consistent with cave interiors that receive little or no daily sunshine. The same thermal profiles were observed for all seven caves in Fig. 3.11. Photo courtesy of NASA/JPL-Caltech/ASU.

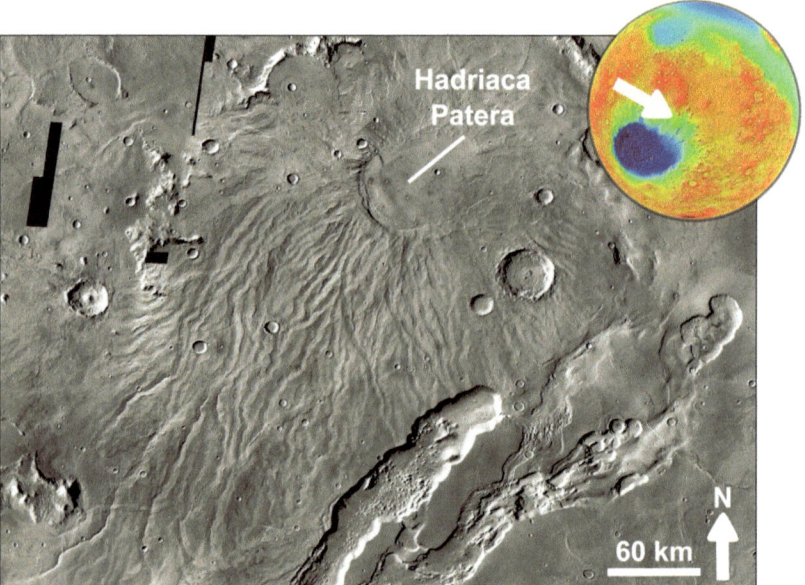

Fig. 3.14 The location of Hadriacus Mons. Photo courtesy of NASA/JPL/ASU THEMIS system.

Although light penetrates the cave "Jeanne" through the hole in its roof, then bounces around the interior, the cave is so large and so deep that almost none of the light that enters it comes out. The hole that we see is only the skylight into a subterranean cavern. At its

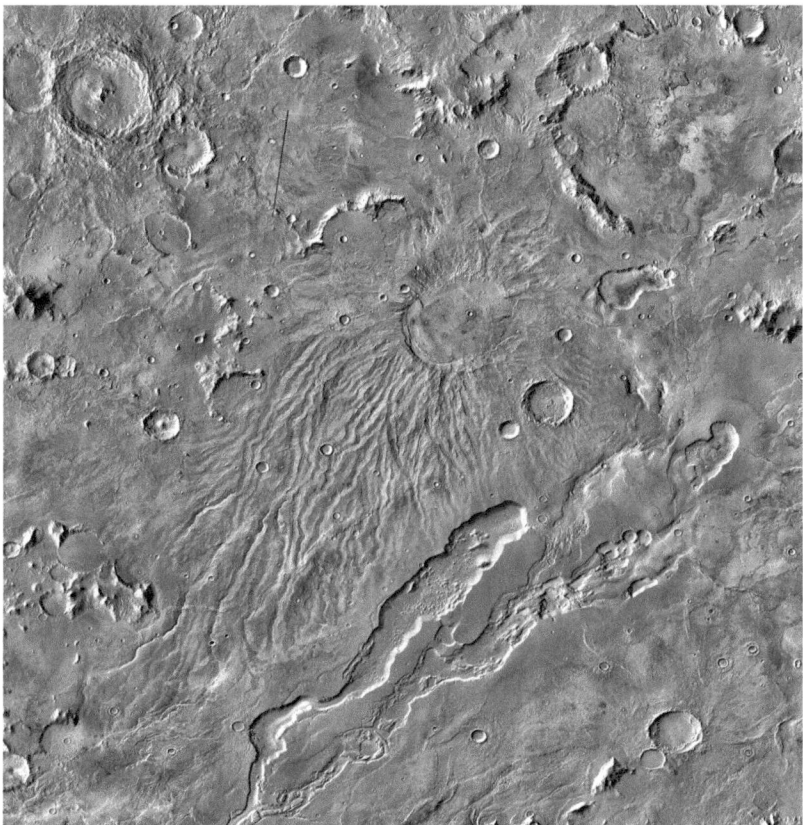

Fig. 3.15 The infrared frames for this mosaic of Hadriaca Mons were taken during the daytime by the THEMIS instrument of the Mars Odyssey orbiter. The upper portion of the image shows relatively smooth terrain in the central caldera, which has been nearly filled in with late-stage lava. The lower half of the image shows lobate (resembling a lobe) lava flows as well as furrows in the ash deposits that make up the volcano's southern flank; these erosional furrows may have formed by surface runoff or sapping (undermining) by groundwater. Just below the center of the image, a few small sinuous troughs are visible, and may be collapsed lava tubes or collapse features related to subsurface water. The number of impact craters on a planetary surface is commonly used as a proxy for its age, an old surface having had time to accumulate more craters than a young surface. The relatively small number of large craters in the image indicates that the surface in this area is younger than the nearby heavily cratered ancient terrains outside the Hellas basin. However, there are more craters on this surface than are found on the average volcanic surface in the Tharsis region. There are some very large old craters on the volcano's flank to the southeast of this image. Photo courtesy of NASA/JPL/ASU THEMIS system.

highest resolution, the HiRISE camera of the NASA Mars Reconnaissance Orbiter reveals the detailed shape of the slightly scalloped edge of the hole, but no amount of image enhancement can bring out any details within. This means that the walls of the cave are overhanging (the cave is larger below the ground than the entrance at the surface) and that it must be very deep. The dusty atmosphere would scatter light sufficiently to illuminate the floor of a shallow cavern. Existing orbital sensors do not have the resolution to reveal what the interior of this cave looks like.

The Thermal Emission Imaging System camera on NASA's Mars Odyssey orbiter took three images of the skylight of the cave named "Annie." This has a diameter about twice the length of a football field. A USGS team led by Glen E. Cushing used both the visible and infrared sensors to determine that the hole is cooler than the surrounding surface in the afternoon and warmer at night. Such thermal behavior is consistent with an opening into an underground cavity, just like caves on Earth. Chapter 6 discusses the optimum characteristics for a cave that would be habitable.

The USGS team has recommended that the next generation of thermal-infrared cameras ought to have a resolution of better than 10 m per pixel (32.8 ft per pixel) in order to better analyze skylights and cave interiors. Although the visible imager subsystem of the THEMIS instrument has a resolution of 19 m per pixel (62 ft per pixel), the infrared subsystem achieves just 100 m per pixel (328 ft per pixel); see their reports in References.

Scientists have reasoned that these sites may not be the best sites to investigate life on Mars. Perhaps older caves that were made by aqueous processes would be more likely to contain evidence of past or even present life. One such example is Hadriacus Mons, situated to the northeast of the vast Hellas Planitia impact basin in the rugged highland terrain of the southern hemisphere.

Mars possesses a handful of low volcanic structures (originally called paterae, after the Latin word for a shallow dish) in the vicinity of the Hellas basin. These show a different style of volcanism to the tall edifices that formed in the Tharsis region. Hadriacus' low slopes, broad structure, and deeply etched flanks imply it is readily eroded volcanic debris known as pyroclastic (from the Greek meaning fire-broken). Evidently, the magma of Hadriaca (and other volcanoes of this type) had significant interaction with subsurface water and produced mostly explosive ash deposits (pyroclastic flows) rather than spread lava across the surface. Nearby sources of water may have included Dao Vallis, which lies on the southern flank of Hadriaca.

4

Terrestrial Cave and Lava Tube Research

4.1 NASA CENTERS AND ORGANIZATIONS

The NASA Centers and their contractors have ongoing multidisciplinary studies and projects involving exploration of the Moon and Mars, some of which look at using caves as shelters. The expertise of each Center is reflected in this research. No single project or Center specifically addresses all of the aspects of adapting a cave or lava tube for habitation. Below is a summary of the general expertise of the various NASA Centers relating to critical technologies that provide the basis for a broad set of human exploration capabilities. Some relate to getting a team of astronauts to the desired site and some relate to assisting them once they are there, specifically: exploring the area, locating the desired cave, assessing its suitability, deploying robots, rovers and base camp equipment, deploying inflatables, setting up communications, using imaging scanners and other instruments, and settling in for the long haul. Note that some of these capabilities have nothing to do with the proposed mission, but are just general spaceflight technologies. Only after further study does one see their relevance to a mission involving caves and lava tubes.

Ames Research Center (ARC):

- Autonomous systems.
- Intelligent Robotics Group.
- Human Exploration Tele-robotics.
- Deployable Aeroshell Concepts and Flexible Thermal Protection Systems.

Armstrong Flight Research Center (AFRC):

- Supersonic flight.
- Vertical flight.
- Noise reduction.
- Low-carbon fuels.

© Springer Nature Switzerland AG 2019
M. von Ehrenfried, *From Cave Man to Cave Martian*,
Springer Praxis Books, https://doi.org/10.1007/978-3-030-05408-3_4

Glenn Research Center (GRC):

- In-space Propulsion Space Power Generation and Storage.
- Nuclear Power Systems.
- Cryogenic Propellant Transfer and Storage.
- Solar Electric Propulsion.
- Structures and Materials.

Goddard Space Flight Center (GSFC):

- Satellite Servicing and Robotics.
- Satellite and Sensor Development.
- Instruments and Tools.

Langley Research Center (LaRC):

- Lightweight Materials and Structures.
- Advanced Radiation Protection.
- Hypersonic Inflatable Aerodynamic Decelerator.
- MSL Entry Descent and Landing Instrumentation.
- In Situ Resource Utilization.
- Habitats.

Johnson Space Center (JSC):

- Human-Robotic Systems.
- Human Research Program.
- Next Generation Life Support.
- Autonomous Landing and Hazard Avoidance Technology (ALHAT).
- Mission Planning and Architecture.
- Space Suit and Portable Life Support Systems Development.
- Habitats.

Jet Propulsion Laboratory (JPL):

- Flight Projects.
- Mission Operations.
- Robotics.
- Spacecraft Design.

Kennedy Space Center (KSC):

- In Situ Resource Utilization (ISRU).
- Launch readiness and operations.

Marshall Space Flight Center (MSFC):

- Composite Cryogenic Propellant Tank.
- ISRU and Habitat Structures.

Here are just examples of some of the work that NASA and their organizations are doing that relates to the subject of this book.

4.1.1   NASA Programs

There are several NASA programs which involve research and contracts that are applicable to a precursor mission that employs caves and/or lava tubes for crew protection in initial phase of exploration.

NIAC

Although the NASA Innovative Advance Concepts (NIAC) program is funded by the Space Technology Mission Directorate (STMD) at NASA Headquarters, it is actually implemented at the Centers. NIAC funds hundreds of papers on various space science and related projects. A good example relating to the subject of this book is "The Caves of Mars Project," which is an early 2000's program to assess the best site to place the research and habitation modules for a human mission to Mars. The final report, published in mid-2004, stated: "This project developed a revolutionary system to exploit the novel idea of extraterrestrial cave use."

The benefits that the study identified for using caves as a foothold in Martian exploration included:

- Atmospheric temperature variations are less experienced in caves than on the surface of Mars.
- Underground caves protect lifeforms and equipment from harmful solar and galactic radiation.
- Underground caves protect crews from dust storms and micrometeoroid impacts.
- Exploring underground caves is a key scientific goal because it makes it easier to study the geology, history, and possible presence of life on Mars without heavy excavation equipment.
- The ability to make caves more habitable for the crews.
- Allows the easy extraction of subsurface materials such as ice and minerals.

The report assigned a Technology Readiness Level to Enabling Technologies previously identified by NASA, including innovations deemed necessary for the use of caves. For example, the innovation "Foamed-in-place Airlocks" is assigned a TRL of 5, while the "Inert Pressurization of Caves" is assigned a TRL of 2. (In this scheme, the higher the level the nearer it is to being made operational.) One key issue was that the amount of gas that would be needed to pressurize a cave would probably not be feasible in the initial phase of exploration. However, one group is investigating a sealant for the walls of the cave which would facilitate pressurization. The insertion of an inflatable seems to makes more sense. Perhaps both would be required, or at least provide additional advantages.

The report described "Essential Tasks" for exploiting a cave:

- Finding extraterrestrial caves.
- Protecting the scientific environment inside of a cave.
- Providing lighting solutions for the interior of the habitats.
- Life support.

The report discussed technology and related subjects.[1] Some are discussed in Chapter 5.

---

[1]The final report entitled "Human Utilization of Subsurface Extraterrestrial Environment" can be accessed at: http://www.niac.usra.edu/files/studies/final_report/710Boston.pdf

Fig. 4.1 Prof. "Red" Whittaker and friends. Photo courtesy of Carnegie Mellon University/ NAI/Dr. Thomas Kieft.

Another example of a NIAC grant which relates to the subject of this book is PERISCOPE: (PERIapsis Subsurface Cave OPtical Explorer) granted to Jeffrey Nosanov and NASA-JPL. This proposes an instrument and mission concept with the goal of investigating and mapping lunar skylights from an orbiting platform by the use of photon time-of-flight imaging. A vehicle in a very low orbit would direct laser pulses into the lunar skylights, detect light returning to the spacecraft after multiple reflections in the cave, and transmit a summary of those data back to the Earth, where a team would process that data to develop a 3D profile of the interior void of the skylight, even though it was at all times beyond the direct line of sight of the spacecraft. Phase I tested the theoretical feasibility of this mission concept using a variety of simulations and analytical tools. In Phase II, the intent was to bring this concept to a level capable of supporting a full mission proposal. Detailed trade studies, analyses, and experiments were then performed using real-world materials as analogs for expected lunar subsurface material.

Another NIAC grant is a report by Professor William "Red" Whittaker called "Technologies Enabling Exploration of Skylights, Lava Tubes and Caves." This discussed robotic exploration of skylights and caves to seek out life, investigate geology and origins, and open the subsurface of other worlds to exploration. The planetary voids present perilous terrain that requires innovative technologies for access, exploration, and modeling. This research reviewed the technologies that would be needed to venture underground and conceived mission architectures for robotic missions that would explore skylights, lava tubes and caves.[2] This work will be discussed further in Chapter 5 Technologies.

---

[2]Professor Whittaker's report can be accessed at:
https://www.nasa.gov/pdf/718393main_Whittaker_2011_PhI_Cave_Exploration.pdf

NASA's Langley Research Center undertook still another NIAC study, entitled "Radiation Shielding Materials Containing Hydrogen, Boron, and Nitrogen: A Systematic Computational and Experimental Study." This is covered more fully in Chapter 5.2 Radiation and Shielding.

SBIR and STTR

Two other NASA programs funded by the Space Technology Mission Directorate and implemented at the Centers are Small Business Innovation Research (SBIR) and Small Business Technology Transfer (STTR). In August 2018, Astrobotic Technology, Inc., located at Carnegie Mellon University, announced $250,000 in new contract awards through these two programs to help its Future Missions and Technology (FM&T) department to develop novel technologies and strategies for the exploration of space and planetary surfaces.

The SBIR contract "Software Defined Reliability for Low Cost Digital Signal Processors on Small Spacecraft" is to enable FM&T to address the requirements of the growing space computing market for the next wave of robotic spaceflight customers with Astrobotic's proprietary "Software Defined Reliability" (ASDR) technology. This research will allow the company to develop advanced robotics capabilities for its Peregrine lunar lander, such as precision landing and hazard avoidance.

A key challenge of high performance computing in space is radiation, which can cause errors in digital electronics and software. The conventional solution is to design systems that are "hardened" to this radiation, but such customized and low volume production is expensive and can introduce significant development delays. Astrobotic intends to address this issue by moving the "hardening" into software with efficient distributed consensus algorithms to carry out flexible error detection and correction using off-the-shelf processors with minimal overhead.

The STTR contract "Mission Coordination and Co-Localization for Planetary Rover Teams" is in partnership with Carnegie Mellon University. The objective of the 13-month contract is to devise strategies to enable accurate localization of multiple planetary exploration rovers, without requiring high fidelity sensing or onboard high-performance computing.

Astrobotic is a nascent lunar logistics company that will deliver payloads to the Moon for companies, governments, universities, non-profiteers, and individuals. It will accommodate multiple customer payloads on a single spacecraft, offering flexibility at an industry-defining low price of $1.2 million per kg. The company is an official partner with NASA through the Lunar CATALYST program, has 26 prior and current NASA contracts, a commercial partnership with Airbus DS, a corporate sponsorship with DHL Logistics, 12 deals for its inaugural mission to the Moon, and 130 customer payloads in the pipeline for further missions. It was founded in 2007 and is headquartered in Pittsburgh, Pennsylvania.

See Chapter 4.3.3 Carnegie Mellon University.

4.1.2 Johnson Space Center (JSC)

JSC is well known for its role in Human Space Flight Research, the astronauts, and Mission Control. While there is no direct research ongoing that specifically addresses caves and lava tubes, it is clear that JSC research applies to astronauts working on the Moon and Mars. Of particular interest is the creation of pressure suits that make it easier to work in such environments as well as all of the EVA tools that would be needed. An ISS

EVA astronaut is provided with a great deal of gear. Astronauts on the surface of the Moon or Mars will probably have a rover full of tools as well as what they carry on their suit.

Much of the JSC activity that applies to our return to the Moon and precursor missions to Mars required coordinating the organizations within that Center. For example, the work on the latest advanced pressure suits such as the Z-2/3 suit in the Advanced Exploration Systems organization was coordinated with the work on the Advanced EVA Portable Life Support System in the Space Technology Mission Directorate. It seems that whatever may be the final designs for such a mission, the suit will likely retain a white outer layer for thermal reasons and its helmet will resemble that of the ISS suit and have lamps.

Similarly, coordination between those working on habitats for the deep space transit phases are working with those focusing upon the habitats for the lunar and Martian surface, who, in turn, are interested in inflatables that would be inserted into caves or lava tubes. Consequently, the JSC organization structure requires a great deal of inter-organizational coordination in order to address the technologies that relate to the mission objectives.

In recent decades, JSC has participated in many space analogs that are related to habitats, rovers, and crew exploration;[3] for example:

- Human Exploration Research Analog (HERA).
- NASA Extreme Environment Mission Operations (NEEMO).
- Desert Research and Technology Studies (Desert RATS).
- Hawaii Space Exploration Analog and Simulation (HI-SEAS).

The attachment of the Bigelow inflatable habitat to the ISS is an operational test and evaluation and proof of concept for the use of such structures. A similar structure could be used on the Moon and Mars. While the work on inflatables in general is being investigated at many NASA Centers for a range of applications, eventually they must be focused on their application in lunar and Martian caves.

4.1.3   JSC's Lunar and Planetary Institute (LPI)

The LPI is managed by the Universities Space Research Association (USRA), an independent non-profit organization, by a cooperative agreement with the NASA Science Mission Directorate. Located near JSC in Houston, its role is to advance understanding of the solar system by providing exceptional science, service, and inspiration to the world. It serves as a scientific forum which attracts world-class visiting scientists, postdoctoral fellows, students, and resident experts. It supports and serves the research community by newsletters, meetings, and other activities; collects and disseminates planetary data while facilitating the community's access to NASA science; engages and educates the public on space science, and invests in the development of future generations of explorers. Hence the research carried out at LPI supports NASA's long-term efforts to explore the solar system.

LPI USRA interdisciplinary activities address biomedicine, planetary science, astrophysics, and engineering, with applications ranging right from fundamental research through to facility management and operations.

---

[3]To read about the analogs NASA is conducting, go to: https://www.nasa.gov/analogs

In 1985, Friedrich Horz led a group that wrote a paper entitled "Lava Tubes: Potential Shelter for Habitats." This said that natural caverns occur on the Moon in the form of lava tubes that are the drained conduits of underground lava rivers. The interior diameters of these tubes can measure tens to hundreds of meters and their roofs are expected to be thicker than 10 m. Consequently, lava tube interiors offer an environment that is naturally protected from the hazards of radiation and modest impacts. Furthermore, constant, relatively benign temperatures of minus 20°C prevail regardless of the extreme temperature variations at the surface. Such an environment would be extremely favorable to human activities and industrial activities, and would therefore deliver significant operational, technological, and economical benefits.

The annual Lunar and Planetary Science Conference (LPSC) organized by the LPI and JSC brings together international specialists in petrology, geochemistry, geophysics, geology, and astronomy to present the latest discoveries in planetary science. Since its start in 1970 as the Apollo 11 Lunar Science Conference, it has been the premier annual gathering of planetary scientists, attracting typically 1800 attendees from dozens of countries.[4]

4.1.4 NASA Ames Astrobiology Institute (NAI)

Astrobiology is the study of the origins, evolution, distribution, and future of life in the universe. This interdisciplinary field requires a comprehensive, integrated study of biological, geological, planetary and cosmic phenomena. It particularly includes the search for habitable environments in our solar system and on planets around other stars; the search for evidence of prebiotic chemistry or life on solar system bodies such as Mars, Jupiter's ice-enshrouded moon Europa, Saturn's ice-enshrouded moon Enceladus, and Saturn's moon Titan; and studies of the origin, early evolution, and diversity of life on Earth. Hence astrobiologists address three fundamental questions: How does life begin and evolve? Is there life elsewhere in the universe? What is the future of life on Earth and beyond?

As part of a concerted effort to address this challenge, NASA established the NASA Astrobiology Institute (NAI) in 1998 as an innovative way to develop the field of astrobiology and provide a scientific framework for flight missions. NAI is a virtual, distributed organization of competitively selected teams that integrate astrobiology research and training in concert with the national and international science communities.

The NAI teams are supported by cooperative agreements between their host institutions and NASA that involve substantial contributions from both NASA and the team. Executive summaries from each team's annual report describe their contributions to astrobiology research. Currently, the NAI has 10 teams including around 500 researchers distributed across approximately 100 institutions. It also has 13 international partner organizations. The Director and administration staff are based at the NASA Ames Research Center in Mountain View, California. The NAI Director and Deputy Director, and the Principal Investigators of the teams form the Executive Council, which considers

---

[4]For LPI data relating to research on caves go to:
https://www.lpi.usra.edu/search/?cx=002803415602668413512%3Acu4craz862y&cof=FORID%3A11&q=cave+research&sa=Search&siteurl=https%3A%2F%2F

Fig. 4.2 Dr. Penelope Boston poses with Naica cave gypsum rock. Photo courtesy of the NAI/Dr. Thomas Kieft.

all matters relating to Institute-wide research, space missions, the development of relevant technologies, and external partnerships.

As Director of the NAI, Dr. Penny Boston leads the scientific activities of its member teams, and also all operations designed to fulfill its mission to perform, support, and catalyze collaborative interdisciplinary astrobiology research; train the next generation of astrobiologists; provide scientific and technical leadership for astrobiology space mission investigations; and develop approaches which will enhance information sharing between widely distributed investigators.

Dr. Boston has written extensively about the mechanisms that form caves, and their ability to preserve indicators of previous planetary conditions. She has also addressed the potential role of caves as habitats for extraterrestrial life as well as their role in limiting the scope of forward and backward contamination, so much of her work applies to the use of caves for future human exploration.[5]

---

[5]Watch Dr. Penny Boston on YouTube at: https://www.youtube.com/watch?v=yioXvqux7_A

The NAI is just one of six elements in the NASA Astrobiology Program. The others are:

- The Exobiology Program. The goal is to understand the origin, evolution, distribution, and future of life in the universe. Research is centered on the origin and early evolution of life, the potential of life to adapt to different environments, and the implications for life elsewhere. This research is in the context of NASA's ongoing exploration of our stellar neighborhood, and the identification of biosignatures appropriate for in situ and remote sensing applications.
- Planetary Science and Technology Through Analog Research (PSTAR). NASA uses analog missions to research the requirement for integrated interdisciplinary field experiments as an integral part of preparation for future human and robotic missions. Future planetary research for solar system exploration requires the development of relevant, miniaturized instrumentation capable of operating in a wide variety of environments throughout the solar system.
- MatiSSE. The Maturation of Instruments for Solar System Exploration Program supports the advanced development of spacecraft instruments likely to be helpful in future planetary missions. The goal is to develop and demonstrate planetary and astrobiology science instruments to the point that they can be proposed in response to future announcements of flight opportunities without a period of additional extensive technology development. The proposed instrument must address specific scientific objectives of likely future planetary science missions.
- PICASSO. The Planetary Instrument Concepts for the Advancement of Solar System Observations Program supports proposals to develop new spacecraft-based instrument systems that show promise for use in future planetary missions. The goal of the program is to conduct planetary and astrobiology science instrument feasibility studies, concept formation, proof of concept instruments, and technology development to the point that new instruments can be proposed in response to the Maturation of Instruments for Solar System Exploration Program.
- The Habitable Worlds Program. The goal is to apply knowledge of the history of our planet and the life upon it as a basis for investigating the general processes and conditions that establish and maintain habitable environments. It also searches for ancient and contemporary habitable environments to explore the possibility of extant life beyond the Earth. The program includes elements of the Astrobiology Program, the Mars Exploration Program, and the Outer Planets Program.

4.1.5 NASA Goddard Robotics Laboratories

The Goddard Space Flight Center (GSFC) in Greenbelt, Maryland, created two special facilities to answer the question: "How will robots work in space?" The Robotic Operations Center and the Servicing Technology Center act as incubators for satellite servicing technologies. GSFC has facilities in which space systems, components and tasks are tested in simulated environments, honed and refined, and finally declared ready for operation in orbit. Although it is directed mostly at satellite servicing, this facility will be able to accommodate robots intended for use on the Moon or Mars. Its capabilities range from simulating the actions of a robotic arm servicing a satellite in space, to practicing the approach of a satellite to an object (such as a spacecraft or perhaps a rotating asteroid) to evaluate how the sloshing of propellants in tanks in microgravity would affect the behavior of the operation. Mission developers use such data to fine-tune systems, controllers, and systems for optimum performance and environmental interaction. Another facility belonging to the West Virginia Robotic Technology Center (WVRTC) has an additional test area for servicing technologies.

In the past, Goddard has used robotic simulations to demonstrate lunar rover modularity and serviceability, tested sensors for optical recognition and relative navigation, dynamically simulated a Lunar Surface Manipulator System concept, and supported preparations for the fourth Servicing Mission to the Hubble Space Telescope.

This research will have to be coupled with activities at other NASA Centers in order to focus on the requirements of a mission to create a shelter in a cave. What kind of robotic assistance will the astronauts require for potentially hazardous or difficult tasks? Which technologies would be applicable? Goddard could develop prototypes and analog mission profiles using robots.

Goddard also has a team that develops the state-of-the-art for instruments and tools which astronauts could operate. For example, hand-held X-ray fluorescence spectrometers, LIDARs, drills, etc., all of which could be used on missions to the Moon and Mars. A wide range of tools were created to enable Apollo astronauts to deploy the ALSEP stations and function as field geologists, and those missions spent only a few days on the lunar surface. (See Appendix A.2.6)

4.1.6   Langley Research Center (LaRC)

The NASA Langley Systems Analysis and Concepts Directorate is taking the lead in studying how In Situ Resource Utilization (ISRU) will enable sustained human pioneering of Mars.

A study in 2014–2015 addressed three objectives in support of that goal:

- Trade the systems and technology requirements for a sustained human presence on the surface of Mars.
- Explore options for an evolvable Earth-Mars transportation architecture that leverages automation and ISRU in order to reduce costs and risks.
- Integrate surface and transportation concepts into a phased build-up that demonstrates a transition path from systems and technologies available today to a sustained human presence on the surface of Mars.

The resulting paper presented an analysis of the impact of ISRU, reusability, and automation on sustaining a human presence on Mars, involving a transition from Earth dependence to Earth independence. The study analyzed the various surface and transportation architectures in terms of the importance of ISRU and reusability. A reusable Mars lander, named Hercules, would eliminate the need to deliver a new descent and ascent stage with each cargo and crew landing on the planet. This would significantly reduce the mass delivered from Earth. As part of an evolvable transportation architecture this is key to achieving continuous human presence on Mars. The extensive use of ISRU reduces the logistics supply chain required to support population growth at Mars. Reliable and autonomous systems, along with robotics, are required to achieve ISRU architectures, because systems must operate and maintain themselves while the crew is not present.

Langley is also working on two types of habitat for use by astronauts on the surface of Mars: a monolithic habitat for initial crews, and pressurized logistics carriers that can be modified after arrival to augment the habitable volume. The monolithic habitat will be delivered as a payload on either a disposable lander or the Hercules Single-Stage Reusable Lander (SSRL). However, it seems that this work has not considered using caves or lava tubes as shelters. Their initial habitat has a dry mass of 20 tons with a habitable volume of 100 m^3, and requires 30 kWe of electrical power.

Their concept of using pressurized logistics carriers seems compatible with the concept of initially using caves for habitats (which would probably provide many times the volume) as well as their concept for providing a monolithic habitat that would be built on Earth. In their design, each pressurized logistics carrier delivers 5.5 tons of cargo, has a dry mass of 3.2 tons, has a diameter of 3.5 m (11.5 ft) and a length of 8.2 m (27 ft), and needs 1 kWe of electrical power. They will supply food, spare parts, and miscellaneous consumables. In conjunction with equipment delivered by subsequent flights from Earth to outfit the facility, and the expansion of ISRU capabilities by later missions, the surface crew would be increased. See the Reference section for the Langley reports.

4.1.7 Marshall Space Flight Center (MSFC)

For long duration missions on other planetary bodies, the use of in situ materials will become increasingly critical. As human presence on these bodies expands, so must the necessary structures: habitats, laboratories, berms, garages, solar storm shelters, greenhouses, etc. The use of in situ materials will reduce the mass and volumes of launches from Earth. The Habitat Structures project being conducted at the Marshall Space Flight Center under the auspices of the In Situ Fabrication and Repair (ISFR) Program is developing materials and construction technologies to support the development of in-situ structures. This research by the Prototype Development Laboratory (PDL) includes the development of extruded concrete and inflatable concrete dome technologies based upon waterless and water-based concretes, and the development of regolith-based blocks with potential radiation shielding binders including polyurethane and polyethylene, pressure regulation systems for inflatable structures, the production of glass fibers and rebar using a molten lunar regolith simulant, the development of regolith bag structures, and a variety of others that include automation design issues.

NASA's 3D-Printed Habitat Challenge encourages teams of citizen inventors to push the state of the art of additive construction to design and build sustainable shelters to enable humans to live on Mars. Previous levels of this challenge have yielded advanced habitat concepts, innovative material compositions, and printing technologies. The current stage of the multi-level contest challenges participants to use Building Information

Modeling (BIM) software tools to describe both the physical and functional characteristics of a "house" on Mars.

Eighteen teams have submitted their designs and judges have selected the top 10 teams to compete for $100,000 in prize money that will be awarded to the best five teams.[6]

4.1.8   Jet Propulsion Laboratory (JPL)

JPL is well known for its planetary missions, not only from a design standpoint but also from television coverage of its mission operations. Of particular interest here are the following:

- The Mars Odyssey mission was launched in 2001 and is still operating.
- The Mars Exploration Rovers, Spirit and Opportunity. They were both landed in 2004 for nominal missions of 90 days. Spirit became bogged down in soft soil in 2009. Opportunity succumbed to loss of power in a severe dust storm on 2018.
- The Mars Reconnaissance Orbiter was launched in 2005 and continues to operate. It has spotted potential areas for caves and collapsed lava tubes.
- The Mars Phoenix lander set down in the north polar region in 2008 on a mission lasting several months.
- The Mars Science Laboratory, also known as the Curiosity rover, landed in 2012 and continues to operate.
- The Mars InSight mission was launched in 2018 to land later that year. It will deploy a seismometer and drill into the surface to emplace heat-flow sensors.
- Set to launch in July 2020, the Mars 2020 mission which will be the most sophisticated planetary rover ever built.

What is interesting about the 2020 mission is that it reuses the proven design of the Curiosity rover and upgrades the technology. It will land at a site selected to optimize the chances of finding evidence of past or present life.

Of particular relevance to the subject of this book is the Robotics Section at JPL, where over 150 engineers are working on a broad set of issues related to both spaceflight and terrestrial applications. Those that relate to cave oriented missions to the Moon and Mars are discussed in Chapter 5 Technologies.

JPL partners with other NASA Centers, contractors and universities to conduct exploration and research involving lava tubes. Two of their reports include "Lava Tube Exploration Robot and Payload Development,"[7] and "Reference Mission Architecture for Lunar Lava Tube Reconnaissance Missions."[8]

---

[6]For more information on NASA's 3D-Printed Habitat Challenge go to:
http://www.nasa.gov/3DPHab

[7]https://www.hou.usra.edu/meetings/2ndcaves2015/pdf/9027.pdf

[8]https://www.lpi.usra.edu/meetings/caves2011/pdf/8013.pdf

4.2 ESA/EUROPEAN PLANETARY SCIENCE CONGRESS

4.2.1 Cave Research

In addition to its space exploration missions and participation in the International Space Station, ESA is also pursuing an "inner space" program. Several projects involve cave science and astronaut training. Their CAVE (Cooperative Adventure for Valuing and Exercising) human behavior and performance skills research has been active for many years, frequently in partnership with Canada, U.S.A., Japan, and Russia.

CAVES uses the Sa Grutta Cave in Supramonte, a mountain range located in central-eastern Sardinia, Italy. Six astronauts spend up to two weeks deep inside caves, isolated in the dark and cold, doing scientific research and daily tasks as a group, as if they were in space. Moving in the cave system is also comparable to spacewalking because it involves the use of harnesses and safety devices.

Spending time underground might not be the most obvious environment to rehearse spaceflight procedures, but Hans Bolender, who heads up the European astronaut training division explains: "There are many similarities to spaceflight such as a lack of day-night cycle, sensory deprivation, minimal hygiene and the necessity to work as a team and solve problems together."

This project is more thoroughly explained in Appendix 2 Projects.

4.2.2 ESA Astronaut Training

ESA also has PANGAEA (Planetary ANalogue GEological and Astrobiological Exercise for Astronauts). Whereas CAVES focuses on teams and the operational aspects of spaceflight, PANGAEA provides the knowledge and skills required for planetary geology and astrobiology. It is organized into three main areas:

- Earth and lunar geology.
- Sedimentary processes and field activities.
- Volcanism and practical geological self-directed traverses with a focus on planetary protection and geomicrobiology.

The course is designed to provide European astronauts with introductory and practical knowledge of Earth and planetary geology in order to prepare them to become effective partners for planetary scientists and engineers in designing the next exploration missions. The course also aims to provide a solid knowledge in the geology of the solar system from leading European scientists. It will include both field work and classroom instruction, so that when astronauts reach the Moon, Mars, or other celestial body and start to collect samples they will know what they are looking at.

The first session for PANGAEA is organized at the premises of the European Astronaut Center, with a field traverse at the Ries Crater, a circular depression in western Bavaria, Germany (as a lunar analog). The second session is held in the Italian Dolomites (a Mars analog for sedimentary processes). The third session is in the Lanzarote Geopark in Spain (a Mars and lunar analog for volcanism). This project is explained further in Appendix 2 Projects.

4.2.3   EPSC

The 2017 European Planetary Science Congress (EPSC) meeting in Riga, Latvia, featured a discussion about the future of planetary exploration, including asteroid mining, Moon villages, and living in lava tubes on Mars. The 2018 meeting was held in Berlin with over a thousand participants from 44 countries.

Some of the papers by European universities were:[9]

- Next generation of lunar orbiters and radars capable of locating lava tubes.
- Systematic comparison of lava tube candidates on the Earth, Moon and Mars based on high resolution Digital Terrain Models (DTM).
- Effects of gravity on lava tube sizes.
- Use of multi-frequency sounding systems to detect lava tubes.

The next EPSC meeting will be held in Geneva, Switzerland, from September 15–20, 2019.

4.3   UNIVERSITIES

The following are just a few representative examples of U.S. university research which relates to the subject of this book. There are also many foreign universities involved in research of Earthly and planetary caves with their sights on lunar and Martian caves.

4.3.1   Arizona State University (ASU)

ASU is world-renowned for heading up major NASA technology, robotics, and planetary/ astronomy science investigations and missions. At ASU, the study of space, the planets, and the origins of life involves expertise spread across many academic units, including the School of Earth and Space Exploration (SESE) and the School of Life Sciences (SoLS). ASU has a strong multidisciplinary faculty that includes astronomers, geologists, engineers, physicists, microbiologists, cell biologists, tissue engineers, immunologists, vaccinologists, and others to create a superior research environment.

Many ASU researchers participate in NASA projects, most notably the Mars Exploration Rovers, Spirit and Opportunity. Professor Philip Christensen led the Mini Thermal Emission Spectrometer (Mini-TES) used by these rovers, and also the Thermal Emission Imaging System (THEMIS) on the Mars Odyssey orbiter. THEMIS was developed by ASU at Tempe in collaboration with Raytheon Santa Barbara Remote Sensing. Lockheed Martin Astronautics, Denver, was the prime contractor for the Odyssey project, and developed and built the orbiter. Mission operations are conducted jointly from Lockheed Martin and JPL.

At the Mars Space Flight Facility of ASU, scientists and researchers are using instruments on spacecraft at Mars to investigate its geology and mineralogy. The facility also houses the ASU Mars Education Program. This provides workshops, field trips, and other opportunities for teachers and students to join with scientists in the excitement of Mars exploration.

---

[9]Abstracts can be accessed at: https://www.epsc2018.eu/programme/abstract_download.html

Fig. 4.3 Dr. Philip Christensen in the field. Photo courtesy of Arizona State University.

The Lava Tube Database of the Ronald Greeley Center for Planetary Studies is a digital version of a hard-copy collection that includes data sheets, maps, pictures and other documentation of lava tube caves in 34 countries on 6 continents and it is expanded as new results come in. It specifies tube length, floor slope, volcanic complex, and various other parameters, and it can be searched, sorted, and filtered to produce general data for lava tubes.[10]

JMARS is a GIS (Geographical Information System) platform containing all of the Mars orbital data that NASA has ever collected. This program is the backbone of the Mars Student Imaging Project and offers students and teachers the ability to ask questions ranging from relatively simple to extremely sophisticated.[11]

4.3.2 Northern Arizona University

A good example of university research into caves is that of Dr. J. Judson Wynne of the Department of Biological Sciences at the Northern Arizona University. He presented a paper, "The Scientific Importance of Caves in Our Solar System," at the 2nd International Planetary Caves Conference in Flagstaff, Arizona, reviewing some of the ongoing work in

---

[10]The Lava Tube Database is located at: https://rpif.asu.edu/index.php/ltdb

[11]JMARS can be accessed by going to: https://jmars.asu.edu/

Fig. 4.4 Dr. Jut Wynne working at the Sierra de las Nieves National Park, Andalusia, Spain. Photo courtesy of NAU/Dr. Jut Wynne.

caves as well as providing a long list of engineering and scientific advantages of cave shelters and habitats.

Dr. Wynne is supportive of using the Moon as a laboratory for exploring caves as habitats for both the Moon and Mars. He has examined existing orbital imagery and proposed new approaches for detecting caves by integrating visible spectrum, thermal infrared, and gravimetry data. His overall goal is to model cave entrance structures and investigate their origins.

His team's work is focusing on what the early missions will require to have in order to select the appropriate caves for exploration, then adaptation as a habitat. This effort will involve 3D computer vision analysis, a communications system and data link, and a form of power generation for an underground habitat that is isolated from solar energy.[12]

4.3.3 Carnegie Mellon University (CMU)

Carnegie Mellon University is a private research university located in Pittsburgh, Pennsylvania. Founded in 1900 by Andrew Carnegie as the Carnegie Technical Schools, the university became the Carnegie Institute of Technology in 1912 and began granting four-year degrees. In 1967, the Carnegie Institute of Technology merged with the Mellon Institute of Industrial Research to form Carnegie Mellon University. It has seven colleges and independent schools:

---

[12]See the References section for his reports. For a video on "Colonizing the Caves of Mars" go to: https://www.youtube.com/watch?v=BgbNzqKYcnQ&t=37s

- College of Engineering.
- College of Fine Arts.
- Dietrich College of Humanities and Social Sciences.
- Mellon College of Science.
- Tepper School of Business.
- H. John Heinz III College of Information Systems and Public Policy.
- School of Computer Science.

The university also has campuses in Silicon Valley, California, and Qatar, with degree-granting programs on six continents.

The Robotics Institute (RI), a division of the School of Computer Science, is a world leader in robotics research. Dr. William L. "Red"Whittaker is the Fredkin Professor of Robotics, Director of the Field Robotics Center, and founder of the National Robotics Engineering Consortium, all of which are at Carnegie Mellon University. He is also Chief Scientist at RedZone Robotics. He has an extensive record of successful developments of robots for craft, labor, and hazardous duty, including field environments such as mines, work sites, and natural terrain. His work includes the development of computer architectures for controlling mobile robots; the modeling and planning of non-repetitive tasks; complex problems of objective sensing in random and dynamic environments; and the integration of complete field robot systems.[13] His work on caves was previously mentioned in Chapter 4.1.1. NIAC.

The Field Robotics Center (FRC) developed the Sandstorm and Highlander robots that finished second and third in the DARPA Grand Challenge, and also Boss, which won the DARPA Urban Challenge. The RI also partnered with the spinoff company Astrobotic Technology to land a Carnegie Mellon University robot on the Moon in pursuit of the Google Lunar XPrize.[14] The robot, known as Andy, was designed to explore lunar pits that might include entrances to caves.

Field robotics is the use of mobile robots in field environments such as work sites and natural terrain. Such robots are required to safeguard themselves while executing non-repetitive tasks and objective sensing, as well as self-navigation in random or dynamic environments. The strategic goal for the FRC is to push this technology to its limits.

The National Robotics Engineering Center (NREC) is an operating unit in the RI that works closely with government and industry clients to develop and mature robotic technologies from concept to commercialization. A typical NREC project includes a rapid proof-of-concept as a preliminary to an in-depth development and test phase that produces a robust prototype with intellectual property for licensing and commercialization. Throughout this process, NREC applies best practices for software development, system integration, and field testing.

NASA's Small Business Technology Transfer (STTR) program funds research, development, and demonstration of innovative technologies which support future missions. Each of the three proposals secured by Astrobotic and Carnegie Mellon University is valued at $125,000. These proposals are:

---

[13]You can watch his briefing on Extreme Robots at:
https://www.youtube.com/watch?v=7X4-jozFVFo

[14]As of 2018, the Google Lunar XPrize remains unclaimed.

- Long-Range Terrain Characterization for Productive Regolith Excavation seeks to develop sensors and software for precisely detecting minerals in lunar regolith. This technology is key to securing the resources required for deep-space exploration and habitats on the Moon and other planetary bodies.
- Perception and Navigation for Exploration of Shadowed Domains seeks to develop imaging technology that combines visual cameras, light detection and ranging (LIDAR), and thermal cameras in order to perceive, localize, and map a robot's surroundings. This technology will produce detailed 3D models which can be used for navigation in shadowed craters at the lunar poles.
- Subsurface Prospecting by Planetary Drones will develop navigation and perception technology to enable flying robots to explore caves, pits, and a variety of other complicated topography without requiring support from Earth.

Astrobotic has developed custom designs for sensors and rovers for planetary surface activities such as exploration, site preparation, and resource extraction. It developed Polaris as an excavation vehicle that could serve as a robotic precursor for human planetary colonization by preparing terrain and mining ice and other volatiles. The company is currently exploring the use of ultra-efficient drivetrain components for enhanced mobility in the rigorous space environment and is also experimenting with sensing packages and software algorithms to enable safe and rapid traverse. The Polaris rover is a test and demonstration platform for planetary exploration, mobility, and regolith manipulation.

The FM&T department is developing a precise and robust robotic navigation system to enable free-flying vehicles to rapidly explore GPS-denied environments such as lunar caves, skylights, and lava tubes. Key challenges include providing accurate position information from a highly dynamic flying platform, transitions between light and dark environments, and robust operation in all flight scenarios. Visual and LIDAR data is tracked by an incremental algorithm for Simultaneous Localization and Mapping (SLAM). Planned improvements will include the use of Field Programmable Gate Arrays (FPGA), integrated circuits that differ from Programmable Read-Only Memory (PROM) chips in that, as their name implies, they can be programmed in the field. The range of applications includes FPGA acceleration, dense modeling, monocular operation, and deployment in a variety of domains such as underwater, at high altitudes, and in space. The development is being funded by a Phase II NASA SBIR technology development contract.[15]

See Chapter 5 for more on robotics.

4.3.4   University of Hawaii (UOH)

The Geology and Geophysics Department of the University of Hawaii includes the NASA Pacific Regional Planetary Data Center. It is involved with planetary analog sites which include pit craters. Their distribution can be interpreted to be the surface expression of dikes in volcanic areas, and can be used to help define the internal structure of a volcano. In addition, pit craters may be skylights into lava tubes, and thus are of interest in studying

---

[15]For more information, visit https://www.astrobotic.com

possible shelters for astronauts on the surface of the Moon or as possible sites for astrobiological activity on Mars.

There are currently three active volcanoes in the Hawaiian Islands. The main island of Hawaii has Mauna Loa and Kīlauea in the Hawaii Volcanoes National Park. Mauna Loa last erupted in 1984 but Kīlauea has been continuously active since 1983. Loihi is located underwater off the southern coast of Hawaii Island. Erupting since 1996, this emerging seamount may eventually break the surface. Other volcanoes that tourists and researchers can hike and explore are Haleakala on Maui, Leahi (a.k.a. Diamond Head) and the National Memorial of the Pacific at Punchbowl on Oahu.

The university has the advantage of having a very active volcano on its own doorstep to study since its establishment in 1907. Over the years, Kīlauea has caused considerable property damage and destroyed a number of towns. The 2018 eruption resulted in the evacuation of some 2000 residents and the loss of 700 homes.

Western contact with the Hawaiian Islands, and thus written history, began in 1778. The first well-documented eruption of Kīlauea was in 1823. Since then, it has erupted repeatedly. Most historical eruptions were either at the summit or in the eastern rift zone, and were prolonged and effusive in character. Nevertheless, the geological record indicates that violent explosive activity was common prior to the arrival of Europeans.

The university has studied many pits and craters over the years, and this work on formation mechanisms and the interior structures has been of interest to other planetary scientists studying pits and craters on other planets, including the ones that were discussed in Chapter 3. Many of the papers mentioned in the Reference section use the data provided by Hawaiian researchers, who frequently team with colleagues from other universities and organizations, notably the USGS, to carry out field trips. Planetary geologists do not (yet) have the luxury of peering into a Mars pit or cave, so they seek out comparable geology here on Earth to serve as analogs.[16]

4.3.5   Kansas State University (KSU)

KSU geologist Dr. Saugata Datta is part of a NASA study to explore microbial life in lava caves on Earth. A four-wheeled rover called CaveR is investigating accessibility issues to assist a future search for life on Mars. "Studying delicate, extreme closed environments on Earth can give insight into the probabilities of finding more such features that relate to life and [one day] its presence on some extraterrestrial planetary surfaces," he explains.

The CaveR rover also will deploy a NASA instrument that will measure how much water is on the cave walls. This instrument will be used on a future NASA mission to seek ice in caves on the Moon. Such an operation will be led from a command center on the surface, several miles away, by researchers who have not actually been inside the cave.

The Biologic and Resource Analog Investigations in Low Light Environments (BRAILLE) project, which is funded by NASA's PSTAR Program, involves two high-visibility field campaigns in the summers of 2018 and 2019 to undertake an astrobiology mission simulation at the Lava Beds National Monument (LABE).

---

[16]For a short video explaining the creation of lava tubes in Hawaii, go to:
https://www.nps.gov/media/video/view.htm?id=FF3C412C-AA80-39D5-ADB4619491F256FB

Lava tubes are a terrestrial analog for the caves that have been detected on the Moon and Mars. Although the lava tubes at LABE are not a perfect analog when compared to the arid surface conditions on Mars, they still represent a reasonable analog for conditions early in the history of Mars. The BRAILLE project is also the first of its kind to combine CaveR technology with remote science operations in an extreme environment like lava tubes, where there is limited light and other resources. By conducting field research inside lava tubes, the findings will assist NASA to develop future missions to explore caves on other planets to search for evidence of past life and perhaps even current life, mitigating the risks associated with planetary exploration.

The question of whether there is life (specifically, microbial life) elsewhere in the solar system is one of the most important questions in astrobiology research. This goes back to the significance of investigating lava tubes, because caves are sheltered from harmful surface conditions that would otherwise have destroyed evidence of life. If microbes are preserved, they will be in minerals in the cave walls, and will provide NASA with a geological record of the types of microbial communities that are present in lava tubes. If we can characterize the presence of microbial life (and other biosignatures) in terrestrial lava tubes, and also refine the procedures for remote science operations using a rover, we can justify support for future expeditions of extraterrestrial caves.

Dr. Datta's research focuses on what sensors will detect when water, rock, and soil chemistry in caves interact to create biomass; living or once living material. This will give the researchers an idea of the type of microbial community that can survive in a system at

Fig. 4.5 BRAILLE CaveR rover in a lava tube. Photo courtesy of Dr. Saugata Datta, Project BRAILLE, KSU.

a certain temperature and level of humidity. Techniques for examining microbial communities in such environments on Earth is an analog for doing that in extreme environments on Mars.

The KSU team is working with co-investigators from the Planetary Systems Branch at the NASA Ames Research Center, the Intelligent Robotics Group, the Flight Instrument Electronics Division, and others from the University of New Mexico, Northwestern University, and the Desert Research Institute.

4.3.6   University of Texas at San Antonio (UTSA)

The Department of Geological Sciences at UTSA combines academic expertise, research excellence, and student success. Its undergraduate program in geology started in 1976, and the graduate program followed in 1986 with its specialism in water resources and environmental geology. It established the Center for Water Research (CWR) in 1986 to promote a multidisciplinary approach to research on water resources in San Antonio and South Texas. It is headed by Dr. Yongli Gao, whose research features karst hydrogeology and geomorphology. In 2004 UTSA approved a program that emphasized Geographic Information Science (GIS) and Remote Sensing tools. This was followed by a Geoinformatics graduate program specializing in geospatial science and technology. The facilities of the department include biogeochemistry, hydrogeology, isotope geochemistry, micropaleontology and stratigraphy, remote sensing/spatial analysis, river science, and studies in sea ice.

UTSA scientists have investigated a number of caves in Texas, including the Robber Baron Cave, Wuzbach Bat Cave, and Bracken Bat Cave. This work was focused on cave atmospheres as planetary analogs. The atmospheres of caves on the Moon and Mars could provide a record of ongoing geological processes and current biological activity. See the Reference section paper called "In Situ Mass Spectrometer Measurement of Cave Atmospheres as an Analogue to Future Planetary Cave Mission."

4.3.7   University of North Dakota (UND)

The University of North Dakota is a public research university located in Grand Forks. Established by the Dakota Territorial Assembly in 1883, six years before the establishment of the State of North Dakota, it is the State's oldest. UND was founded with a strong liberal arts emphasis, then expanded to include scientific research. It offers a variety of professional and specialized programs, including the State's only schools of law and medicine. Its best known college may be the John D. Odegard School of Aerospace Sciences. This was the first in the U.S.A. to offer a degree in unmanned aircraft systems operations. A number of national research institutions are present on the university's campus, including the Energy and Environmental Research Center, the School of Medicine and Health Sciences, and the USDA Human Nutrition Research Center.

Within the John D. Odegard School is the Department of Space Studies, and within that is the Human Spaceflight Laboratory. It provides pertinent real-world experiences to students from all across the world, formal involvement in graduate and undergraduate research and NASA projects and other activities that relate to human spaceflight. Its particular focus

Fig. 4.6 The Human Spaceflight Laboratory suit and habitat team. Photo courtesy of Dr. Pablo de León.

is the design and production of space suit and habitat prototypes. The North Dakota Experimental-1 (NDX-1) space suit is designed for operating on the surface of Mars, and the NDX-2 suit is meant for testing in lunar simulations. UND is also involved in the design, construction, and testing of an Inflatable Lunar Habitat (ILH), with an attached electric rover. The ILH consists of a frame surrounded by a bladder which isolates the habitat from the ambient atmosphere. The electric rover has a pair of NDX-2 planetary suits attached to it. The combined habitat, rover, and planetary suits are meant to test a concept for operating on either the Moon or Mars. This concept is also compatible with precursor missions to explore caves and lava tubes. In early 2009, a team led by Dr. Pablo de León, a professor in the area of extravehicular activities and space suit design, was awarded a 3-year NASA grant to develop, design, construct, and test advanced inflatable habitat architecture concepts that could be adapted for the surfaces of the Moon and Mars.[17]

The Planetary Exploration Initiative consists of an Inflatable Lunar and Mars Analog Habitat (ILMAH), Pressurized Electric Rover (PER), and two space suits connected externally to the rover via suitports. The main elements are connected to allow the inhabitants of the ILMAH to move into and out of the rover without having to venture "outside." Completed in 2012, the pressurized ILMAH is 40 ft long, 10 ft wide, and 8 ft high. It consists of a rigid frame covered by an inflatable bladder, much like the innermost layers on space suits, to protect the habitat from the surrounding atmosphere. The design allows both the tensile and compressive loads to be transferred from the soft fabric to the

---

[17]For a 5 minute video of Dr. de Leon discussing the pressure suit work go to:
https://video.search.yahoo.com/search/video?fr=tightropetb&p=Pablo+de+Leon#id=3&vid=98b571
8169df7bbf2dd4249e74a9c236&action=click

Fig. 4.7 The NDX-2 prototype suit. Photo courtesy of UND Department of Space Sciences.

rigid frame, avoiding punctures or penetrations. The inflatable material is malleable and retains strength during folding. It is also lightweight and stows in a significantly smaller volume, which is an important feature for a long-term spaceflight to Mars. This expandable soft goods structure offers a lower mass solution with increased volume in contrast to the use of metal or rigid composite materials. The habitat can support a crew of four people for up to thirty days. There are four bedrooms, a galley, dining area, bathroom with shower and toilet, and a lab area for scientific work. The ILMAH is outfitted with extensive internal climate control to maintain its crew in relative comfort.

4.3.8 University of New Mexico (UNM)

The University of New Mexico is a public research university in Albuquerque. Founded in 1889, it provides bachelors, masters, doctoral, and other professional degree programs in a wide variety of fields. In addition to the 600-acre campus in Albuquerque, there are branch campuses in Gallup, Los Alamos, Rio Rancho, Taos, and Los Lunas. UNM offers a variety of academic programs through twelve Colleges and Schools. The academic options include in excess of 215 degree and certificate programs.

The Arts and Sciences Department is the largest single research unit on main campus. It attracts roughly $50 million in new and continuing external research funding per annum from agencies that include the National Institutes of Health, the National Science Foundation, the National Endowment for the Humanities, the Department of Energy, the Department of the Interior, and the State of New Mexico. In addition, many private foundations and non-governmental agencies provide faculty and student fellowships.

There are several research organizations:

- Sevilleta National Wild Life Refuge/Long-Term Ecological Research.
- U.S. Long-Term Ecological Research Network.
- Earth Data Analysis Center (EDAC).
- Established Program to Stimulate Competitive Research (EPSCoR).
- Resource Geographic Information System (RGIS).
- The Statistics Clinic.

The UNM Department of Biology is one of the biggest academic units in the State. Collectively, research activities generate $6–13 million per annum and the department hosts some of the university's most prestigious academic programs, particularly in ecology and evolution, embracing diverse research in ecosystems ecology, plant population biology, behavioral ecology, metabolic ecology, and a number of collections-based studies with the Museum of Southwestern Biology. In addition, the department has fostered the concept of a single large interactive department spanning the spectrum of modern biology, one that blurs traditional boundaries and favors collaborative and multidisciplinary approaches.

Dr. Diana E. Northup is an Associate Professor of the Department of Biology, College of University Libraries and Learning Sciences. She and her colleagues have been studying life in caves since 1984, particularly the geomicrobiology of cave ferromanganese deposits, calcium carbonate formations called pool fingers, the microbial diversity of sulfur caves, and lava caves worldwide. Dr. Northup's teams have been studying Subsurface Life In Mineral Environments (SLIME), seeking to answer:

- How microbes help form the colorful ferromanganese deposits that coat the walls of Lechuguilla and Spider Cave in Carlsbad Caverns National Park.
- How these deposits compare to surface desert and rock varnish coatings.
- How microbes participate in the precipitation of the calcium carbonate formations called pool fingers and the microbial diversity located in the Cueva de las Sardinas hydrogen sulfide cave in Tabasco, Mexico.

These projects are being funded by the National Science Foundation Life in Extreme Environments Program and the NSF Geosciences Directorate.

Dr. Northrup has recently been honored by having her research featured on the NOVA popular science television series and by being asked to serve as a Guest Editor for a special issue of *Geomicrobiology Journal* on "Geomicrobiology of Caves." In addition, she is a recent recipient of a grant from the Charles A. and Anne Morrow Lindbergh

Fig. 4.8 Dr. Diana Northrup in the field. Photo courtesy of UNM

Foundation to study the impact of human activities on the microbial populations in caves.[18]

Another group of UNM researchers joined an international, multidisciplinary team to create a new paleoclimate rainfall record that highlights the contribution of man-made industrial emissions to reduced rainfall in the northern tropics.

The team, which included UNM Professor Yemane Asmerom, Senior Research Scientist Victor Polyak in the Department of Earth and Planetary Sciences, and Associate Professor Keith Prufer and graduate student Valorie Aquino from the Department of Anthropology, reconstructed a rainfall record stretching back 450 years from unprecedented speleothem samples in the Yok Balum cave in Belize using uranium thorium dating from the chemical composition in the speleothems containing aragonite, which is high in uranium. Their results, "Aerosol forcing of the position of the intertropical convergence zone since AD 1550," published by *Nature Geoscience* in 2015, identified the effects of from man-made use of fossil fuels. In particular, they noted a significant drying trend that started around 1850 and paralleled the steady rise in sulfate aerosols in the atmosphere resulting from the industrial boom in Europe and North America.

4.3.9 New Mexico Tech/NCKRI

New Mexico Tech has over a dozen research divisions which work with private industry, government agencies, and other universities to employ undergraduates, generate opportunities for graduates, and contribute ground-breaking research to the scientific canon and the world of industry, science and engineering.

---

[18]Watch Dr. Diana E. Northup on YouTube at:
https://www.youtube.com/watch?v=XuIYzZcvW7w&t=20s

The university has affiliated science and engineering centers, one of which is the National Cave and Karst Research Institute (NCKRI) established by the U.S. Congress in 1998 as the national authority on caves and karst with a mission to:

- Further the science of speleology.
- Centralize and standardize speleological information.
- Foster interdisciplinary cooperation in cave and karst research programs.
- Promote public education.
- Promote national and international cooperation in protecting the environment for the benefit of cave and karst landforms.
- Promote and develop environmentally sound and sustainable resource management practices.

NCKRI was created in partnership with the federal government through the National Park Service, the City of Carlsbad, New Mexico (the city in which it is headquartered), and the State of New Mexico through the New Mexico Institute of Mining and Technology (a.k.a. New Mexico Tech; NMT), which administers NCKRI. New Mexico Tech in Socorro, New Mexico, has jokingly been called a research institution that happens to have a university; in fact, that is not far from the truth. NCKRI is one of more than a dozen research divisions at NMT which partner with private industry, government agencies, and other universities. Nearly all professors in every academic department have active research projects which involve students working on undergraduate and a range of advanced degrees.

NCKRI was initially an institute within the U.S. National Park Service, but in 2006 it was reformed as a non-profit 501(c)(3) corporation in order to increase its flexibility in pursuing its mission. Its work addresses hydrogeology, geophysics, diverse environmental management issues, and geomicrobiology. It also carries out public and academic education programs, and hosts a variety of conferences. Its Karst Information Portal (www.karstportal.org) is a free, on-line international cave and karst reference library.[19]

4.3.10   St. Mary's University

St. Mary's University is a Catholic and Marianist liberal arts institution in San Antonio, Texas, with a highly regarded master's level school. Its cave research has been conducted by the Department of Physics and Environmental Science, whose staff include a number of leading scientists in the field.

Dr. Evelynn Mitchell was awarded a doctorate by the Environmental Science and Engineering program at the University of Texas at San Antonio. Before that she taught physics at the University of the Incarnate Word and gained experience in engineering and applied physics working as a manufacturing engineer at Sony Semiconductor from 2000 to 2003. She studied the Edwards Aquifer in Texas as part of her dissertation research, combining hydrogeology and geophysical skills to determine specific storage values using the compression of earthquake waves. She has also used seismic and resistivity applications in search of the underlying geology and water table in the Texas Hill Country and in Jalisco,

---

[19]For more information on NCKRI visit www.nckri.org

Mexico. As a professor at St. Mary's, she has worked with students to analyze levels of carbon dioxide in local caves, using ground penetrating radar to detect voids, and using electrical resistivity to investigate subsurface locations. Mitchell also applies her knowledge of hydrogeology by representing St. Mary's on the West Side Creeks Oversight Committee.

Mitchell has teamed with her husband, Joseph N. Mitchell, at the Southwest Research Institute, headquartered in San Antonio, to conduct research on carbon dioxide cycles in caves. They have also collaborated on the use of the Cave Mass Spectrometer to detect atmospheric constituents that might have applications for detecting microbial biological life on other planetary bodies. In addition, she has used ground penetrating radar and resistivity data to detect cave passages.

Melissa Karlin was an aviation environmental science compliance specialist for the county government in south Florida and an environmental scientist for a private firm. She also worked as a grant writer and served as an adjunct faculty member who taught introductory environmental science courses. She gained her doctorate through the Infrastructure and Environmental Systems program at the University of North Carolina at Charlotte. In her dissertation Karlin reported on the spatial ecology of the endangered red wolf (Canis rufus), incorporating both geographical information system and remote sensing science to study dispersal, interspecific interactions, and habitat use by the red wolf in North Carolina. She has also studied disease transmission in the threatened gopher tortoise (Gopherus polyphemus) in southeastern Florida, as well as home range patterns and burrow distribution of this species.

At St. Mary's, Karlin worked with Dr. Evelynn Mitchell on the Robber Baron Cave study on seasonal variations in species and the impact of human visitors to the cave. In teaching ecology, wildlife research, geographic information systems and experimental design she incorporates extensive field work undertaken by her undergraduate students.

See the Reference section for papers related to cave research from St. Mary's University.

4.4   MARS INSTITUTE AND SETI INSTITUTE

4.4.1   Mars Institute

The Mars Institute is a non-profit research organization dedicated to the scientific exploration and public understanding of Mars. Planetary scientist Dr. Pascal Lee co-founded and chairs the Mars Institute.

Research at the Mars Institute focuses on Mars and other planetary destinations that could serve as stepping stones to Mars, in particular our Moon, the two small moons of Mars, and near-Earth objects. It investigates technologies and strategies that will facilitate and optimize the future human exploration of Mars. It operates the Haughton-Mars Project (HMP) Research Station on Devon Island, Nunavut, Canada, in the High Arctic. This is currently the largest privately operated polar research station in the world, and the leading field research facility for planetary science and exploration.

In relation to cave exploration, the Mars Institute is a leader in research which applies new technologies in the advancement of cave exploration on the Earth, the Moon, and Mars; in particular using Unmanned Aerial Vehicles (UAV) or drones. Together with

NASA, the SETI Institute, and industry partners such as Astrobotic and FYBR,[20] the Mars Institute has been investigating the utility of UAVs in the context of both human and robotic exploration. The functions for drones include scouting, surveying and sampling, and these are being actively researched through field tests at analog sites.

For the exploration of the Moon and high-altitude sites on Mars where there is insufficient atmosphere to operate airfoiled drones, Mars Institute scientists are investigating the use of rocket-thrustered UAVs. If kept small and nimble, these could be used to explore lava tubes and other caves. Most lava tubes, crater pits, and skylights identified on Mars thus far are located on the flanks of Mars' giant volcanoes, at altitudes where the air is very tenuous, making the rocket-thrustered drone approach more practical than an airfoiled one.

4.4.2   SETI Institute

The SETI Institute is a 501(c)(3) non-profit scientific research institute based in Mountain View, California. It is a key research contractor to NASA and also the National Science Foundation, and collaborates widely with industry partners. It was founded in 1984 and currently employs more than 130 scientists, educators, and administrative staff. Its work is anchored primarily at three centers: the Carl Sagan Center for the Study of Life in the Universe, the Center for Education and the Center for Outreach.

There are six Research Divisions:

- Astrobiology.
- Astronomy & Astrophysics.
- Climate and Biogeoscience.
- Exoplanets.
- Planetary Exploration.
- SETI.

Several of these themes relate to caves and lava tubes and the possibility of biology within them. There is also the Frontier Development Laboratory (FDL), which is an applied artificial intelligence research accelerator and public/private partnership between the SETI Institute and the NASA Ames Research Center.

In 2018, SETI Institute planetary scientist Pascal Lee reported the discovery of candidate impact-melt lava tubes and skylights on the floor of Philolaus Crater near the north pole of the Moon. The discovery was reported following a survey of hundreds of images of the polar regions acquired by the Lunar Reconnaissance Orbiter. If verified, the lava tubes in Philolaus would be the highest latitude caves found on the Moon to date. In 2018, Dr. Lee was also awarded a grant from the Dubai Future Foundation to begin a systematic survey of what is known about the geotechnical characteristics of lava tubes on Earth and the conditions under which some harbor ice, plus their potential implications for lava tubes on the Moon and Mars. As a case study, and with additional funding support from NASA, Dr. Lee led a team to the 3500 year-oldLofthellir Lava Cave complex in Iceland, one of the relatively few ice-rich lava tube systems known on our planet. They made the

---

[20]FYBR Solutions is a geospatial data company.

Fig. 4.9 Entrance skylight of the ice-rich Lofthellir Lava Cave system in Iceland. Photo courtesy of Pascal Lee.

Fig. 4.10 The team that studied the Lofthellir Lava Tube/Ice Cave in Iceland in 2018. L-R: Pascal Lee (Mars Institute, SETI Institute), Julian Watson (WAG-TV), Eirik Kommedal (Mars Institute, SETI Institute), Kerry Snyder and Erik Amoroso (Astrobiotic), Tony Mews and Daddi Bjarnason (WAG-TV) and Anton F. Birgisson, the guide. Photo courtesy of Tom Mews..

Fig. 4.11 3D mapping of the ice-rich Lofthellir Lava Cave, Iceland, using Astrobotic's lidar-equipped drone. Photo courtesy of Pascal Lee.

first 3D map of a lava tube using a lidar-equipped drone developed by the space exploration company Astrobotic. Lessons learned and insights gained are helping to create a rigorous framework for assessing the habitability of lava tubes on the Moon and Mars, and the resources they might offer.

Other aspects of planetary habitability are also being investigated at the SETI Institute. For instance, scientists Nathalie A. Cabrol and Edmond A. Grin, have edited the book "From Habitability to Life on Mars," published by Elsevier in 2018. Its authors were directly involved in past, current, and forthcoming Mars missions and they explore the current state of knowledge and questions about the past habitability of Mars, and the role that rapid environmental changes might have played in the transition of prebiotic chemistry to life. The preservation of biosignatures in the geological record will help to determine future exploration strategies. The authors point out that investigations of terrestrial analogs to early Martian habitats under various climates and environmental extremes will provide critical clues to the understanding of where, what, and how we should search for biosignatures on Mars.

4.5 PLANETARY SOCIETY

The Planetary Society is an American internationally active, non-governmental, non-profit foundation which undertakes research, public outreach, and political advocacy for engineering projects related to astronomy, planetary science, and space exploration. Their support for Mars exploration is well known, as is their influence on governmental and public opinion. It is hoped that this book will be received by them as a form of public opinion advocating the use of caves on the early missions to the Moon and Mars.

One of the Society's principles is to prioritize human spaceflight technology development in areas that sustain human psychological and physical health for long duration spaceflight. Another principle is to place emphasis on technology development to ensure the future success of deep space exploration, including a landing on Mars. They are concerned that many key technologies necessary for the long duration survival of humans in space have not yet been demonstrated. In particular, we ought to prioritize development of technologies that relate to human physical and psychological health.

Many people have their sights set on the ultimate program goal of landing on Mars and exploring the planet, but few people consider other options for the first few missions where the use of caves, lava tubes, or other geological features will provide the protection that astronauts will need without having to develop all the infrastructure currently envisioned for surface habitats. If the initial use of caves proves satisfactory, perhaps the introduction of surface habitats could be put off until a later phase of Mars exploration.

In late 2018 the Society emphasized the need for an orderly transition from the International Space Station toward future deep space exploration, particularly the need to provide adequate budgets in order not to delay a mission to Mars. The ISS represents a significant annual cost of approximately $4 billion to operate, supply, and crew. NASA and its partners do not currently have the budget to sustain the ISS and also to develop a robust human deep space exploration program. If new funding is not forthcoming, then the nation's priority must be on the deep space exploration effort and NASA must transition away from its primary funding for, and management of, the ISS. This would free up critical resources to be spent on projects that directly support efforts to send and sustain human missions beyond low Earth orbit.

Certainly, the use of caves as shelters would significantly reduce the cost of those first missions back to the Moon and on to Mars.

4.6 U.S. GEOLOGICAL SURVEY

The U.S. Geological Survey's Astrogeology Science Center was established in 1963 in Flagstaff, Arizona, by Eugene Merle Shoemaker to provide geological mapping of the Moon and to assist in the training of Apollo astronauts. It is the home of astronaut Harrison H. "Jack" Schmitt, as well as many of the scientists who trained the Apollo astronauts such as Keith A. Howard, Don Wilhelms, and Raymond Batson. The Astrogeology Science Center continues to provide support for past, present, and future space missions. It serves the nation, the international science community, and the American public in the pursuit of knowledge about our solar system.

The USGS has worked with NASA and other space agencies to lead scientific investigations, select rover landing sites, create geological maps and cartographic products for numerous missions across the solar system, including many to Mars. It possesses the largest library for Earth sciences in the world.

Fig. 4.12 Michael Dunn, Director of 4th Planet Logistics. Photo courtesy of 4th Planet Logistics.

A good example of the contribution of the USGS to the subject of this book is the work of Glen E. Cushing entitled "Candidate Cave Entrances on Mars." This used images from the visible-wavelength instrument of Mars Odyssey's Thermal Emission Imaging System (THEMIS) and the Context Camera (CTX) carried by Mars Reconnaissance Orbiter. Cave candidates were then examined by the much more powerful High-Resolution Imaging Science Experiment (HiRISE) on Mars Reconnaissance Orbiter. Another example is the Astrogeology Science Center's support for analog research, such as the DRATS (NASA's Desert Rats) training operations and their research on Hawaiian volcanoes by Laszlo Kestay and Greg Vaughan.

The main science component of the Cartography and Imaging Sciences Node (IMG) of the NASA Planetary Data System (PDS) is located at the Astrogeology Science Center. Engineering support and technical management for the IMG is provided by a partner facility at JPL. The IMG is a PDS science discipline node which archives over 500 terabytes in a growing collection of image and ancillary data provided by dozens of past and present space missions. It serves the NASA planetary science community with digital image archives, necessary ancillary data sets, software tools, and the technical expertise to fully utilize the vast collection of digital planetary imagery.[21]

Wherever NASA chooses to go in the solar system, the USGS provides basic information to the planning committees and plays a major role in the mission.

---

[21]The Planetary Data System can be accessed by going to:
https://pds-imaging.jpl.nasa.gov/index.html

4.7 4TH PLANET LOGISTICS

4th Planet Logistics is an Illinois registered limited liability company which was founded for the purpose of designing, building, and evaluating human habitats and related support components for utilization on the Moon, Mars, and beyond. Their Mars Lava Tube Pressurization Project (MLTPP) is seeking to develop a range of technologies and strategies to enable human colonies and bases on the Moon and Mars exploiting existing ancient lava tubes. Their goal is to create and test various habitats using terrestrial lava tubes as analogs. Using existing caves will reduce the amount of construction materials that must be launched from Earth and landed on the surface of the target. MLTPP's goal is to evaluate the feasibility of pressurizing terrestrial lava tubes directly by creating atmospheric barriers that use the lava tube's naturally occurring regime as a primary or secondary structural shell. This will involve the use of 3D printable, locally derived materials and the appropriate robotic assembly techniques.

The MTLLP intends to utilize the Stefanshellir cave in Iceland for feasibility studies of turning lava tubes into shelters. It is one of the largest caves in Iceland, situated in a large flood basalt lava field called Hallmundarhraun about 144 km (90 mi) from Reykjavík. The team will carry out detailed surveys and model the lava tubes to obtain the information to develop systems that can employ additive manufacturing and robotic assembly methods to demonstrate making a barrier that would be capable of pressurizing and insulating a habitat.[22]

---

[22]For a 27 minute video of Mike Dunn's talking about the conversion of lava tubes to shelters, go to: https://www.youtube.com/watch?v=A9CvkwXbWJk

5

Technologies

5.1 INTRODUCTION

All of the work undertaken by NASA Centers is reflected in the NASA Strategic Technology Investment Plan (STIP), which is goal driven. This in turn, provides guiding principles for investment in the technologies that are documented in the NASA Technology Roadmaps. The updated roadmaps are organized into 15 key technology areas and a cross-technology one. The roadmaps list the technologies that the Agency could develop. This portfolio includes technology development programs and projects from each mission directorate and office involved in the development of technology, and spans the entire technology maturity life cycle, including a variety of early-stage conceptual studies which are pursuing entirely new technologies.

Technology Readiness Level (TRL) is a scale used to provide a measure of technology maturity:

- Level 1: Basic principles observed and reported.
- Level 2: Technology concept and/or application formulated.
- Level 3: Analytical and experimental critical function and/or characteristic proof of concept.
- Level 4: Component and/or breadboard validation in a laboratory environment.
- Level 5: Component and/or breadboard validation in a relevant environment.
- Level 6: System/subsystem model or prototype demonstration in a relevant environment (ground or space).
- Level 7: System prototype demonstration in an operational (space) environment.
- Level 8: Actual system completed and (flight) qualified through test and demonstration (ground and space).
- Level 9: Actual system (flight) proven by successful mission operations.

The portfolio encompasses both near and long-term developments, and enables the discovery and advancement of technologies that could fundamentally change the way we live and explore our world and the universe. However, the roadmaps catalog many more

© Springer Nature Switzerland AG 2019
M. von Ehrenfried, *From Cave Man to Cave Martian*,
Springer Praxis Books, https://doi.org/10.1007/978-3-030-05408-3_5

technologies than can be funded in the technology portfolio. The Agency must therefore understand how each technology might be of benefit to national goals and specific space missions, and carefully allocate its technology investments.

Currently, there is no defined mission that specifically describes an exploration mission to the Moon or Mars which includes exploring caves and converting them into habitats. Only by reviewing the ongoing work in the various technology areas does the scope for this application become apparent. For example, the technology area In-Situ Resource Utilization discusses infrastructure, and the one on Human Mobility Systems discusses rovers and tools. Likewise, Habitat Systems contains a section on "smart" habitats and habitat evolution, although most of this work is for in-space and surface habitats. It is envisioned that the work would most likely be applicable to subsurface habitats such as caves and lava tubes.

This book will extract the technology details from the NASA work that apply to a mission which involves the exploration of lunar and Mars caves for habitation underground. It is clear from the ongoing research that nearly all the technology elements are in work, although not necessarily focused on the first missions that might involve a cave or lava tube. The Technology Roadmaps focus on the most advanced missions. The following are just a few examples of recent and ongoing research that could be directly applied to living in caves on the Moon and Mars.

5.2 RADIATION AND SHIELDING

5.2.1 Introduction

Space radiation exposure has clinically relevant implications for the lifetime of a crew in space. It is currently estimated that long-term missions could increase the likelihood of fatal cancer anywhere from 5 to 21 per cent above the baseline risk. Radiation health and performance risks include carcinogenesis, acute syndromes, acute and late Central Nervous System (CNS) effects, and degeneration of tissue. The challenge for crew safety requires "risk limits" to be sufficiently understood, defined, and properly implemented. The purpose of the radiation sub-goals is to ensure that the radiation risk to astronauts is reduced to As-Low-As-Reasonably-Achievable (ALARA). The focus is on developing technologies to increase crew mission duration in the free-space environment while upholding the Permissible Exposure Limits (PEL) for radiation. NASA's development objectives center on reducing uncertainty in identifying the risk of death due to exposure to radiation, extending the number of safe days in space using either shielding or Biological Countermeasures (BCM), improving the ability to predict future space weather events and their duration in order to prepare and protect a crew, and developing small, low-mass, low-power, crew-friendly sensors for monitoring the radiation environment. It is widely accepted that extending crew mission duration in free space will require improved understanding of risks, better ways of predicting and monitoring radiation, and using a combination of shielding and BCM.

For many years, it has been suggested that lava tubes on the Moon and Mars could not only shelter a human base from natural hazards such as radiation and meteoritic material, but also eliminate the diurnal extremes of temperature on the exposed surface by maintaining an almost constant temperature. An analysis of radiation safety issues for lunar lava tubes has been carried out for the influx of Galactic Cosmic Rays (GCR) and Solar Particle Events (SPE), the latter being a transient increase in energetic protons levels, and large Coronal Mass Ejections (CME) that follow on from solar flares. There is no current ability to predict the onset, intensity, and evolution of a SPE or the arrival time of the material from a CME. Although SPE exposure can, in theory, be mitigated by passive shielding, the mass constraints on a spacecraft will limit this mitigation strategy for a long duration mission. It is therefore essential to develop tools capable of forecasting the occurrence and magnitude of such events to enable Mission Control to alert deep space exploration crews to impending danger. The forecasting tools would have to project ahead for time windows ranging from days to weeks, and provide a high degree confidence and a low likelihood of false alarms. While the risks of radiation are intrinsic to space travel, they need not be continued on the surface if it is possible to go underground.

In order to tackle all of the technology challenges with regard to spaceflight, NASA created Technology Roadmaps for all of its projected exploration targets. Technology Area 6, Human Health, Life Support and Habitation Systems, looks at what NASA and many of their contractors have been doing in this area. Many NASA Centers have an interest in space radiation, but the Johnson Space Center in Houston, Texas, has the prime role in those areas affecting the crew.

This section addresses areas that are related to the theme of this book, namely a "precursor" mission whose objective is to seek subsurface shelter from the hostile environment.

The Technology Roadmap for Human Health, Life Support and Habitation Systems identified five topics:

- Environmental Control and Life Support Systems and Habitation Systems.
- Extravehicular Activity Systems.
- Human Health and Performance.
- Environmental Monitoring, Safety and Emergency Response.
- Radiation.

These were further broken down into many sub-areas, each of which had many sub-goals. As a result, there are almost 200 desired capabilities at different levels of technology readiness. Some of these areas are generic to crew safety, but some relate directly to the shielding of astronauts and mitigating their risks of radiation, both in transit to and on the surface of the Moon or Mars. There are some that are directly applicable to protecting the crew by exploration of caves, lava tubes, and other potentially suitable features for shelters. Not all of the areas can possibly be covered in this book. The following is just one example that is being investigated and has interesting new technology applications.

5.2.2 Radiation

The radiation area is focused on developing technologies to increase crew mission duration (100 to 1000 days, depending on the mission) in the free-space radiation environment while not exceeding the space radiation Permissible Exposure Limits (PEL). It is generally accepted that an integrated, optimized approach that utilizes shielding options, biological countermeasures, an improved understanding of the risks, and tools to improve prediction and monitoring of radiation will be required in order to achieve the sub-goal of extending crew mission duration. The primary radiation technologies in need of advancement to achieve this sub-goal include:

- Risk Assessment Modeling: The focus is to develop tools which enable, quantify, and reduce uncertainty in assessing astronaut risk due to space radiation exposure and to improve mission planning, mission operations, and system design for missions in low Earth orbit and into deep space to the Moon and Mars.
- Radiation Mitigation and Biological Countermeasures (BCM): The focus is to develop BCMs that can minimize or prevent physical, cognitive and behavioral disorders due to space radiation without incurring adverse side effects and loss of life.
- Protection Systems: The focus is to advance the design of new, integrated radiation protection shielding technologies. That is, to provide passive or active shielding through design advances, advanced materials, lightweight structures, and in-situ resources. Specifically, the objective is to provide reasonable (mass and power) shielding together with countermeasures for 1000 safe days during Mars missions.
- Space Weather Prediction: The focus is to advance improvements in Solar Particle Event forecasting and associated alert systems that will minimize operational constraints for missions outside the protection afforded by the Earth's magnetic field.
- Monitoring Technology: The focus is to prototype and mature advanced, miniaturized technologies for radiation measurement, and to demonstrate them as integrated vehicle systems using available platforms.

5.2.3   Protection Systems

Since we are interested in caves and lava tubes for protection, let's examine one of the above items a little further, namely Protection Systems. While the crew is exploring the Moon or Mars they will be exposed unless they are in a rover that has been designed to protect them. Chapter 5.4 of this book (Robotics) discusses rovers; for example, the updated Space Exploration Vehicle (SEV) and the more advanced Multi-Mission Space Exploration Vehicle (MMSEV) concept vehicle. The final vehicle will most assuredly have a lot of radiation shielding because it will be used as a temporary habitat until more lasting protection can be found; in this case a cave, lava tube, or other suitable geological structure.

The technology areas being developed include the following:

- Develop and advance structural performance of high-hydrogen-content materials and other material systems in order to supersede the traditional materials for the primary and secondary spacecraft structure. This area is also being studied by those in the material technologies area. The goal is to replace existing mass with better mass for radiation protection which also meets structural requirements. High-hydrogen materials can include polymer matrix composites, where the polymer and/or fibers are high in hydrogen content.

- Develop vehicle equipment and components using radiation-protective materials. This in-situ passive shielding technology will employ stowed hydrogen-rich logistics (such as food, water sources, other supplies, and waste) as multipurpose shielding. Develop multi-purpose containers for bio-materials that can make use of human waste without affecting crew (odor, leakage, and handling transfer).
- Develop in-situ passive shielding using planetary surface materials; for example regolith manipulation; the processing of building materials; the characterization of regolith for utility as shielding; the transformation of regolith into shields, structures and tunnels; and additive manufacturing using regolith material. Regolith can be consolidated by sintering using conventional, solar, microwave or laser heating. It can also be combined with polymer-matrix materials for an increased hydrogen content. (Note that this can be avoided by using caves and lava tubes.)
- Develop high-temperature superconductor technology and performance for active shielding. Structural components will require technologies to keep the magnet at its cold operating temperature and to prevent it from flexing. Other developments are high-temperature superconductors and splicing technologies. (This can also be avoided by using caves and lava tubes.)

5.2.4   Shielding Material

Shielding will be required in transit to the Moon and Mars, while on the surface exploring for sites, while in pressure suits, while in a habitat, and even while in a cave or lava tube.

There are many ideas for radiation shielding, some applying to the spacecraft and others to surface vehicles and habitats. Some such as creating a magnetic or electrostatic field around the crew are unrealistic any time soon. In terms of the surface and subsurface, it appears that one particular concept is more likely than any others to be effective and efficient from a cost and mass standpoint, especially in the context of using caves and lava tubes for protection.

One material currently in development at NASA's Langley Research Center is hydrogenated boron nitride nanotubes, known as hydrogenated BNNTs, or even super-hydrogenated BNNTs. These tiny, nanotubes made of carbon, boron, and nitrogen have hydrogen interspersed throughout the interstices between the tubes. Boron is also an excellent absorber of secondary neutrons, making hydrogenated BNNTs ideal for making shielding. Neutrons are released as secondary radiation when the GCR and SPE interact with the walls of a structure (spacecraft, lander, rover or habitat) and also with regolith material. A neutron flux poses a threat to humans because it can cause radiogenic cancers. Hence there is immense scope for using BNNT for multifunctional radiation shielding structural materials for a space exploration architecture.

In 2013 the NASA Langley Research Center, Jefferson Sciences Association (JSA) and the National Institute of Aerospace (NIA), cooperatively synthesized long, highly crystalline boron nitride nanotubes (BNNT) using a novel pressure and vapor condensation method. This synthesis technique was then licensed by BNNT, LLC, which built a factory in Newport News, Virginia, to start creating commercial products. In 2016, this team won the Government Invention of the Year Award for "Boron Nitride Nanotubes."

BNNT material is made up entirely of low atomic number (Z) atoms. Boron (Z = 5) and nitrogen (Z = 7) are larger than hydrogen (Z = 1) but are still small, and certainly much smaller than aluminum (Z = 13). It can theoretically be processed into structural BNNT, which is thermally stable up to 800°C (1472°F) in air. In addition to being stable at high temperatures, it has extraordinary strength. This enables it to be employed for load bearing structures. Furthermore, its molecules can be incorporated into high hydrogen polymers and the result used as a matrix resin for structural composites. The BNNT molecular structure is attractive for hydrogen storage. And, as noted, boron has one of the largest neutron absorption cross sections of all the elements in the Periodic Table, and nitrogen has a larger neutron absorption cross section than carbon. Incorporating the boron-10 isotope into the boron nitride would provide even better protection against neutrons.

A NIAC study conducted at the Langley Research Center listed the benefits of hydrogenated BNNT for space applications as follows:

- Hydrogen is effective at fragmenting heavy ions found in Galactic Cosmic Radiation (GCR).
- Stopping protons from Solar Particle Events (SPE).
- Slowing down the neutrons which are issued as secondary radiation when GCR and SPE interact with matter.
- Hydrogen is not a structural material. Polyethylene (CH_2) contains a lot of hydrogen and is a solid material, but it does not possess sufficient strength for load bearing aerospace structural applications.
- Researchers have successfully created yarn out of BNNT which is flexible enough to be woven into fabrics.

The NIAC report implied that the aerospace industry appears to be stuck with aluminum alloys for primary structures retrofitted with polyethylene or water for radiation shielding, but that the application of BNNT technology, processed into composites, fabrics, yarn, and film forms, may solve the problem of radiation as well as many other space problems. These materials may well be applicable to space suits, surface vehicles, and habitats.[1]

5.3 INFLATABLES AND OTHER HABITATS

5.3.1 Introduction

The concepts and work on inflatable and expandable structures dates back more than half a century. Wernher von Braun had concepts for inflatable space stations in the 1950's. German architect Frei Otto published a book *Tensile Structures* in 1962 in which he

---

[1]Watch the following videos for more information:

(a) Dr. Sheila Thibeault on NASA Radiation Shielding Materials Containing Hydrogen, Boron, and Nitrogen https://www.youtube.com/watch?v=ADA-FtQ_Vno

(b) Dr. Catharine Fay on Boron Nitride Nanotube – Tiny Tube With Great Potential https://www.youtube.com/watch?v=CoHSNiZqwEY

(c) Dr. Catharine Fay on Boron Nitride Nanotube Development https://www.youtube.com/watch?v=r25RMceegKM

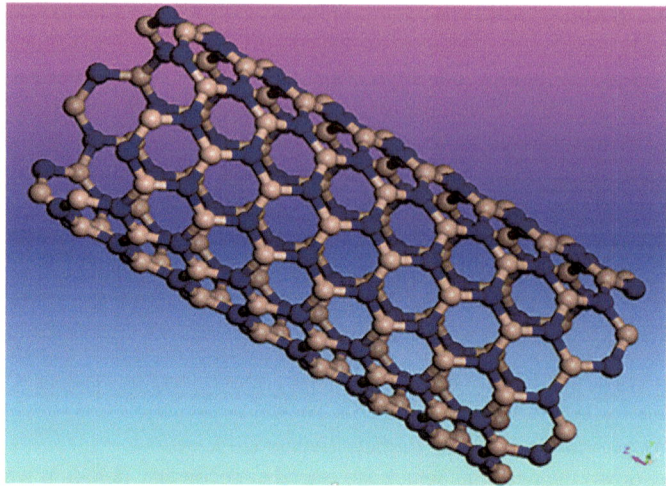

Fig. 5.1 The BNNT structure. Photo courtesy of NASA.

Fig. 5.2 A BNNT "puff ball." Photo courtesy of BNNT, LLC

described the attributes and design consideration of pneumatic and air supported structures. Expandable structures contributed to the early years of the communications satellite
era in the form of passive relay balloons: Echo 1 being orbited on August 12, 1960 and

then Echo 2 on January 25, 1964. Also in 1964, F. W. Forbes of the Air Force published *Expandable Structures for Space Applications*. In the Apollo era, the 1970 World Exposition in Osaka, Japan, had many pavilions that were air supported.

With the ground work having been laid, it is now time to apply these concepts to the Moon and Mars using the latest technologies.

Firstly, it should be noted that the words inflatable and expandable can possess different meanings. A balloon expands as it is being inflated. A volume can be expanded mechanically and then pressurized, a phased process that may assist in attaining the full expansion and also make the volume useable or livable, as was the case for the BEAM module on the International Space Station.

Designing for space habitation is tightly constrained by the dimensions of the launch vehicle payload bay. Thus the high cost of space transportation drives the subsystem designer to utilize every cubic centimeter available. The result of this process is often a highly efficient, extraordinarily expensive, and unique piece of hardware. If this volume constraint could be relaxed, allowing greater use of off-the-shelf technology, the result could well be cheaper systems for humans use in space. Cost savings could be realized in launch operations as well, by allowing a more flexible manifest with fewer vehicle-dependent payloads. The largest single element of a space station or lunar base is habitation. This includes the habitat as well as the crew habitation needs. If this element could be separated into its basic components, then that would increase the launch options. For example, it may be launched using a single large rocket, if one were available, or on a series of less powerful ones. This flexibility in launch operations can be attained by launching the pressure vessel separately from the habitat components, then integrating them in place. This method requires that the pressure vessel be collapsible, so that it can be packaged efficiently (a rigid vessel has the same volume whether it is empty or full).

A fabric structure is the simplest way to achieve a collapsible pressure vessel. Some advantage may be claimed for a rigid module which carries its air with it, whereas the air to pressurize an inflatable must be supplied separately, requiring space on the launch vehicle. However, the transportation of air will be part of the routine logistics flow for a base on the Moon or Mars. It must be supplied to the base continuously to make up for losses (chiefly through the airlocks, which lose 10 per cent of their air by volume each time they are used). There must also be a supply on hand at all times for emergency repressurization.

An efficient system for the transportation of volatiles to the Moon or Mars will be required, perhaps using cryogenic technology in order to carry them as liquids. The development of such a system is particularly likely if the vehicles themselves use cryogenic propellants. The advantage to the program of getting the first load of air "free" inside a habitation module will therefore be minimal for a base on the airless Moon. The atmosphere of Mars is 95 per cent carbon dioxide, 2.7 per cent nitrogen, 1.6 per cent argon, and traces of other gases. Some equipment would be needed to enable the crew to capture the atmosphere and process the gases into a useable habitat atmosphere. If water ice is found, this could also be broken down into hydrogen and oxygen for use as propellants, and astronauts could breathe the oxygen in combination with nitrogen which is either transported from Earth or is derived from the ambient atmosphere.

The long range plans for habitats on the Moon and Mars are able to draw on experience with the facilities being used successfully on the International Space Station. Current concepts for the Deep Space Habitat (DSH), the Multi-Mission Space Exploration Vehicle (MMSEV) and the design for a surface rover, will all involve a major effort to get them into low Earth orbit, let alone to the Moon or Mars.

What is proposed in this book is just for the early missions back to the Moon and on to Mars. Those habitats will most likely be rather primitive and for short durations, and will not provide the relative luxuries of the state-of-the-art. They will most likely be in the "bivouac" or "camp" category. They will suffice until more time and cash becomes available to send more capable and longer lasting equipment. Bear in mind that while the Moon is close by, Mars is not. But there are trajectory options available for Mars that do not require spending a year and a half on the surface. The Short-Stay (Opposition Class) option involves spending only 40–60 days on the surface until the trajectories are correctly aligned for the trip home. Perhaps a temporary inflatable habitat would be adequate for a Short-Stay whose task was to undertake exploration that would allow engineers on Earth several more years to build and launch more permanent habitats.

Once the first crew lands on the Moon or Mars, they must find an acceptable cave or lava tube for shelter which can be transformed into a place to temporarily live and work. They cannot merely pressurize a giant opening. Even if they were provided with a huge source of gas for pressurization, it would leak out through the porous rock. One company is working on how to seal a cave to allow it to be pressurized. The answer is to insert an inflatable, expandable habitat that can be pressurized and made suitable for living. The cave floor would have to be either relatively smooth or at least not littered with boulders. There are robots that can assist humans to prepare a site. The habitat would need to be anchored, and there are robots that can to do that too. Those first few missions would need a volume for at least four people, and later missions perhaps double that. This initial facility would need airlocks to enter and exit without losing any gas/air, or at least have replacement gas. The inflatable habitat could possess compartments that would be out-fitted for sleeping quarters, a galley, and working area. It would also include the life support and power systems, some of which could be outside the habitat with interfaces to the interior. The following sections describe the possibilities.

5.3.2   Relevant Technologies

To tackle the technology challenges that face spaceflight, NASA has developed roadmaps for each of the planned exploration targets. In the case of habitats, this is partially covered in Technology Area 12 on Materials, Structures, Mechanical Systems and Manufacturing. It is also covered in Area 6 on Human Health, Life Support and Habitation Systems. Area 6 is more concerned with crew health and safety. Area 12 relates to the physical structures and it is these that are discussed in this section.

Many of the NASA Centers are involved in accordance with their particular charter and areas of interest, but the Johnson Space Center has the prime role in areas that affect astronauts.

The NASA technology-focused report covers the entire spectrum of NASA's missions, and involves all of its Centers. The Roadmap characterizes the areas as follows (with habitat comments):

- Materials are the enablers behind the structures, devices, vehicles, power, life support, propulsion, entry, and many other systems; for example, the habitat material.

- Structures represent the design and analysis content that applies materials in a manner which results in certification for the intended environments. Certification and sustainability can be among the most costly and time-consuming aspects of spacecraft development. Many of the advancements in structures technology are critical enablers for sending humans into deep space. Inflatable, expandable habitats will most likely incorporate interior structures.
- Mechanism systems are essential to performing the functions required at almost every stage of spaceflight operations. They must be designed to be robust, long-lived, and capable of performing in harsh environments. The habitats would require inflation mechanisms and pressure controls.
- Advanced manufacturing will concentrate on the highest value innovative opportunities and will integrate new tools into the evolving manufacturing arena. Manufacturing technological advancements are essential in order to bridge the gap of cross-disciplinary advances and guarantee that advanced manufacturing capabilities become available for significant improvements in cost, schedule, and overall performance. Consider the manufacturing of high strength, leak resistance materials which exploit the latest concepts in radiation shielding fabrics and materials, such as BNNT.

This section will address those areas related to the subject of this book, namely "precursor" missions which seek subsurface shelter from the hostile environment.

The Technology Roadmap for Technology Area 12, identified five areas as follows:

- Materials.
- Structures.
- Mechanical Systems.
- Manufacturing.
- Crosscutting.

These five areas were divided into 22 sub-areas, each of which yielded many sub-goals. Thus there are over a hundred desired capabilities, each at a different level of technology readiness. Some of these areas are generic to crew safety in general. Most of the other areas are related to spacecraft, but some are specific to structural habitats. Some are directly applicable both to habitats made of metals and composites and also to inflatables. This book cannot cover them all, but the following sections review the technologies which are currently in inflatables and could protect a "precursor" mission against radiation, temperature extremes, and micrometeoroids on the Moon and Mars.

Materials

NASA's work in materials science covers all aspects of spaceflight and goes far beyond the concept of applying them to inflatable habitats. Broadly speaking, material properties and capabilities provide the form and function to structures, sensors, thermal, and other protection and management systems, safety and life support systems, power generation and energy storage systems, and many other systems. The issue of vehicle mass is always of importance, therefore materials must provide the desired functions and have a low overall mass. Inflatables fall nicely into this category. Developing materials which have improved properties directly aimed at forthcoming mission requirements is critical to the success of future missions. The range of material types and applications is so wide that it is necessary to ensure that NASA develops materials that will enable or reduce the risk involved in forthcoming missions. The materials section of the Technology Roadmaps seeks to link materials development needs with the requirements for human exploration missions, science, and aeronautics missions. This section is subdivided to address the needs of lightweight structures, computational design materials, flexible material systems, environment, and specialist materials. This book will only address those that relate to inflatables. Bear in mind though, that NASA is developing inflatables for much more demanding applications such as deployable heat shields and space telescopes.

The sub-areas under the general category of Materials are:

- Lightweight Structural Materials.
- Computationally Designed Materials.
- Flexible Material Systems.
- Materials for Extreme Environments.
- Special Materials.

Many of these relate not only to inflatable habitats but also to other systems or components meant for much more extreme environments ranging from cryogenic temperatures to reentry heating. But as noted in the previous section, for habitats we will desire materials such as boron nitride nanotubes for radiation protection. Perhaps that technology is in the lower to mid-levels of technology readiness. We desire a very lightweight, flexible material having sufficient strength to withstand pressurization and provide appropriate structural stiffness. A cave or lava tube on the Moon or Mars may have a rough interior, hence the inflatable fabric must not be susceptible to tearing, scuffing, or being punctured. The temperature extremes should be relatively mild underground, but an inflatable for humans will need to sustain a significant pressure differential once inflated, particularly on the airless Moon. If the habitat is to be inflated on the exposed surface, greater temperature variations must be taken into account, and in the case of Mars it must be capable of surviving dust storms.

Fabrics

The materials employed for habitats will include fabrics produced by a range of methods and possess the special characteristics required by that application. Many will incorporate aramid fibers. These heat-resistant, strong synthetic fibers have many uses in aerospace and military applications, including ballistic-rated body armor fabric and ballistic composites, bicycle tires, marine cordage, marine hull reinforcement, and as a substitute for asbestos. The following examples serve to illustrate their value.

The most common ones are carbon based aramids coupled with special resins. The advantages of carbon fibers include high stiffness, high tensile strength, low weight, high chemical resistance, tolerance of high temperatures, and low rate of thermal expansion. Such properties are attractive in aerospace, civil engineering, military, and motorsports, along with other competition sports. Carbon fibers are usually combined with other materials to create a composite. When impregnated with a plastic resin and baked, it forms a carbon-fiber-reinforced polymer with a very high strength-to-weight ratio. Although it is extremely rigid it is somewhat brittle. Carbon fibers are also composited with other materials, such as graphite, to make reinforced carbon-carbon composites that have a very high tolerance of heat, an excellent space application being the nose cone and wing leading edges of the Space Shuttle.

Kevlar, which is considered to be a para-aramid, is a heat-resistant and strong synthetic fiber that is related to aramids such as Nomex and Technora. Developed by DuPont in 1965, this high-strength material was first commercially used in the early 1970s as a replacement for steel in racing tires. Typically it is spun to form ropes or fabric sheets which can be used as such or as an ingredient in composite material components. The applications of Kevlar include bicycle tires and racing sails and bulletproof vests. Because of its high tensile strength-to-weight ratio, it is five times stronger than steel. It also enables modern marching drumheads to withstand high impact. When used as a woven material, it is suitable for mooring lines and other underwater applications.

Several grades of Kevlar are available:

- Kevlar K-29 is used in industrial applications, such as cables, asbestos replacement, brake linings, and body/vehicle armor.
- Kevlar K49 provides a high modulus for applications such as cable and rope.
- Kevlar K100 is a colored version of Kevlar.
- Kevlar K119 is a higher-elongation, flexible and more fatigue resistant version.
- Kevlar K129 has higher tenacity for ballistic applications.
- Kevlar AP has 15 per cent higher tensile strength than K-29.
- Kevlar KM2 offers enhanced ballistic resistance for armor applications.
- Kevlar XP is a lightweight version of KM2 for softer body armor.

Although Kevlar will maintain both strength and resilience down to cryogenic temperatures, its tensile strength is diminished at higher temperatures. When at a temperature of 160°C (320°F), 10 per cent reduction in strength occurs after about 500 hours. At 260° C (500°F) 50 per cent strength reduction occurs after 70 hours. Ultraviolet degrades and decomposes Kevlar, so it is rarely used outdoors without protection against sunlight.

Vectran is manufactured by Kuraray Company Ltd., in Japan, which acquired the business from Celanese Advanced Materials, Inc., in 2005. Vectran is a high-performance multifilament yarn which is spun from a liquid crystal polymer. Its fiber exhibits exceptional strength and rigidity. Weight for weight it is five times stronger than steel and ten times stronger than aluminum. Its "golden fibers" are noted for their thermal stability at high temperatures, high strength and modulus, low creep, and good chemical stability. They are moisture-resistant and stable in hostile environments. A polyester coating is often used around a Vectran core to improve abrasion resistance and serve as a water barrier. Vectran has a melting point of 330°C (626°F) with progressive reduction in strength starting at 220°C (428°F). Like Kevlar, Vectran's low tolerance of ultraviolet limits its exposure to sunlight. Its fibers can be used as a reinforcing matrix for ropes, electrical cables, sailcloth, and advanced composite materials, professional bike tires, and certain electronics applications. It is used as one of the layers in the soft goods structure of the Extravehicular Mobility Unit manufactured by ILC Dover for NASA, and was the fabric used for all of the airbag landings made on Mars: Mars Pathfinder in 1997 and on the twin Mars Exploration Rovers Spirit and Opportunity in 2004. It was used again on NASA's 2011 Mars Science Laboratory in the bridle cables for lowering the Curiosity rover down onto the surface of Mars. Vectran is also a key component of all the inflatable spacecraft of Bigelow Aerospace: Genesis I, Genesis II and BEAM, the latter currently on the International Space Station. In addition, Vectran is to be used for the much larger B330 module.

5.3.3   NASA's Prior Inflatable Habitat Research

Research into inflatables for space applications has been going on for decades. Here are two representative examples of the design and operational issues which remain valid to this day.

1988 JSC's Inflatable Lunar Base

This was a conceptual design conducted at the Johnson Space Center that built upon research from the previous 20 years, notably that at the Langley Research Center and the Goodyear Aerospace Corporation in the Apollo era. It discussed the cost savings, mass and volume savings, and many operational and habitability advantages. It explored all the different possible forms and geometries associated with stresses and also different structural designs as well as materials. It took into account radiation, temperature extremes, micrometeoroids, leakage, and puncture and abrasion resistance, and defined a candidate inflatable habitat.

Their design concept was a sphere 16 m (52 ft) in diameter containing 2145 m^3 (75,750 ft^3) inflated to a pressure of one atmosphere. It was for 8 to 12 astronauts who would live and work on four floors. It would be made of Kevlar-29, and have a mass of 2200 kg (4850 lbs). In the JSC concept, the inflatable was to be covered over by 3 m (9.8 ft) of regolith in sand bags. (I can't see the crew filling sand bags and putting them on top of the habitat.)

The study concluded that inflatable habitation holds great promise for human presence on the Moon and in space.

1998 TransHab (Transit Habitat)

A decade later, the TransHab concept was pursued by NASA in order to develop the technology required for expandable habitats which would be inflated in space. This reflected the original intention to design an interplanetary vehicle to transfer humans to Mars. Specifically, TransHab was envisioned as a replacement for the planned rigid ISS Habitation Module (which was later cancelled). When deflated, an inflatable module is a compact package that is somewhat easier to launch than a metal shell. When fully inflated, TransHab would expand to a diameter of 8.2 m (27 ft) as compared to the 4.4 m (14.4 ft) of the Columbus science module that is on the ISS.

Considerable controversy arose during the TransHab development effort as a result of delays and rising costs of the ISS program. In 1999, the National Space Society issued a policy statement which urged NASA to continue research and development of inflatable technologies but curtail development of the TransHabmodule for the ISS. Finally, in 2000, despite objections from the White House, House Resolution #1654 prevented NASA from undertaking further work on the TransHab. Nevertheless, the bill catered for the possibility of leasing an inflatable habitat from private industry (hence Bigelow Aerospace's interest, see below).

The project concluded that the technologies required to design, fabricate, and use an inflatable module in space applications had been proven by the TransHab team in the

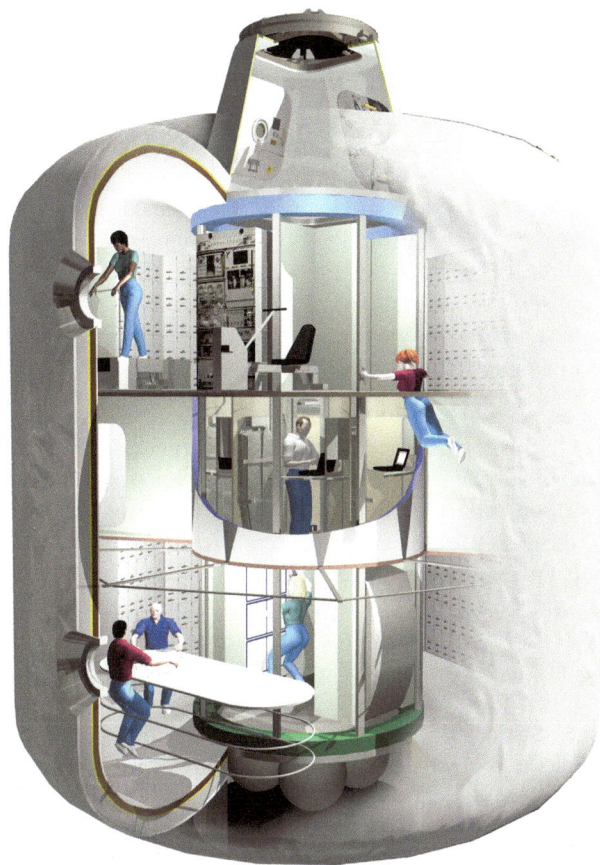

Fig. 5.3 TransHab cutaway, 1999. Photo courtesy of NASA.

development phase of that program. A number of issues of inflatable space structures had been addressed by way of testing and hands-on development, including ease of manufacturing, structural integrity, micrometeoroid protection, folding, and vacuum deployment. The TransHab inflatable technology program had not only proven that inflatable structures are viable, but also that they offer significant advantages over conventional metal structures.

5.3.4 Industries Habitat Research

The following are contracts for space habitats, not surface habitats, but some of the concepts should be transferrable. This work is sponsored by a program called Next Space Technologies for Exploration Partnerships (NextSTEP) that is being managed from the Advanced Exploration Systems (AES) Division of the NASA Human Exploration and Operations Mission Directorate. NextSTEP establishes public-private partnerships to advance commercial development of space systems while advancing capabilities for deep space exploration, so that the Moon can be used as a proving ground for Mars.

In 2016, NASA chose six U.S. companies to expand knowledge, commercial capabilities and opportunities in space by the development of full-sized ground prototypes and concepts for deep space habitats under the second Broad Agency Announcement called NextSTEP-2.

An effective space habitat comprises a pressurized volume plus an integrated array of complex systems and components which include environmental control and life support systems (ECLSS), a facility for docking, logistics management, radiation mitigation and monitoring, fire safety, and crew health. Prototypes will investigate the configurations of the habitat and how the various systems interact with one another and with other capabilities such as propulsion modules and the use of airlocks. They will provide platforms to ensure the standards and common interfaces being considered are well thought out.

The six selected partners are:

- Bigelow Aerospace LLC of North Las Vegas, is to develop and test a prototype of the Expandable Bigelow Advanced Station Enhancement (XBASE). The testing on this platform will advance methods for deep space missions and serve as a basis for commercialization in low Earth orbit. With a volume of 330 m^3, the XBASE is based upon the B-330 expandable spacecraft for the mission-specific purpose of mating at the International Space Station as a visiting vehicle. XBASE leverages the results of the 16 m^3 Bigelow Expandable Activity Module (BEAM) that was attached to the ISS in June 2016 for long-term evaluation.

- Boeing of Houston, Texas, is developing a modular habitat system that leverages experience in designing, developing, assembling on-orbit and operating the International Space Station for more than 15 years. It will include the production of a full-scale habitat to provide design analysis and high-fidelity demonstration and test capability in order to simulate how humans might safely live and work in deep space while conducting long missions. This ground demonstrator will test and validate interface standards, systems functionality, and critical exploration technologies.

- Lockheed Martin of Denver, Colorado, is to refurbish a surplus Multi-Purpose Logistics Module of the type that enabled the Space Shuttle to carry equipment and supplies to/from the International Space Station. It will become a full-scale habitat prototype, with integrated avionics and ECLSS. The high-fidelity ECLSS prototype will provide risk reduction and permit form-and-fit testing. The avionics prototype will prove data communications between the habitat and the new Orion spacecraft, and demonstrate crew interfaces between an Orion and a deep space habitat. The company will also exploit virtual prototyping to validate the habitat module's form, fit, and function.

- Orbital ATK of Dulles, Virginia, will increase the maturity factor of the mission architecture and design of their "initial cislunar habitat" concept which is based upon the Cygnus spacecraft that is currently being used to supply the International Space Station. The prototype will support testing of critical interfaces with the Orion spacecraft and other modules. It will mature the Cygnus-derived habitat design for long-term operation in deep space. This work will establish a proposed roadmap designed to facilitate Mars exploration. (Orbital ATK was purchased by Northrop Grumman in 2018, and became Northrop Grumman Innovation Systems.)

- Sierra Nevada Corporation (SNC) Space Systems of Louisville, Colorado, is to refine the concept of operations for a flexible architecture that could serve as a deep space habitat. This will leverage three to four commercial launches to construct a modular long duration habitat. The prototype is to be based on the Dream Chaser cargo module as a foundation for the SNC NextSTEP-2 proposal, and will allow the company to assess its ability to meet the criteria for each operation phase and identify risks. After launch from the Dream Chaser spacecraft, the SNC NextSTEP-2 module will be combined with a large inflatable fabric environment module, an ECLSS system, and a propulsion system. The design and prototype will confirm the proof-of-concept and ensure critical subsystems seamlessly integrate together.
- NanoRacks of Webster, Texas, working in conjunction with its partners, Space Systems Loral and the United Launch Alliance, collectively called the Ixion Team, will conduct a comprehensive feasibility study regarding the conversion of an existing launch vehicle's upper stage, or propellant segment, into a pressurized habitable volume in space. This will provide insight into an innovative and low-cost approach that can be used for any rocket system, including the new Space Launch System (SLS).

The NextSTEP awards are to inform the acquisition and deployment approach for the next phase of flight systems for deep space missions, including important aspects such as standards and interfaces, module configurations, and options for deployment using SLS and Orion and commercial vehicles. Of course, NASA is also open to collaborative opportunities with international partners to develop a robust fully operational deep space habitation capability.

The following sections elaborate on the Bigelow Aerospace work, which may be transferrable to a surface habitat, and on the work pursued by ILC Dover with the Langley Research Center on surface habitats.

Bigelow Aerospace
In 1999 the private company Bigelow Aerospace purchased the patent rights for the TransHab developed by NASA, then set out to design a private space station. The company launched two successful subscale modules in the Genesis series in 2006 and 2007 aboard Dnepr rockets (former intercontinental ballistic missiles) fired from a Russian Air Force Base. These were essentially test and evaluation proof-of-concept missions. On April 8, 2016, the Bigelow Expandable Activity Module (BEAM) was launched aboard the SpaceX CRS-8 Dragon cargo vehicle that berthed at the International Space Station. On April 16, British astronaut Tim Peake used the ISS's robotic arm to retrieve BEAM from the unpressurized trunk of the spacecraft and attached it to the Tranquility node. The 1400 kg (3086 lbs) module was then inflated in May to a length of 4 m (13 ft), a diameter of 3.2 m (10.5 ft), and a volume of 16 m^3 (565 ft^3).[2,3]

---

[2]For a 3 minute video of the astronauts entering BEAM go to:
https://www.youtube.com/watch?v=5kZZdp727ek
[3]For a 6 minute video explaining more about the BEAM operations go to:
https://www.youtube.com/watch?v=gARj5wmlFKg

The rationale for mounting BEAM on the ISS was to collect performance data on expandable habitat technologies, in particular its structural integrity, leak rate, radiation dosage, and thermal responses. After a year, it was to be jettisoned. But in October 2017 NASA announced that BEAM would remain in place for another 3 years, with options for two further extensions of 1 year.

After NASA and Bigelow successfully completed collaborative analyses on the BEAM life extension and stowage feasibility, astronauts gained additional storage capability aboard the station by removing the hardware used to inflate the module and converting the sensors which monitor the BEAM environment from wireless mode to wired mode in order to prevent interference from future stowage items on transmission of sensor data. Next, they installed air ducting and nets to define the stowage volume. In this condition, BEAM can hold 130 Cargo Transfer Bags to accommodate a wide variety of items. It is likely that a power and data interface will be added to undertake further technology demonstrations using BEAM within the partnership agreement.

The in-space use of BEAM as part of a human-rated system allowed Bigelow Aerospace to increase the maturity of its technology for commercial activities in low Earth orbit. Initial studies indicated that soft materials can perform as well as rigid materials for habitation volumes, and that BEAM has performed as designed in terms of resistance to impacting space debris.

A suite of sensors installed by the crew automatically take measurements and monitor BEAM's performance to help inform designs for future habitat systems. This extension will deepen NASA's understanding of expandable space systems. It was the excellent performance of the module during its first year that convinced NASA to retrofit BEAM as a storage facility. Astronauts have entered the module more than a dozen times since it was inflated.

ILC Dover

Since the start of project Apollo, ILC has been the designer and producer of the space suit pressure garments for NASA. They developed and supplied the Space Suit Assembly as part of the Extravehicular Mobility Unit that is used by astronauts during assembly and maintenance of the International Space Station. ILC has also carried out work on Inflatable Impact Attenuation Systems (a.k.a. airbags) not for cars, but for the Moon and Mars. Its vented airbag technology can land extremely heavy payloads exceeding 20 tons. For the Mars Pathfinder and Mars Exploration Rover missions, the company designed and produced a unique all-encompassing airbag system to safety deliver the Sojourner, Spirit, and Opportunity rovers onto Mars.[4]

Also not as well-known as the pressure suits is ILC Dover's work to develop and manufacture a variety of inflatable habitats, airlocks, and shelters for use in Earth orbit and for exploration of the Moon and Mars. The company's projects include the X-Hab Lunar Habitat, the InFlex Lunar Habitat, the Toroidal Lunar Habitat, and the Expandable Lunar Habitat. It also participated in the Antarctic Habitat Planetary Analog Study, the Lawrence Livermore Inflatable Space Station, and the Minimum Function Habitat. In 2007, ILC

---

[4]Watch this short video of the ILC's landing bags for rovers.
https://www.youtube.com/watch?v=KyktvC7w7Js

Fig. 5.4 Full scale mockup of the BEAM. Photo courtesy of NASA.

teamed up with the NASA Langley Research Center to build a prototype surface habitat for test and evaluation of the concept and technology.

ILC has produced a variety of inflatable structures for military and aerospace applications. Inflatable products having space applications include ballutes and decelerators; inflatable and deployable antennas; sunshields, solar sails and solar arrays; radiation shields; and planetary balloons.

5.3.5 Mobile Surface Habitats

Ten years ago, the Space Exploration Vehicle (SEV) was conceived as both a space and surface vehicle. Years later, it evolved into the Multi-Mission Space Exploration Vehicle (MMSEV). Following the cancellation of the Constellation program, it was reconfigured as a free-flying scout vehicle for the exploration of an asteroid. Then it was transformed from a two person to a four person descent and ascent vehicle that would be attached to the Deep Space Habitat and flown down to the lunar surface. In this configuration, it resembles a larger version of the Apollo Lunar Module in that it would serve as a launch vehicle for returning to the "mother ship" for the ride home.

The technology for MMSEV would clearly be directly applicable to a mobile surface habitat, but the landed MMSEV should not be risked to drive around the Moon or Mars. A separate wheeled variant would be created to use the technology and systems of the spacecraft when appropriate, plus the other systems needed for surface operations. One could reason that a redundant version should be available for backup, and that it should also be used for dual exploration. One early study was of a version that enabled the two vehicles to dock. In that case, a crew could transfer from one vehicle to the other in the event of a breakdown of one vehicle, or simply to temporarily increase the habitability.

As the above mentioned studies by the six companies are deep space driven, rather than surface driven, it is not clear how much of their work is transferrable. Certainly there are commonalities in support systems. Maybe that should be the subject of the next study. However, bear in mind that there is another category of habitat, namely the one in which the habitat is inserted into a cave or lava tube.

5.3.6   Universities Habitat Research

An annual eXploration Systems and Habitation (X-Hab) Academic Innovation Challenge seeks to develop strategic partnerships and collaborations involving universities. It has been organized to help bridge strategic knowledge gaps and to increase knowledge in capabilities and minimize the technology risks for NASA's visions and missions. In 2016, the scope of the Challenge was extended beyond habitation to address additional aspects of exploration systems. The competition is intended to link up with senior and graduate level design curricula that emphasize hands-on design, research, development, and then manufacturing of the functional prototypical subsystems that will assist space habitats and deep-space exploration missions. NASA will directly benefit by sponsoring the development by academic institutions of innovative concepts and technologies seeking solutions that can be applied to exploration strategies.

The Advanced Exploration Systems (AES) Division of the NASA Human Exploration and Operations Mission Directorate is offering multiple awards of $15,000 to $50,000 to design and produce studies and functional products that are of interest to AES as proposed by universities in accordance with their expertise. Institutions interested in participating submitted X-Hab proposals for review by teams of technical experts. Universities were able to form collaborations to act as a single distributed team. Promising concepts will then be integrated into existing NASA-built operational prototypes for further investigation.

Each annual Challenge stimulates another generation of talented engineers to explore new ideas, with concepts being investigated during the academic year and functional prototypes presented to NASA for evaluation.

Four university teams in 2018 undertook studies and developed partial system mockups featuring design commonalities among Mars transit and surface habitats:

- The California Polytechnic State University developed several concepts with design commonality between in-space and surface habitats. These layouts were used during the transit to Mars and then reconfigured after landing to better accommodate the partial gravity of that environment. A concept inspired by the "roly-poly" (woodlouse) bug was selected for a demonstration by a combination of virtual reality and a physical mockup. This showcases a unique strategy for mobility and transformation of the surface habitat.
- The Pratt Institute, Brooklyn, New York, designed a large habitat system and created a ground mockup of a key element of the architecture. Called EDEN, it is a Transit Hub to provide a simulated Earth gravity (one-G) environment; a

sustainable, fresh food source; and an environment that includes nature and natural materials to soothe the rigors of space travel. The one-G environment achieved by the rotation of the habitat system mitigates detrimental health impacts of microgravity. To illustrate the viability of the expandable torus design, a scale segment was built on campus.

- The University of Maryland developed a full concept design for Multi-mission Artificial Gravity Reusable Habitat (MARSH). Their program included mission design, subsystem analysis, and virtual reality testing. The resulting habitat addressed all aspects of spacecraft design and also major concerns resulting from partial gravity, including Coriolis effects and maintaining pointing of radio dishes and solar arrays. Furthermore, first-year engineering students designed, created and evaluated various partial gravity staircase designs in a neutral buoyancy facility.
- The University of Michigan developed the Argo concept. This is a dual-purpose habitat architecture suitable for deep space transit as well as the surface of Mars. It addressed the conflicting requirements of combining partial gravity with microgravity by using artificial gravity to ensure that the crew experience consistent physical conditions. The team developed virtual reality models for the transit and surface configurations, and also physical mockups of an example node, a partial torus (part of a rotating system to simulate gravity), and an airlock for a concept demonstration.

The 2019 X-Hab Challenge addresses eight topic areas, some of which relate to NASA's new Lunar Orbital Platform-Gateway. One of those topics relates to the planned precursor mission. The goal of the Inflatable/Deployable Crew-Lock to Enhance Gateway Ground Test and Evaluation is to design and fabricate a full-scale, low-pressure inflatable/ deployable structure that demonstrates a crew-lock for use in gateway ground test efforts.

5.3.7 The Future of Inflatables and Expandables

It seems clear that many, but not all, of the technologies required for a habitat on, or beneath the surface of the Moon and Mars have achieved a high level of flight readiness. While some of the materials and other technologies have been tested, it is the nature of NASA and the aerospace industry to continue to conduct research seeking improvements, application to other needs, or simply to lower the costs of manufacturing.[5] As a human return to the Moon is probably not going to happen until the 2020's and missions to Mars for another decade after that, there is plenty of time to advance the associated technologies and increase the readiness level of others. Also, since NASA has not focused on habitats for caves and lava tubes for the early missions to either body, there is time to add a "precursor" of this type to the overall project architecture and focus on the requisite technologies.

The Langley/ILC habitat in Fig. 5.5 is an historical reference point for surface habitat development. The prototype was an inflatable structure with a diameter of 12 ft (3.65 m)

---

[5]For an example of an advanced concept for an inflatable habitat that involves 3D printing, see this video by a team at Northwestern University, Evanston, Illinois at:
https://www.youtube.com/watch?v=mxzoO9ADqOE&feature=youtu.be

Fig. 5.5 Langley Project Leader Karen Whitley stands before the ILC habitat module. Photo courtesy of NASA/LaRC/Jeff Caplan.

that was made of multilayer fabric for ground-based evaluation of emerging technologies such as flexible structural health monitoring systems, self-healing materials, and radiation protective materials. Attached to the structure is a smaller inflatable which demonstrates an airlock. Both are essentially pressurized cylinders and are connected by an airtight door. It was used to evaluate materials, lightweight structure technologies, crew interfaces, dust mitigation methods, and compatibility with robotics and other surface equipment. Testing and evaluation of this type has now gone on at NASA and in industry for over a decade. NASA, the National Science Foundation, and ILC Dover built an inflatable structure for tests in the Antarctic. Although Antarctica is not as harsh an environment as the Moon or Mars, this exercise provided valuable lessons.

As an example of a "crosscutting" technology, there is another project which also uses advanced materials technology and engineering that relates to habitats for the Moon and Mars. NASA is also using inflatables for heat shields! In 2012, Inflatable Reentry Vehicle Experiment (IRVE)-3 was reentered into the Earth's atmosphere at Mach 10. It successfully survived temperatures peaking at 538°C (1000°F) and peak deceleration of 20 g's. The IRVE was done for the Hypersonic Inflatable Aerodynamic Decelerator (HIAD) Project within the Game Changing Development Program, itself part of NASA's Space Technology Program. The Langley Research Center led the effort and built two of the four segments of the IRVE-3 payload. Wallops Island provided its rocket expertise and built the other two payload segments. Airborne Systems in Santa Ana, California, provided the inflatable structure and thermal blanket. If inflatables can survive in that extreme environment then they will be able to be used for habitats in caves and lava tubes!

NASA engineers are studying different designs and materials. The inflatable structure has to be made of incredibly strong, yet flexible fabric in order to keep its shape and withstand the force of rushing through an atmosphere at as much as 25,000 miles per hour (about 42,300 km per hour). The current design emerged from a partnership with the private company called HDT Global. It is a series of inflated rings that are stacked together and made of braided Kevlar, the material that is used for bulletproof vests. Each ring is lined with silicon on the inside. The scientific name for the donut-shaped ring is a torus. The Kevlar gives each torus its strength and the silicon liner confines the compressed gas inside, in much the same way as a bicycle tube and tire. Kevlar straps keep the rings attached to each other and to the payload hardware.

Inflatable entry shields will significantly benefit missions to Mars, where it is possible to use the "aerocapture" technique in which a vehicle arriving from deep space penetrates the upper atmosphere on a trajectory that emerges into space for insertion into planetary orbit, rather than using rocket propulsion for the braking maneuver. This will enable missions to efficiently put much larger payloads into orbit. It may also be possible to use inflatable shields to deliver heavy modules to the surface.

The evolution of inflatable habitats has seen an increase in the acceptability of the concept for several applications including on-orbit, deep space, the surface of the Moon and Mars, and the proposed subsurface mission advocated by this book. Inflatable habitats have been used for years in analog research, and will continue as such for several years to come. At some point, a final design for a surface and subsurface application will need to be specified for the Moon and for Mars. Since these two environments are different, there will be some fundamental differences in the designs but not for their interior objective of protecting crews and providing them with a habitable environment for extended periods.

5.4   ROBOTICS

5.4.1   Introduction

Human exploration of the Moon and Mars will definitely use different kinds of robots. When using the word "robot," one really means a fairly sophisticated AI software algorithm which controls hardware. When you conceive of missions that explore caves and lava tubes and a wide variety of terrains on the Moon and Mars, you are visualizing a multitude of robots, each designed to perform given types of task. Some robots may operate autonomously, others will respond to their human handlers, and some will be able to think for themselves. I see humans and robots working together as a team.

Our Earthly realm is already replete with many types of robots. The ones we think of most are those in factories that excel at repetitive tasks such as welding or painting car parts. We even see them in hazardous areas such as mines, deep sea oil pipe lines, and nuclear power plants. They are appearing in hospitals, assisting surgeons. But these are all special purpose robots that carry out very specific tasks programmed by their designers with inputs from the user/operator.

When I was young, the first robot was introduced at the 1939 World's Fair. It was named Elektro, and was created by the Westinghouse Electric Corporation in Mansfield, Ohio, in the late 1930's. It had a humanoid appearance, was 2.1 m (7 ft) tall, weighed

Fig. 5.6 Elektro and Sparko. Photo courtesy of the Senator John Heinz History Center.

120.2 kg (265 lbs), could respond to voice command and speak about 700 words (using a 78-rpm record player), move its head and arms, smoke cigarettes, and inflate balloons. Its body consisted of a steel gear, cam and motor skeleton covered by an aluminum skin. Its photoelectric "eyes" could distinguish red and green light. In 1940, it reappeared with Sparko, a robot dog which could bark, sit, and beg to humans.

The Robot Hall of Fame was established in 2003 by the School of Computer Science at Carnegie Mellon University. This has my favorite robot, Gort, one of the most memorable pop culture images from the Cold War era. The 1951 movie *The Day the Earth Stood Still* features a mysterious spaceman named Klaatu who comes to Earth in a flying saucer on a mission of peace, but is accompanied by a self-aware robot bodyguard that is capable of awesome destructive power. Even in black and white, the 8-ft-tall robot is a stirring presence as he emerges from the spacecraft.

Fig. 5.7 Gort and Klaatu. Photo courtesy of 20th Century Fox Studios.

If, indeed, we soon return to the Moon and go on to reach Mars by 2039, it will have been 100 years since Elektro, or about four generations of humans.[6]

5.4.2 NASA's Robotics Research

The NASA Technology Roadmaps define a viable path forward for research and development. Technology Area 4 deals with Robotics and Autonomous Systems. Many NASA Centers are involved in robotics in accordance with their particular charter and areas of interest. The roadmap covers the entire range of the Agency's missions.

This section will address those areas which relate to the subject of this book, namely the "precursor" missions to the Moon and Mars that seek to explore the subsurface for shelter from the hostile environment. It is presumed that the first crews will have to fully leverage intelligent and versatile robots for many tasks, including exploring long

---

[6]You can read about all your robot friends like Robby, the Robot, R2-D2 and C-3PO at: http://www.robothalloffame.org/about.html

distances, surveying potential sites, deploying sensors and other equipment, preparing a site, deploying a habitat, setting up scientific equipment, and undertaking operations potentially too hazardous for astronauts to perform themselves. In essence, the robots will be part of the crew.

The Technology Roadmap for Robotics and Autonomous Systems identified seven areas as follows:

- Sensing and Perception.
- Mobility.
- Manipulation.
- Human-System Interaction.
- System Level Autonomy.
- Autonomous Rendezvous and Docking.
- Systems Engineering.

These areas were further broken down into 46 sub-areas. Each of these sub-areas had many sub-goals. Consequently, there are hundreds of desired robotic capabilities that are at a various levels of technology readiness. Some areas are generic, such as software and engineering methods. Others relate to making any robot functional. A number are directly applicable to assisting astronauts in the exploration of caves, lava tubes and other potentially suitable features, and their conversion to shelters. Not all of the areas can possibly be covered in this book. The following are just examples.

Mobility

Mobility pertains to moving from one place to another in an environment. This is distinct from intentionally modifying the environment. For example, mobility on, into, and above a planetary surface. This can involve various tasks, such as flying, walking, climbing, rappelling, tunneling, swimming, sailing, and thrusting. As an illustration, think of astronauts who have just landed near a potential site, be it on the Moon or Mars. Perhaps a rover has been positioned nearby (possibly we will have moved on from delivering the rover as cargo on the lander, as in the case of the Lunar Rover of the Apollo program). What new technologies might the crew require to fetch the rover, which may be some distance away? How will the crew then drive to candidate sites? How will they enter sites? How will they make use of a site?

NASA grouped mobility technologies into the following general categories:

- Extreme-Terrain Mobility. This provides mobility across terrains with challenging topographies and challenging regolith properties for bodies with substantial gravity (hence not small moons, asteroids, or comets).
- Below-Surface Mobility. This provides access to and mobility below a solid or liquid surface.
- Above-Surface Mobility. This provides coverage of, access to, and mobility above planetary surfaces.
- Small-Body and Microgravity Mobility. This provides mobility across surfaces of small bodies or microgravity environments without surface contact.
- Surface Mobility. This provides efficient mobility across non-extreme terrains or liquid surfaces.

- Robot Navigation. This provides autonomous or supervised mobility for surface, above-surface, and extreme terrains.

Since we are seeking caves and lava tubes for protection, let us examine just two of the above categories: Surface Mobility and Below-Surface Mobility. These describe what is required after landing on the surface of the Moon and Mars, then exploring a given cave or other chosen protected feature. Bear in mind that when developing these capabilities NASA was considering landing on various planetary bodies. Also, some of the stated capabilities do not necessarily apply to the Moon and Mars.

Surface Mobility

The state of the art in surface mobility includes six-wheeled, passive-suspension, rocker-bogie mechanisms with front and back steerable wheels. A prime example is the new Mars 2020 rover, which is based on the proven Curiosity configuration.

The rover for transporting astronauts could well resemble the Multi-Mission Spacecraft Exploration Vehicle (MMSEV) but given how the field of robotics is maturing it might look something like Fig. 5.10.

In 2017, the Kennedy Space Center released its latest Mars Rover concept. It was designed and build by Parker Brothers Concepts, is nearly 4 m (13 ft) wide, 3.35 m (11 ft) tall, and 7.3 m (24 ft) long. It employs six 50-inch airless wheels. (They didn't say what it might cost or weigh, and what it would take to get it to Mars.) It is said that "What Man Can Conceive, Man Can Achieve," but I have to wonder at what cost![7]

Back in reality, the NASA rover technology must demonstrate mobility across tens of kilometers of relatively flat terrain with low-grade slopes (shallower than 25°), widely-separated positive and negative obstacles, and other terrain hazards such as sand-dune slippage. The speeds of robotic Mars rovers have generally not exceeded 5 cm per second (0.11 miles per hour), yet they have demonstrated their ability to climb obstacles with diameters of a wheel-radius. The challenges to be overcome included mobility across a large variety of terrains with limited power, computation, and control of wheels to minimize energy and wear and tear on the vehicle.

In the case of human surface exploration, the challenges will include efficient mobility for crew and payloads across natural terrain. The latter will include the transportation of in-situ resource processing equipment, habitats, science analysis facilities, and other assets such as cranes, haulers, and davits. The objective is to transport payloads, equipment, and other surface assets at much higher traverse speeds for both human and other missions and also to increase the robustness of onboard sensing, control, and navigation software. This will include addressing issues relating to safety of crew on (or near) vehicles operating at relatively high speeds (these are covered in the technology area for Safety and Trust).

Human drivers possess a remarkable ability to perceive terrain hazards at long range and to pilot safely on dynamic trajectories. Despite the limitations of human sensing and cognition, it is generally true that an experienced driver can pilot at speeds near the limits set by physical laws (frictional coefficients, tip-over, and other vehicle terrain kinematic and dynamic failures). This is remarkable, given the huge computational throughput

---

[7]For a short video on the Mars Rover "concept car" go to:
https://www.youtube.com/watch?v=fZddqJ7y5TI

Fig. 5.8 An artist's rendering of the JPL-designed Mars 2020 rover. Rendering courtesy of NASA/JPL-Caltech.

requirements needed to quickly assess subtle terrain geometric and non-geometric properties (for example, visually estimating the properties of soft soil) at long range sufficiently rapidly to maintain a speed at or near the vehicle limits. This ability is lacking in today's best obstacle detection and hazard avoidance systems. A further focus will be on improving the payload mass fraction, specific power, speed, and endurance in terms of time and distance of traverse.

The "off-world" speed record for humans on a rover was set by Gene Cernan at 18 km per hour (11.2 miles per hour) on Apollo 17. If you watch that video and observe the regolith being kicked up and the wheels bouncing off the surface, you will realize he was "pushing the envelope" of the vehicle. In the future, astronauts might want to reach a pre-landed package, a cave, or other destination that is in sight, but recognize that high speed might not be the best choice. Go to YouTube and watch the videos of John Young and Gene Cernan exercising Apollo rovers. Irrespective of how robustly rovers are made, they are unlikely to be capable of rapid "cruising" on the Moon or Mars. The Apollo astronauts were careful not to turn or brake abruptly, and avoided making sharp turns lest their rover should tip over.

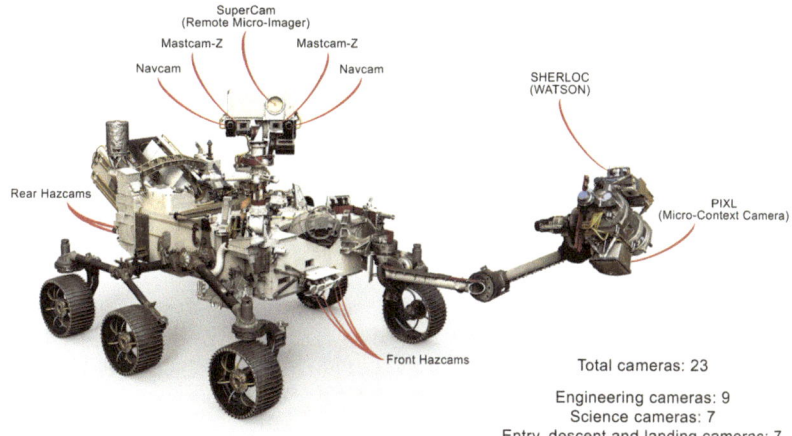

Fig. 5.9 An annotated drawing of the Mars 2020 rover. Drawing courtesy of NASA/JPL-Caltech.

Fig. 5.10 A concept for a crewed Mars Rover. Photo courtesy of NASA/KSC.

The undulatory nature of the surface produced a "wild ride." What kind of autonomous system could do what the human instinctively does. Yet, one of the technology goals is to traverse the Moon or Mars at a much higher speed than is currently achieved using automated rovers. If that comes to pass, it will be quite a ride, even at 1/6 or 3/8 g!

Challenges for surface rovers include the development of suspension systems and compliant wheels with performance similar to pneumatic tires on Earth, high-performance actuators, energy storage, thermal control, passive and active spring-damper systems, and so on. One can also imagine robotic rovers that can be given a command to go somewhere and pick up a package, or to enter a risky-looking area and take 3D images that are relayed back to the crew for analysis (or even to Earth for more sophisticated analysis).

There are other technologies related to surface mobility that are too detailed for this discussion. In summary, they include:

- Adaptive Autonomous Surface Navigation.
- Autonomous Navigation for Tethered Systems.
- Low-Altitude Above-Surface Navigation.
- Below-Surface Navigation.
- Small-Body/Microgravity Navigation.

The challenges for all above-surface mobility platforms include adaptation to extreme heat or cold, extremely high or low pressures (air density), and also the chemical composition of the atmosphere (if any). Additional challenges include power, communication, energy collection and storage, weight of structures and avionics, resilience of materials, controllability and autonomy (including high-speed mobility), and reusability of engines in the case of dynamic-lift systems. Moreover, validating system-level capabilities in relevant environments can be challenging, if not impossible, and when possible, in many cases it can only be done in parts. One concept for surface (and possibly below-surface mobility) is the use of a robot as a stand-in for a human.

This was first demonstrated in 2011 by the Desert Research and Technology Studies (RATS) analog in Arizona. Also that year, Robonaut 2 was installed on the International Space Station and proved itself.[8] It was later returned to Earth for an upgrade. But the pace of development has been rapid since that time, and the technology in more modern robots (such as Atlas, made by Boston Dynamics; see below) will be available when we return to the Moon and venture to Mars. By then, the state-of-the-art for robust robots capable of traversing difficult terrain will be considerably advanced, and exploration will have become a collaboration between humans and robots.

Below Surface Mobility

This pertains to access through any subsurface such as naturally occurring terrain cavities, particularly caves, lava tubes, and overhangs. The action of intentionally modifying the medium to generate cavities or holes, either by deep drilling or by excavation, is addressed by another technology area, and because it is beyond the capability of the first few missions to the Moon or Mars it is not included here.

Below-surface mobility has not yet been used on other planetary surfaces. On Earth, it has been used for underwater and underground activities. The latter was primarily driven by the oil, gas, and mining industries by way of deep-directional drilling and underground mining operations. We have no problem in thinking of ways to bore holes in the Earth and transfer materials and people through tunnels. Now we must devise ways to gain access to a cave on another planet, and it is not as easy as you might think.

---

[8]For a briefing on Robonaut go to:
https://www.youtube.com/watch?annotation_id=annotation_1062681065&feature=iv&src_vid=g3u48T4Vx7k&v=ePWjFlSdB4U

Fig. 5.11 Robonaut and a rover go exploring. Photo courtesy of NASA.

Below-surface mobility is made particularly difficult by, and must account for, the absence of direct sunlight, the lack of direct line-of-sight communication, and the nature of the medium (for example, abrasiveness of regolith while burrowing). While the Technology Roadmap discussed burrowing, this lies beyond the scope of this book, which envisages that the first few missions will not require much in the way of burrowing activity, their goal will be to find a cave that is big enough to accommodate an inflatable habitat.

Another technology that is being developed is robots that will be able to assist astronauts in exploring caves or lava tubes. These could be free-floating (flying) robots that are capable of self-positioning and self-orientation in microgravity for sensing and operations. A "Moon Bat" or "Mars Bat" could assist by entering the cavity first, to evaluate the situation. Is the cave worth further exploration, or is it clearly unsuitable? Another technology is a class of hopping or tumbling surface robots that provide mobility on the surface while sensing their surroundings. Still another class of robot is an anchoring robot that can both anchor and de-anchor an object to a surface such as the wall or floor of a cave. This could be helpful in securing an inflatable habitat. Another class of robot is the Track & Wheel Hybrid Robot that integrates the two forms of locomotion for enhanced mobility.[9]

---

[9]See for example:
https://www.researchgate.net/publication/321997018_Kylin_A_transformable_track-wheel_hybrid_robot

Fig. 5.12 Valkyrie, a next-generation robot built upon the Robonaut experience. Photo courtesy of NASA/JSC.

Fig. 5.13 Damage to the Curiosity rover wheels after 1000 sols on Mars. As of September 2018 it had gone 19.6 km (12.2 mi). Photo courtesy of NASA/JPL-Caltech/Science Magazine.

Mobility Components

Some mobility components relate to both surface and below surface components, be they robotic, semi-robotic, or just hardware. They include what you currently see on terrestrial prototypes and on rovers like the half-century-old LRVs on the Moon and modern rovers like Curiosity on Mars. They are tractive elements, such as wheels, tracks, anchors, and footpads; long-life actuators tailored for mobility in terms of speed and torque range and ability to withstand temperature extremes. There are also special-purpose elements, such as terrain-properties sensors and the field programmable gate arrays whose logic can be specified by a customer.

The wheels of terrestrial vehicles are almost always compliant by means of pneumatic pressurization, but this is generally unsuitable for space environments where there are extreme temperature variations. On the airless Moon the tires of the LRV were made of a metal mesh, but these are not believed to be suitable for very long-range missions. To date, Mars rovers have had rigid wheel rims which do not conform to the terrain. Tracked vehicles distribute the load across a much larger area and are frequently used on Earth for soft terrain, but the proclivity of rocks to become entrained in the running gear means either a very high torque in order to crush the rocks, or elasticity to make "throwing a track" unlikely.

A good example of the lessons learned from operating the Curiosity rover is the redesign of the wheels for the NASA Mars 2020 rover to avoid the wear and tear observed on the Curiosity wheels by driving over sharp, pointed rocks. The 2020 rover will be improved by having thicker, more durable aluminum wheels of reduced width and a slightly increased 52.5 cm (20.7 in) diameter than the 50 cm (20 in) wheels of Curiosity.

Fig. 5.14 The redesigned wheel for the Mars 2020 rover. Photo courtesy of NASA/ JPL-Caltech.

The aluminum wheels will be covered with cleats to enhance traction and curved titanium spokes to provide springy support.[10]

The footpads for landers can be made very lightweight, but they have not yet been applied to walking vehicles because of their low power efficiency. Mobility actuators are generally at the extremities of a vehicle and need to operate under a wide range of temperatures that dip to cryogenic temperatures, and are therefore difficult to thermally protect. Due to the chilly night-time temperatures on Mars, the rovers there have required significant power to heat the wheels and steering actuators prior to starting in the morning. Dissipating waste heat can also be an extreme challenge. For example, motors for lunar cargo vehicles will require to sustain thermal cycles lasting 29.5 Earth days, enduring a tremendous range of temperatures. The LRV for Apollo was obliged to operate only during the fairly benign "early morning" conditions. A rover capable of operating "all day" will probably require a heat-pipe or similar technology to shed excess heat during the hottest part of the day and then "disconnect" thermally from the environment to prevent heat loss during the frigid night.

Existing Mars rover wheel actuators are designed to have a rim thrust equal to half the weight of the vehicle in order to allow the vehicle to extricate itself from holes, even though the peak-power operating point is typically associated with a rim thrust less than

---

[10]For more information on the Mars 2020 rover go to:
https://mars.nasa.gov/mars2020/mission/rover/body

5 per cent of the vehicle weight. In terrestrial vehicles this is accomplished by a transmission capable of multiple speeds, but for rovers it has been achieved by significant design compromises in order to preserve simplicity. Future legged systems could incorporate some sort of bio-inspired elastic energy storage into their actuators to achieve a reasonable efficiency, but this may pose some control challenges. Prototypes for rovers have demonstrated continuous and "fast" autonomous navigation by incorporating field programmable gate arrays. In addition to speeding up perception (stereovision, terrain classification, state/pose estimation), they are utilized to assess hazards and evaluate the safety of paths to prevent flip overs, high centering, sinkage, and other factors relevant to mobility. All of these elements will not only require to be developed to a sufficient level of technology readiness but also to be flight qualified.

Some of these components are:

- Wheels for Planetary Surfaces.
- Actuators for Mobile Robots.
- Terrain Adhesion.
- Sensing Terramechanical Properties.

Ames Research Center

Since 1990, the Intelligent Robotics Group (IRG) at the Ames Research Center has led NASA in the development of robotics technology to reinvent planetary exploration. Ames develops and integrates new technologies into autonomous systems for flight missions and terrestrial demonstrations. It undertakes applied research in a broad range of areas which include computer vision, geospatial data systems, planetary mapping, human/robot interaction, software architectures for robots, and interactive 3D visualization.

Areas of research and development include:

- Adaptive control technologies.
- Control agent architectures.
- Embedded decision systems.
- Evolvable systems.
- Intelligent robotics.
- Adjustable autonomy.
- Distributed and multi-agent systems.
- Goal-level commanding.
- Planning and scheduling.

See also Chapter 4.1.4.

Johnson Space Center

In addition to the robotics work for the Technology Roadmap mentioned above, the Human Research Program (HRP) at NASA is dedicated to advancing the best methods and technologies to support safe, productive human space travel. Part of this charter is the use of robots to assist astronauts and to assure their safety. The Dexterous Robotics Laboratory (DRL) was created by the Johnson Space Center as a reconfigurable facility to create advanced robotics. Its Robonaut 2 was used on board the International Space Station. The R5 robot, nicknamed Valkyrie, then expanded upon the Robonaut experience.

The Valkyrie team also partnered with the Florida Institute for Human and Machine Cognition (IHMC) to implement a walking algorithm on NASA hardware in readiness for participating in the Space Robotics Challenge, part of NASA's Game Changing Development Program and Centennial Challenges.[11]

The laboratory includes:

- Dexterous Manipulator Testbed.
- Multi-Use Remote Manipulator Development.
- Systems Engineering Simulator.
- Virtual Reality Laboratory.

The laboratory's work also integrates with their activities in other technology areas.

A spinoff from the Robonaut 2 project was the X1 robotic exoskeleton. NASA and the IHMC in Pensacola, Florida, assisted by engineers at Oceaneering Space Systems in Houston, Texas, have jointly developed the robotic exoskeleton. This 26 kg (57 lbs) device is a robot which a human can wear "over" their body, either to assist or inhibit movement in leg joints. The motivation was to help astronauts to maintain their physical conditioning; operating in its "inhibit" mode, it is an in-space exercise machine that supplies resistance against leg movements and build up muscle tone. In its "assist" mode it offers the additional benefit of being able to help paraplegics to walk in Earth gravity. Worn over the legs using a harness that runs up the spine and around the shoulders, X1 has 10 degrees of freedom using four motorized joints positioned at the hips and the knees, six passive joints for sidestepping, turning and pointing, plus a flexible foot. There also are adjustment points to enable the X1 to be used in many different ways.

X1 has been in research and development for several years, with the primary focus being on design, evaluation, and improvement of the technology. NASA is examining the potential for the X1 as an exercise device to improve the health of astronauts on future long duration missions, particularly to Mars. Without taking up valuable mass or volume during missions, X1 could replicate common crew exercises that are vital to keeping astronauts healthy in microgravity. It will also measure, record, and real-time-stream data to Earth to enable doctors to monitor the impact of the exercise regimen.

Coupled with a space suit, a future version of X1 could improve the ability of astronauts to walk in reduced gravity fields and supply the extra force needed to carry out "heavy work" in caves and lava tubes on the Moon and Mars.

Here on Earth, IHMC is interested in developing and using X1 as an assistive walking device. With IHMC walking algorithms, it has the potential to produce high torques to assist walking in varied terrain, and climbing stairs. Preliminary studies are already under way. The potential of X1 extends to other applications, including rehabilitation, gait modification, and offloading weight from the user. Preliminary studies by IHMC have shown X1 to be more comfortable, easier to adjust, and easier to put on than previous exoskeleton devices. Researchers plan on improving on the design, adding more active joints to areas such as the ankle and hip which will increase the potential uses for the device.[12]

---

[11]For a two minute video on NASA's Space Robotics Challenge go to:
https://www.youtube.com/watch?v=aTpDj5hDO6s

[12]For a one minute video of X-1 go to: https://www.youtube.com/watch?v=ldl-V1n7Efw

Jet Propulsion Laboratory

The Robotics Section at the Jet Propulsion Laboratory (known as JPL Robotics) addresses a range of problems from basic research through to implementation of space missions. It provides unique capabilities for development of systems for in-situ robotic exploration, including visual perception of environment structure and properties, mobility in natural terrains, contact operations for science instrument placement and sampling the environment, operator interfaces, and modeling and simulation of robots and their surroundings.

About half of JPL Robotics is directly supporting spaceflight projects, and the other half is split between NASA and non-NASA funded robotics research. All research is designed to solve technical problems that address immediate research program needs and also contribute to the technology infusion for a range of future missions. Its activities include:

- Provide operators and systems analysts for Mars rovers and landers.
- Assist in the design of the Mars Helicopter technology demonstrator.
- Assist in the design of the Mars 2020 sample acquisition and caching system.
- Assist in the design of a future Sample Return Rover for Mars.
- Assist in the design of technology for climbing, grasping, gripping, and hopping robotic systems.
- Develop descending and climbing robots, such as Axel and Lemur.
- Assist in the design of a hopping robot, such as Hedgehog.
- Assist in the design of robots that can be used in orbital construction activities, such as RoboSimian.[13]
- Develop modeling, simulation and visualization for a wide variety of applications.
- Develop robotic designs for terrestrial land, sea, and air applications.

Kennedy Space Center

The In-Situ Resource Utilization (ISRU) Laboratory established at the Kennedy Space Center is a multifaceted group that conducts applied and basic research in chemical processing systems in support of Advanced Exploration Systems (AES) and the Space Technology Mission Directorate.

The laboratory conducts the following research:

- Designs, builds, and tests reactors and systems for the chemical processes required for outposts on the Moon and Mars.
- Designs, builds, and tests systems to remove the contaminants produced during regolith reduction in ISRU reactors.
- Characterizes contaminants in regolith simulants and also contaminants produced during the reaction, and develops the process that will remove contaminants in the gas phase as well as liquid water to a level safe for water electrolysis.
- Develops inert anodes to be used in Molten Regolith Electrolysis.
- Tests reactor materials for hydrogen that permeates at high temperatures (up to 1000°C).

---

[13]To watch a video of RoboSimian go to: https://www.youtube.com/watch?v=3HFXO_qx5ZY

- Testing and designing a system to produce methane from food wrappers and left-over food of an outpost on the Moon.

KSC also hosts the annual NASA Robotic Mining Competition (RMC) which involves projects submitted by university students.

Langley Research Center
The Langley Structures and Materials Test Laboratory manages the Commercial Infrastructure for Robotic Assembly and Services (CIRAS) project. This focuses on issues regarding the construction of large space structures, including:

- The Tension Actuated in Space MANipulator (TALISMAN) robotic arm for the deployment of a solar array.[14]
- The Strut Assembly, Manufacturing, Utility & Robotic Aid (SAMURAI).
- The NASA Intelligent Jigging and Assembly Robot (NINJAR) that is to demonstrate autonomous truss building.

The objective of CIRAS is to advance technologies to enable the autonomous construction of large platforms in space. It is a collaboration with industry and is seeking to develop a "toolbox" of capabilities for use in servicing, refueling, and ultimately the construction of assets on orbit. It works in collaboration with other NASA Centers and industry partners. The project is part of the In-Space Robotic Manufacturing and Assembly project portfolio managed by NASA's Technology Demonstration Missions Program, and sponsored by NASA's Space Technology Mission Directorate.

Other areas of Langley's robotics research include:

- Lightweight Sensing and Control Systems.
- Amorphous Surface Robots.
- Safe2Ditch Technology.
- In-Situ Characterization and Inspection of Additive Manufacturing Deposits using Transient Infrared Thermography.

See also Chapter 4.1.6.

5.4.3   Industry Robotics Research

The robotics industry comprises a multitude of companies targeting a very large number of applications. Some are now quite popular, such as driverless cars and drones. You might have a floor sweeper in your home, or a robotic lawn mower. Would you believe there is even a robot that not only does your laundry but will also fold the items according to the type of garment or person and then put them into separate drawers! While some tedious, boring, or hazardous human jobs are now being done by robots, in some cases even the service dog is being replaced by a robot. When you see a blind man walking in the street with a "seeing eye" robot instead of a German Shepard, you will appreciate that the world has really changed.

---

[14]To watch a one minute time lapsed video of TALISMAN go to:
https://www.youtube.com/watch?v=fLKqvAsHfss

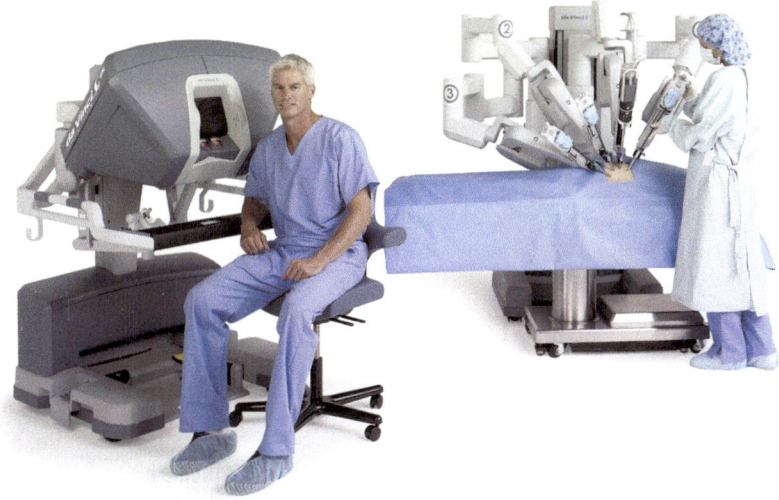

Fig. 5.15 The da Vinci Surgical System. Photo courtesy of Intuitive Surgical and Lourdes Health.

Although not all robotic companies are focused upon space applications, their underlying technologies may be applicable; for example, voice commands, 3D vision, AI software, and even computer chips. However, technologies for space applications such as robotic satellite servicing might not be directly applicable to working in caves and lava tubes on the Moon and Mars. Nevertheless, as these technologies evolve or are adapted they might become relevant.

There are well over a hundred robotics oriented companies around the globe, including in Australia, Canada, Croatia, Estonia, France, Germany, Greece, India, Latvia, Pakistan, Poland, Portugal, Russia, Singapore, Spain, Turkey, and the UK.

Some of the applications in automation and artificial intelligence include the following:

- Wherever there is repetitive or menial labor.
- Warehouse Logistics Management.
- Advanced prosthetics.
- Automated storage and retrieval in warehouses.
- Robotic surgery such as the da Vinci Surgical System.
- Solar power systems.
- Sea patrols.
- Whole Farm Management.
- Automated landing systems.
- Driverless cars and trucks.
- Security.
- Home companion.

Here are a few robotic companies that are representative of the industry:[15]

- Boston Dynamics started as a spin-off from MIT, where they developed the first robots that ran and maneuvered like animals. It is now owned by the SoftBank Group, a Japanese multinational holding conglomerate that has its headquarters in Tokyo. One of Boston Dynamics' many robots is Atlas, which was designed for search and rescue work but has been very useful in other roles.[16]
- Intuitive Surgical, Inc., is a California based corporation which develops, manufactures and markets robotic products designed to improve clinical outcomes of patients through minimally invasive surgery. By providing a surgeon with superior visualization, enhanced dexterity, greater precision and ergonomic comfort, its da Vinci Surgical System makes it simpler for surgeons to undertake minimally invasive procedures involving complex dissection or reconstruction.
- The iRobot Corporation was founded in 1990 by three MIT graduates. It focuses on consumer robots for inside and outside chores with the award-winning Roomba vacuuming robot and the Braava mopping robot. As the leading consumer robot company, it is involved in building smart homes. In addition, it is also seeking to develop robots for space exploration and military defense.
- GreyOrange was founded in 2011 and has its headquarters in Singapore, but the business extends to India, Hong Kong, Japan, Germany and UAE. It has designed and manufactured a range of advanced robotic system for logistics and supply chain management including warehouse automation and fulfilment centers.
- EPSON Robots is the robotics design and manufacturing department of Japanese corporation Seiko Epson, the brand-name watch and computer printer producer. It has created Cartesian, SCARA, and 6-axis industrial robots for factory automation products and solutions.
- Rethink Robotics started a boom in the arena of collaborative robots. A "cobot" or "co-robot" is a robot which is intended to physically interact with humans in a shared workspace. Before their first co-robot, Baxter, was launched in 2012 the robots were handled by skilled technicians.
- Waymo emerged from the Google "driverless car project" and became independent in 2016. Both companies are subsidiaries of Alphabet, Inc. The project started in 2009. The first successful test drive was in 2017. Meanwhile, the company has made corporate acquisitions intended to boost its portfolio in robotics.
- DJI is a Chinese company specializing in drones and unmanned aerial vehicles with stabilization camera systems. Its Geospatial Environment Online (GEO) is a best-in-class information system providing its users with up-to-date guidance where normal flights may be limited by safety concerns or regulations.
- Locus Robotics is another company which has specialized in warehouse automation. It supports the evolving needs of the E-commerce industry, and offers autonomous, collaborative robots to work alongside humans.

---

[15]Watch an interesting video about several types of robots at:
https://www.youtube.com/watch?v=8vIT2da6N_o
[16]Watch Atlas on the following video: https://www.bostondynamics.com/atlas

- SCHUNK started as a mechanical workshop by Friedrich Schunk. It is a world leader in gripping and clamping systems and hydraulic expansion technology.
- Vex Robotics makes tools for educators and mentors to enable problem-solving in topics across the range of Science, Technology, Engineering and Mathematics (STEM).
- Autonomous Solutions provide vendor-independent automation systems for ground vehicles. Its clients include Anglo American, Rio Tinto, Ford Motor Company, Luke Air Force Base and government agencies such as Los Angeles Police Department. The company has automated around 75 vehicle types in nine industries.
- Hanson Robotics is an AI company devoted to research, robotics and the development of the world's most remarkable humanoids. It was founded in 2013 and achieved fame with the robot Sophia, which is (thus far) the only robot to be awarded citizenship of another country (Saudi Arabia).[17]
- SoftBank Robotics (formerly Aldebaran Robotics) is a French company with its headquarters in Paris but owned by the Japanese group SoftBank. It specializes in autonomous humanoid robots and has produced a "family of robots" which address a number of objectives: Nao is geared towards programming, teaching and research, Romeo is intended to assist people, and Pepper is for customer relations.

5.4.4 Robots for the 21st Century

Robotics is now such an integral part of the fabric of world-wide society that the competition is extreme. Indeed, the future is so bright that companies are buying each other up to gain a market share or even a new market. This new technology is disrupting markets and bringing game changing innovation to business, health care, and education. An excellent place to see this at work is at the International Consumer Electronics Show (CES), an annual trade show held by the Consumer Technology Association in January at the Las Vegas Convention Center in Las Vegas, Nevada. Other evidence is that Exchange Traded Funds (ETF) of robotic companies on the stock exchange are worth billions of dollars.

"There's never been a more exciting time especially in robotics, automation, and artificial intelligence, where the rate of change and innovation has become exponential and is poised to surpass anything we've ever seen before. Perhaps it's because of the sheer pace of change that headlines about driverless cars and self-flying delivery drones don't turn heads these days." So says William Studebaker, President & CIO of ROBO Global in summarizing the key points about the future of robotics. He made a number of specific observations. Note that while they were commercial observations, the software and technologies are applicable to space in many cases.

Studebaker's observations included the following points (some of which have been paraphrased and shortened; his points about investing have been omitted):

---

[17]Watch her and other robots on the Jimmy Kimmel show at:
http://www.hansonrobotics.com/robot/sophia

- Robotic Automation and Artificial Intelligence (RAAI) is accelerating at lightning speed. It is a rare day that news of yet another new development is not in the spotlight – no matter how amazing that may be. A hand-held translator that allows you to understand and communicate in more than 80 languages? A humanoid robot that is granted citizenship in Saudi Arabia? Suddenly the world of wonder that we envisioned "back in the day" is no longer a galaxy far, far away, but one we've begun to take for granted in our everyday lives.
- Robotics and AI are already an integral part of our lives. In 2017 Amazon routinely delivered over 5 billion Prime items within two days or less; an accomplishment which hinged on robotics-driven factory automation and AI reasoning. Siri and Alexa help us do everything from texting without a keyboard, to picking the perfect playlist, to discovering a great new place to eat and then avoiding traffic to arrive there on time. Just as the Internet radically transformed how we work and communicate, robotics with AI is revolutionizing how we live, work, and play.
- Big data, AI, and e-commerce will dictate the new global giants. Big data is now the #1 driver of success; the power to process vast amounts of data to provide unique solutions which are customized to the individual in near real-time. The winners are robotics-driven retailers, sensor suppliers, and automation solution providers right across the supply chain.
- The leaders of tomorrow are relying more than ever on machine learning to process, trend, and analyze the massive amounts of data that tell them who is buying, what they are buying, and when they are buying; and how to deliver quickly and cost-effectively. The "must haves" are now AI and machine learning.
- AI will give predictive marketers what they have been seeking for more than 20 years: the power to process vast amounts of data. Every big tech company that is serious about its AI ambitions will attempt to develop its own AI chips to gain an edge in computing power, thereby echoing steps already taken to build specialized chips for vital tasks such as image and speech recognition.
- Robotics innovation and self-guided vehicle technologies are powering collaborative robots to complement human labor and therefore increase productivity. Autonomous robots that can navigate around unstructured warehouses spaces to autonomously pick and place stock keeping units (SKU) and collect inventory data are suddenly enabling new players to scale up automation to keep pace with Amazon and other large retailers without incurring significant downtime.
- Robots will replace humans in service roles, enabling humans to offload ever more service-based tasks to robots. It has been estimated that some 85 per cent of customer interactions will be managed by AI by 2020. The winners are financial services, agriculture, telecommunications, retailers, and robotics providers.
- Healthcare as we know it is set to vanish. Advancements in analytics, AI, and the Internet of Things (IoT) will completely transform how patients are diagnosed and treated, how medical facilities operate, and how health issues are predicted and prevented. Wearable diagnostics and the ability to stream a patient's vital signs directly to their medical records will greatly improve patient outcomes by reducing errors and time to treatment. The use of wearable patches and other devices that sit on the skin will enable continuous biometric monitoring and drug delivery. And

robot-assisted surgeries will continue to improve surgical outcomes and thereby extend the career span of the most experienced surgeons. Block chaining, which has been used extensively in financial services, will allow medical record interoperability throughout the world. These changes will simultaneously reduce costs throughout the healthcare system and also enable doctors to deliver a higher level of care to increasing numbers of patients every day. The winners are patients, healthcare providers, robotics and AI providers.

- Computers will learn to think, and think to learn. Computers will be able to think and learn like humans on factory floors, offices, and the cars we drive. Already semiconductor chip makers are refining machine learning to make truly autonomous cars, virtual assistants, and an unending list of automated processes. Robots can be taught new processes and then, once they have mastered that new knowledge, share it with "co-workers" that share a network, enabling computers to independently optimize logistics, deliver efficiencies, and increase returns. The winners are semiconductor companies, manufacturers and robotics providers.

- Augmented and Virtual Reality will redefine what we view as "real." As new products and capabilities hit the market the interest and investments in augmented reality, virtual reality and mixed reality will escalate. AR is already being adopted in digital transformation for applications to support construction and on-site repair and VR workstations are being powered by everyday eyeglasses. As this technology gets more powerful, and finds its way into more aspects of our lives, it will start to merge with autonomous vehicles, machine learning, and edge computing technologies. Just as the release of iPhone X marked a watershed moment for broadly viable facial recognition, 2018 will see many AR uses move from mere speculation to reality, with VR breakthroughs not far behind. The winners are AR and VR technology providers, automakers, manufacturers, and 3D "additive" printing.

- Voice-assisted AI and voice computing will be fully embraced. Amazon delayed the launch of its Alexa by a full year to be sure the device could respond to the user in less than one second, rather than two. The reason: the company wanted to produce a human interface, not a computer. That diligence has paved the way for an era of voice computing. Hence 2018 will bring voice-assisted AI and voice computing into the mainstream as businesses and consumers seek the power of these technologies. In fact, voice-assisted AI is already being used in everyday tasks that range from scheduling office meetings to trading stocks, to ordering groceries in real time. The winners are AI technology providers and consumers.

It is apparent that these same advances and many more have been made in the aerospace industry as every contractor that makes an airplane or a spacecraft uses advanced robotics. Each NASA contractor, be it prime, subprime or even end-of-chain vendor, makes use of robotics.[18]

---

[18]For a 20 minute video of many of the above mentioned robots go to:
https://www.youtube.com/watch?v=kbaDdg4LA9k

5.5 POWER SYSTEMS

5.5.1 Introduction

It is clear that human missions back to the Moon and on to Mars cannot use the current state-of-the-art power systems, namely the solar panels of the Opportunity rover or the radioisotope thermal generator of the Curiosity rover.

NASA recognized this situation in Area 3 of its Technology Roadmaps. Many power technologies are not even close to a high readiness level, but the Agency moved one project, known as Kilopower, into the Game Changing Development program with a greater priority. The following sections describe the desired power system technologies, then discuss the power system of Curiosity, which represents the state-of-the-art, and the future in the shape of the Kilopower project that can be developed to supply various levels of power for different missions.

For a "precursor" mission to the Moon or Mars of the type advocated by this book, the mission planners still have to determine the power requirements but this will probably be in 10 KWe range and be scalable to higher levels. It is also clear that as the initial base expands, the power systems will be increased or augmented using the latest advances in solar power and the storage batteries.

5.5.2 Relevant Technologies

In order to tackle all of the technology challenges pertinent to spaceflight, NASA has produced roadmaps to all of its likely exploration targets. Technology Area 3: Space Power and Energy Storage is further divided into (1) Power Generation, (2) Energy Storage, (3) Power Management and Distribution, and (4) Crosscutting Technologies.

The NASA technology-focused report covers the entire spectrum of missions, and all NASA Centers are involved as appropriate to their particular charter and areas of interest; for example, the Glenn Research Center has been investigating the Kilopower system.

The Roadmap generally characterizes the power system areas as follows (with comments):

- Power generation subsystems include solar arrays, radioisotope thermal generators, fission and fusion nuclear reactors, and fuel cells. This area includes methods of generating power from chemical, solar and nuclear sources, along with their energy conversion and harvesting technologies. (Note that although these systems are part of the overall NASA program, including the "dream" of fusion reactors, none of them are applicable to the kind of "precursor" mission describe in this book.)
- The electric energy storage options for space missions include batteries, regenerative fuel cells, and flywheels. This area considers these ways of storing energy after that energy has been generated from solar, chemical, and nuclear sources when it is not to be consumed immediately. (Further comments will be focused on the missions describe in the book.)
- Power management and distribution describes the technologies that are needed to manage and control electrical power generated from a source. These technologies have limited ability to tolerate deep-space radiation. The range of operating

temperatures of components are so narrow as to require substantial thermal management and control schemes with some degree of human intervention.

- Crosscutting Technology areas are those discussed in other technology roadmap areas, such as Analytical Tools, Processing, Science Modeling Frameworks, Languages, Tools and Standards, and Nanotechnology. In some cases they relate to Aeronautics rather than space. And some areas relate to deep space electrical power systems.

These four areas were further broken down into 19 sub-areas, each of which had many sub-goals. As a result, there are over a hundred desired capabilities at different levels of technology readiness. Some of these areas are generic to crew support in general, some are specific to powering their vehicles, robots, habitats, and support apparatus such as tools for erecting a habitat. Some of the goals are related to powering spacecraft.

Some technologies are directly applicable to a mission whose objective is to explore for an acceptable cave site and use that for protection. Not all of the areas can possibly be covered in this book. The following discussion is related to those select technologies that support such a mission. A power supply will be required for the various vehicles, habitats, robots, tools and experiments that the crew will employ.

5.5.3 NASA's Power Research

Once the crew has landed on the Moon or Mars, the one obvious technology that they will need is the primary source of power. All of the previous spacecraft sent to these destinations were powered by solar or nuclear radioisotope systems and batteries, but much more power will be required for future spacecraft, rovers, and robots.

The problem was (and remains) that the NASA mission directorates will not include a fission power system in their solicitations until such a system has been flight qualified, and science teams won't propose new missions that require more power than is currently proven and available. An attempt to break this "chicken and egg" situation is the Kilopower project, the goal of which is to advance that technology to a level sufficient for a flight development effort, and so encourage scientists to propose new ideas for higher power missions.

For perspective, the following is the current state-of-the-art as represented by the Mars Science Laboratory's Curiosity rover.

The Current Power System
The Curiosity rover now exploring Gale Crater on Mars draws its power from a radioisotope thermoelectric generator. In it, thermocouples convert the heat from the radioactive decay of plutonium into electricity. Such systems have been used for decades. The current model, the Multi-Mission Radioisotope Thermoelectric Generator (MMRTG), has a mass of 45 kg (99 lbs) and its heat source is 4.8 kg (11 lbs) of plutonium-238 dioxide. It provided some 110 watts of electricity and 2000 watts of thermal power at launch. In everyday terms, this was sufficient to light up one bright bulb! Two 42 ampere-hour lithium-ion rechargeable batteries meet the peak demands of rover activities when the demand temporarily exceeds the output of the generator. The operational life of the MMRTG is 14 years, with little decrease across that time. It was supplied to NASA by the U.S. Department of Energy. Unlike the solar powered Spirit and

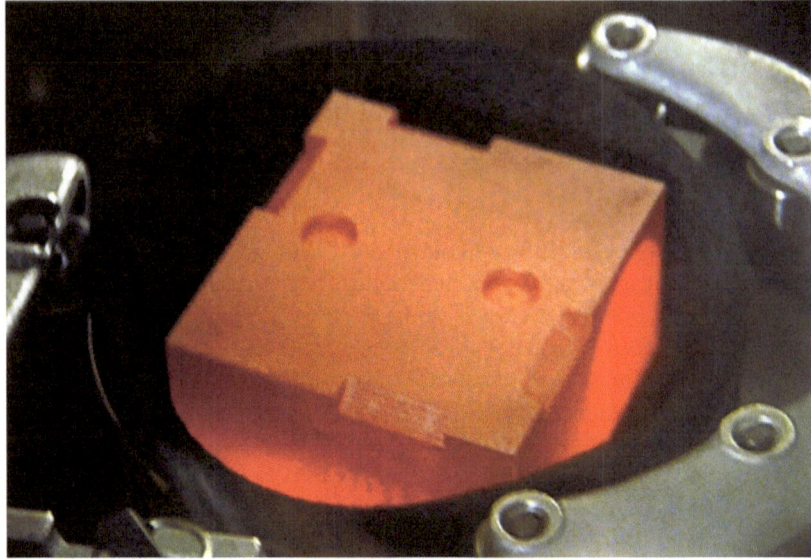

Fig. 5.16 Curiosity's source of power. Photo courtesy of the Idaho National Laboratory.

Fig. 5.17 The MMRTG after a fit check with Curiosity in 2011. Photo courtesy of NASA/ KSC/Kim Shiflett.

Opportunity rovers, the MMRTG allows Curiosity's engineers significant flexibility in operating instruments, even at night, during dust storms, and through the winter season. The Mars 2020 rover will have a similar system.

But obviously a human mission back to the Moon (let alone one to Mars) will need more electrical power than can be provided by a radioisotope thermoelectric generator!

The Future Power Systems

This Technology Area describes subsystems that use nuclear fission as an energy source and convert the waste heat into electricity. The fission power reactors that were launched into space between 1965 and 1987 were operated at coolant outlet temperatures and thermal powers comparable to those required by a 21st century concept for a 40-kWe system.

Space reactor programs have succeeded in developing high-temperature, high performance fuels, materials, and heat transport systems. However, this hardware has not been flown in space. The experience gained from nearly seven decades of terrestrial fission systems can benefit the design and development of future space fission systems. Fuel and materials technologies developed for terrestrial systems (e.g., the Fast Fission Test Facility and the Experimental Breeder Reactor-II, both designed by Argonne National Laboratory in Idaho) will be applicable, especially for first generation space fission systems. Smaller fission systems with outputs in the 1–10 kWe range have potential as alternatives to radioisotope thermoelectric generators for science missions and will be a viable option for human outposts on Mars. Furthermore, high-temperature (~ 800 Kelvin or 980°F) fuels developed for terrestrial applications and test reactors could be useful for ultra-compact, high-specific-power space systems.

Note that this discussion does not describe the systems that would be required for the "grander" missions to the Moon and Mars, only those that are relevant to "precursor" missions. NASA's Technology Roadmap looks much farther into the future.

To satisfy the requirements of current mission architectures, the development of fission power should focus on three different power classes:

- A 1–10 kWe power module common to both a science mission bus power and human exploration surface power.
- A 10–100 kWe "workhorse" system for human exploration surface power and for the electric propulsion on "flexible path" missions; for example, a human missions to an asteroid.
- A 1–5 MWe low-specific-mass system for human missions to Mars. The current human exploration mission concepts may be vastly enhanced by developing technologies to enable very-high-power (exceeding 5 MWe), very-low specific mass (less than 5 kg/kWe) fission power systems for electric propulsion. The top technical challenges for fission systems are application specific. (This includes powering the entire "stack" of transit vehicles as well as the vehicles that operate on the surface.)

This discussion will focus on what the crews will require once they are on the surface in order to explore caves and lava tubes in search of a suitable location to set up their habitat.

A 1–10 kWe fission system would require high-uranium-density fuel; simple, light-weight core-to-power conversion heat transfer; low mass power conversion (at low power); and a design for safety, reliability, and minimum mass. Existing (or near-term) materials, fuels, power conversion, and technologies for rejection of waste heat could be applied. Simply put, the technologies exist for developing near-term, mission-enabling space fission systems. The main challenge for these initial systems is integrating the technologies into a safe, reliable, and affordable system.

The main challenge for second generation space fission systems (and beyond) is developing technologies that further improve performance: specifically, high-temperature reactor fuels and materials; high temperature, high-efficiency power conversion; and lightweight, high-temperature radiators. For example, for space fission power systems exceeding 100 kWe, their performance would benefit from better fuels, advanced power conversion, and lightweight radiator technologies. Innovative designs for the reactor would also improve performance. Some of the technologies needed to provide multi-megawatt systems that have a specific mass below 5 kWe/kg include the development of high-temperature (1800 Kelvin) fuels and advanced power conversion technologies with efficiencies greater than 30 per cent. Lightweight radiators capable of operating at temperatures between 500 and 1000 Kelvin would also raise the performance of the integrated system.

The System to Solve the Problem
The Kilopower project started in 2015 as a near-term technology effort aimed at developing the preliminary concepts and technologies required by an affordable fission nuclear power system that would enable long duration stays on planetary surfaces. Many NASA Centers and other government organizations participated in the effort.

The prototype power system was designed and developed by NASA's Glenn Research Center in collaboration with NASA's Marshall Space Flight Center and the Los Alamos National Laboratory. The reactor core was manufactured by the DOE's Y12 National Security Complex at Oak Ridge, Tennessee. The project is part of the Game Changing Development (GCD) program within NASA's Space Technology Mission Directorate, and is managed by NASA's Langley Research Center. The project will remain in the GCD program until transitioning across to the Technology Demonstration Mission program in Fiscal Year 2020.

The primary goal of Kilopower was a full power test at the Device Assembly Facility at the National Security Site in Nevada within 3 years. This test, known as KRUSTY for Kilopower Reactor Using Stirling TechnologY, had three main objectives:

- Operate the reactor at steady state with a thermal power output of 4 kWt at a temperature of 800°C (1472°F).
- Verify the stability and load following characteristics of the reactor during nominal and off-nominal conditions.
- Benchmark the nuclear codes and material cross sections on the basis of the test data.

The KRUSTY test hardware was specifically designed to be as representative of a flight prototype as possible in order to validate the performance of the full scale system and components. This would maximize the applicability of the data to the flight system. Hence the core, reflectors, heat pipes, Stirling simulators and shield structure were all full scale components for the 1-kWe Kilopower system.

In September 2017, the U.S. Department of Energy's Y12 National Security Complex completed all of its quality assurance checks and received approval for packing and shipping of the Kilopower experiment's highly enriched uranium core to the National Security Site in Nevada. After that, the Kilopower project team successfully completed a non-nuclear dry run of the experiment assembly and equipment hardware at NASA Glenn ahead of its shipment to Nevada.

The flight prototype test reactor is about 1.9 m (6.5 ft) tall and uses a highly enriched solid, cast uranium-235 reactor core which is about the size of a paper towel roll. Reactor heat is transferred via passive sodium heat pipes made of a nickel-chromium-tungsten-molybdenum alloy, and then converted into electricity by Stirling engines. It was designed for an electrical output of up to 1 kW. The first tests employed a depleted uranium core manufactured by the Y12 National Security Complex; exactly the same as the regular highly enriched uranium core apart from the level of enrichment. The KRUSTY test marked the first time the U.S. had conducted ground testing using any space reactor since the SNAP-10A experimental reactor that was eventually flown in 1965, over a half century ago.

The KRUSTY reactor was operated at full power starting on March 20, 2018 during a 28-hour test with a 30 kg (62 lbs) core of 93 per cent uranium-235. An average core temperature of 800°C (1472°F) was achieved, producing about 4.0 kW of steady state fission thermal power. The test evaluated simulated mission failure scenarios that included shutting down the Stirling engines, adjusting the control rod, thermal cycling, and disabling the system that removed excess heat. The experiment was concluded by a SCRAM test. By demonstrating the reactor technology in a relevant environment, it achieved a readiness level (TRL) of 5.

Fig. 5.18 KRUSTY non-nuclear tests at the Glenn Research Center. Photo courtesy of NASA/GRC.

Fig. 5.19 The KRUSTY team takes her up to full power. Shown in the foreground from left to right: Marc Gibson (GRC/NASA) and David Poston (LANL/NNSA). In the background: Georgie McKenzie (LANL/NNSA) and Joetta Goda (LANL/NNSA). Photo courtesy of NASA.

After finishing the Kilopower Reactor Using Stirling Technology (KRUSTY) experiment, the Kilopower project team set about the task of developing mission concepts and performing additional risk reduction activities in preparation for a possible future flight demonstration to pave the way for future systems to power human outposts on the Moon and Mars. This project promises to be a major step towards enabling mission operations in harsh environments, missions that rely on In-situ Resource Utilization to make propellants and other materials, and missions that seek to explore caves and lava tubes in which to establish habitats.

5.5.4   Future Applications

A variety of science and human missions using fission power sources have been independently examined, with positive results. Although scientists have refrained from proposing kilowatt-class missions owing to their non-existence over the past half century, the successful Kilopower reactor demonstration should change this situation.

Specific interests in Nuclear Electric Propulsion (NEP) that use fission have acknowledged the fact that their power requirements are realistically outpacing the radioisotope fuel availability and production. Two decadal survey missions using NEP systems were studied by a Glenn Research Center multidisciplinary collaborative engineering team called COMPASS, whose primary purpose is to perform integrated vehicle systems analyses. Through these analyses, the team conducted trades studies and provided system designs for both exploration and science missions with the objective of delivering an orbiter to the Centaur class object Chiron and a Kuiper Belt Object (KBO). Both studies were able to meet the mission objectives with a 7–10 kWe Kilopower reactor. These missions are well suited for space reactors because the power levels are readily achieved with the abundance of uranium fuel. It may well be possible to enhance many of the decadal survey missions by using nuclear reactors. The prospects will be further investigated as development of the Kilopower technology continues.

The human exploration of Mars will undoubtedly be the greatest achievement of the century, if not actually the millennium. It is rapidly becoming a near term possibility.

The independent studies referred to in this book have pointed out some of the advantages of nuclear surface power, and how the Kilopower reactor can reduce several risks associated with the Martian environment that pose difficulties for a solar powered mission. Both the ISRU and crew phases of an early Mars mission could be achieved using several 10-kWe Kilopower reactors. Such a system won the mass and power trades for human missions by a factor of two, even for sites favorable for solar power. The advantages of nuclear systems are even greater for sites at high latitudes.

The Kilopower reactor is well on its way to surpassing the technology barriers that have prohibited nuclear power for the last half century. With a successful full scale ground test, the technical and programmatic risks for nuclear power in space are significantly reduced.[19] To put things into perspective, this pioneering 10 kWe space fission power system could run several average households continuously for at least a decade. Four Kilopower units would be sufficient to establish an outpost on the Moon and Mars.

5.6 SCIENTIFIC INSTRUMENTS AND SENSORS

5.6.1 Introduction

Unlike areas such as power, this area may be of a higher readiness level because the development of advanced sensors and instruments has been pioneered by the unmanned space program and the International Space Station program for many years. The two technologies covered in this section are lasers and spectrometers, both with over a half

---

[19]For more information about the Kilopower project, visit:
https://www.nasa.gov/directorates/spacetech/kilopower
For a two minute video from Los Alamos on Kilowatt go to:
https://www.popularmechanics.com/space/moon-mars/a15497211/nasa-los-alamos-mini-nuclear-reactor-power-outposts-moon-mars/
For a short YouTube video on KRUSTY go to:
https://www.youtube.com/watch?v=6K8SEkr9I3o
For a 13 minute video on The Problems of Power in Space go to:
https://www.youtube.com/watch?v=m2IiI4UVZP8

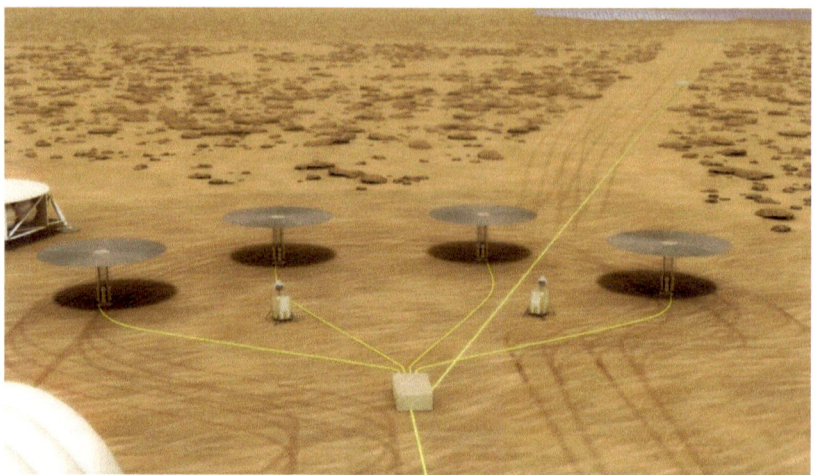

Fig. 5.20 An artist's rendering of an array of four Kilopower systems deployed on Mars. Photo courtesy of NASA.

century of proven flight experience. The new challenge for NASA is primarily to apply the latest technology to reduce the size and weight of the equipment, apply the latest in nanotechnology, and couple this with the latest in computing power and microminiaturization to lower the cost. Even those first flights back to the Moon and on to Mars will not occur for at least another decade. It is not envisioned that the flight rate will be as high as it was in the Apollo era; the costs are now much higher and, in the case of Mars, the target is much farther away. This will provide ample time to further advance the state-of-the-art in all of the hardware and software areas.

5.6.2   Relevant Technologies

To tackle all of the technology challenges with regard to spaceflight, NASA has developed roadmaps to all the planned exploration targets. With regard to sensor systems, this is Area 8: Science Instruments, Observatories, and Sensor Systems. That is further broken down into three areas: (1) Remote Sensing Instruments and Sensors; (2) Observatories; and (3) In-Situ Instruments and Sensors. The specific technologies in this area are applicable to missions from very small to large. The type of "precursor" mission discussed in this book requires technology to assist the crews with locating and exploring caves or lava tubes that could protect their habitat, and then, once that is established, support the mission. The observatories sub-area will not be covered here, nor will those technologies that are focused on the elaborate follow-on missions involving considerably greater resources and more flights.

The NASA technology-focused report covers the entire spectrum of missions and all NASA Centers are involved as appropriate to their particular charter and areas of interest; for example, the Johnson Space Center is responsible for sensors that are related to the spacecraft environment and the health of the crew, and the Los Alamos National

Laboratory may be involved in the development of sensors and instruments. Other Centers may be involved in testing or systems integration.

The Roadmap generally characterizes the science and sensor areas as follows (with comments):

- The technologies for remote sensing instruments and sensors include the sensors and instruments sensitive to electromagnetic radiation, numerous types of particle (i.e. charged, neutral, dust), electromagnetic fields, both direct current (DC) and alternating current (AC), acoustic energy, seismic energy, and whatever physical phenomena the science requires (at least as relevant to the early missions of the program).
- The observatory technologies are necessary to design, manufacture, test, and operate space telescopes and antennas which collect, concentrate, or transmit photons. They enable, or enhance large aperture monolithic and also segmented single apertures as well as structurally connected or free-flying sparse and inter-ferometric apertures. Applications span across the electromagnetic spectrum. (This area doesn't immediately apply to crews exploring the Moon and Mars.)
- The technologies for in-situ instruments and sensors include components, sensors, and instruments sensitive to fields and particles that can perform in-situ charac-terization of the space environment and planetary samples. They enable or enhance a broad range of planned and potential missions for the next several decades. In-situ instruments and sensors will support deep space missions to the Moon, Mars, and other targets such as comets and asteroids. (In our case, we are interested in technologies needed for a cave or lava tube that protects a habitat, but not all the others that discuss making fuel from the regolith or processing local materials to construct a permanent base).

These three areas were further broken down into 12 sub-areas, each of which had many sub-goals. There are approximately 86 desired capabilities at different levels of technology readiness. Some of these will be generic to any missions to the Moon and Mars because remote sensing is a fundamental tool for conducting science and mission operations. Others relate to remote sensing of other celestial bodies or to instruments that are landed on such bodies.

There are some technologies that are directly applicable to missions to explore the surface of the Moon and Mars for an acceptable cave site for protection. Some are applicable to searching for and finding the desired site and others that assist in taking science data once the facility has been established. Of course, not all of the identified areas can be covered in this book. The following discussion is therefore related to two cate-gories of instruments and sensors that can support a mission of the type advocated here.

5.6.3   NASA's Sensor and Instrument Research

The state-of-the art for flight-proven payload technologies is defined in terms of the instrument suites and associated technologies, such as sampling hardware on NASA's planetary landers and rovers. Similar instruments are flown in orbit. The state of the technology for in-situ instrumentation includes the payloads currently under development for forthcoming lander, rover, and sample return missions, an example being the suite of instruments for the Mars 2020 rover. This work is the task of the Jet Propulsion Laboratory. Other examples are the remote sensors for the International Space Station that were sponsored by the Johnson Space Center to continually monitor the crew's air.

Future in-situ technology needs and challenges are mission specific due to the broad range of radiation, thermal, atmospheric, and compositional environments encountered in planetary bodies across the solar system. It is envisioned that the technology needs of the early missions back to the Moon and on to Mars will not be as challenging as the elaborate follow-ons that are now being planned, but will selectively utilize the technologies that are now being developed.

Instead of reviewing the vast array of possible future in-situ technology needs, Area 8 focuses primarily on the key technology challenges that must be solved in order to facilitate the next logical steps in planetary exploration envisioned in the 2013–2022 Planetary Decadal Survey. Among the recommended future missions are ones that could be accomplished using currently available technologies. Other, more adventurous missions will require (or would significantly benefit from) new technologies. However, Area 8 does not provide detailed specifications for the other in-situ technology challenges that face future missions.

Among the broad range of planetary missions envisioned over the next two decades, the most technically challenging will require in-situ drilling, sampling, and analysis capabilities. These will benefit from NASA's sustained investments in technologies for in-situ exploration. Already flown missions and those that are currently under development will enable planetary scientists to leverage existing capabilities developed for in-situ atmospheric, organic, mineralogical, elemental, and geophysical studies. It is envisioned that the crews of the first missions back to the Moon and on to Mars will employ human-rated tools to collect samples for return to Earth. It is likely that in addition to seeking a protected site to establish their habitat, the first crews will deploy a modern equivalent of the Apollo Lunar Science Experiment Package.

Lasers

In addition to the lasers used by the descent modules to measure altitude, once on the surface the crews will need lasers of various kinds during surface exploration and operations; e.g. for measuring the range of an object, 3D mapping a cave or lava tube, and then determining the composition of its rocks or the quality of the interior environment.

The Los Alamos National Laboratory is collaborating with the French Center for the Study of Radiation in Space (CESR) to develop the SuperCam instrument to be carried by the NASA Mars 2020 rover. This will be a significant upgrade to the ChemCam (Chemistry and Camera) instrument of the Curiosity rover. This is just one type of laser, configured to enable researchers to sample rocks and other targets from a distance. A brief description follows that represents what might be available for the "precursor" missions to the Moon and Mars. Their rovers could be crewed, and their sensors used for site selection and determination in addition to scientific purposes. The Mars 2020 technology may be very close to that for the 2025–2035 window, by which time it will have been flight proven.

The ChemCam instrument package has two remote sensing instruments: the Laser-Induced Breakdown Spectrometer (LIBS) and the Remote Micro-Imager (RMI). The LIBS measures the elemental composition of samples and the RMI places that data

Fig. 5.21 The Cold Atom Laboratory payload. Photo courtesy of NASA/JPL-Caltech.

into a geomorphological context. Together, they help to determine which rock and soil targets in the vicinity of the rover are of sufficient interest to use the contact and analytical laboratory instruments for further characterization. ChemCam can also analyze a considerably larger number of samples than can be studied using the contact and analytical laboratory instruments. For example, the ChemCam team anticipates making daily analyses of the soil at the rover location in order to understand variations in the soils, both locally and regionally. What is more, it can analyze samples that are inaccessible to other instruments, notably vertical outcrops where LIBS can target individual strata using its submillimeter beam diameter. ChemCam has the capability to obtain passive spectroscopy data of rocks and soils on Mars. The fact that the spectral range covered by LIBS isn't typical of passive spectroscopy instruments makes it difficult to appreciate what information is useful in its spectra. On the other hand, passive spectroscopy does not have the range limitation of the LIBS, which focuses a powerful laser pulses on a small spot on target to ablate atoms and ions in electronically excited states, from which they decay and generate light-emitting plasma. The resulting light is focused by a telescope onto the end of a fiber optic cable and into the instrument for three dispersive spectrometers to record the spectra. However, the target has to be within 7 m (23 ft) of the rover.

Lasers of all kinds have been proposed for space missions.[20] For example, the Cold Atom Laboratory currently on board the International Space Station is able to study atoms

---

[20]To demonstrate the extremes to which they can be used for science go to: https://coldatomlab.jpl.nasa.gov/ It includes a short video.

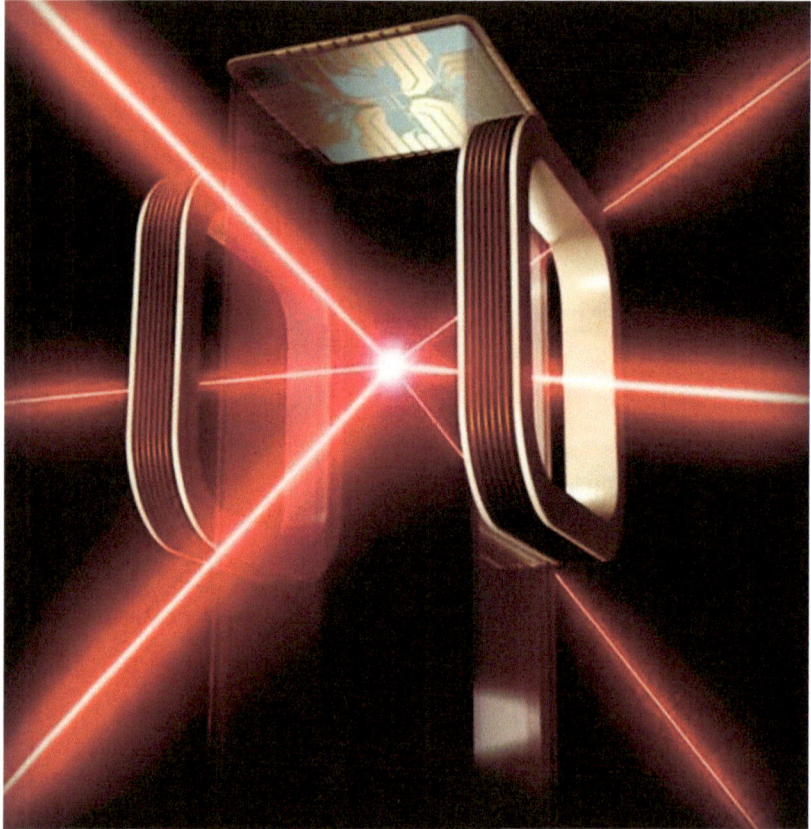

Fig. 5.22 The ISS Cold Atom Laboratory laser. Photo courtesy of NASA/JPL-Caltech.

at the coldest temperatures ever achieved (much colder than deep space) and is enabling scientists to study relativistic physics.

Mass Spectrometers

As a standard method for analysis, mass spectrometers (MS) have already reached other planets and moons including:

- The two Viking landers carried them to the surface of Mars.
- The Huygens probe that was released by the Cassini spacecraft carried a specialized gas chromatograph-mass spectrometer instrument to analyze the atmosphere of Saturn's largest moon, Titan.
- The Cassini spacecraft itself carried a mass spectrometer which analyzed the composition of the plumes emanating from the moon Enceladus.
- The Mars Phoenix Lander had a thermal and evolved gas analyzer mass spectrometer.
- The Cassini spacecraft carried a plasma spectrometer that measured the mass of ions circulating in Saturn's magnetosphere.

The wealth of knowledge developed over years as a result of studying cabin atmospheres, coupled with the clear need for routine monitoring to support long-term human presence in space, has prompted the development of an ambitious plan to implement multiple strategies for monitoring trace contaminants on board the International Space Station (and indeed future spacecraft). This includes using Grab Sample Container (GSC) and Solid Sorbent Air Sampler (SSAS) devices which archive samples for subsequent ground-based analyses, kits to monitor the presence of formaldehyde, and specific analyzers to monitor combustion products and highly toxic hydrazines. It will also (for the first time) use instruments that combine the advantages of a fast Gas Chromatograph (GC) and the sensitivity of an Ion Mobility Spectrometer (IMS) to perform in-situ analysis of a broad range of volatile organics.

With missions of the type that would be required in establishing a base on the Moon or Mars still in the preliminary planning stage, the requirements for future instruments will become even more demanding. A search has already started for a second-generation sensor for volatile organic analysis which is much smaller, less complex and more portable than is presently available, yet has equivalent or even improved analytical performance.

Technologies currently under consideration for the later operational phases of the International Space Station include Fourier-Transform Infrared Spectroscopy (FTIR), which is a technique used to obtain an infrared spectrum of absorption or emission of a solid, liquid, or gas. An FTIR spectrometer simultaneously collects high-spectral-resolution data across a broad spectral range.

Perhaps the two most important areas to make progress are further reductions in size (weight and volume) and the development of more powerful software for instrument control, data analysis, and self-optimization and diagnosis. Apparatus meant for life support monitoring must be capable of autonomous operation, and the user interface must allow varying levels of control.

One example of an instrument that combines robotic technology with a mass spectrometer is the ISS Robotic External Leak Locator (IRELL). This could help mission operators to detect the location of an external leak and rapidly confirm a successful repair. It was developed jointly between NASA's JSC and GSFC. Two instruments working in sync give the IRELL the ability to detect ammonia leaks on the International Space Station. One is a small mass spectrometer, the other is an ion vacuum pressure gauge which measures total pressure in space – it cannot distinguish between different gas molecules, but it can sniff for a large leak up close and determine the position of the leak to within several inches.[21] Canada's Dextre robot, which is fully controlled by ground operators, can pick up the tool and point it toward the station's cooling lines while engineers on Earth monitor the data.[22]

---

[21]For a tutorial on mass spectrometers watch this NASA Goddard YouTube video:
https://www.youtube.com/watch?v=_L4U6ImYSj0
[22]To watch a short video of Dextre using the IRELL go to:
https://www.youtube.com/watch?v=nNcRDBK8zxY

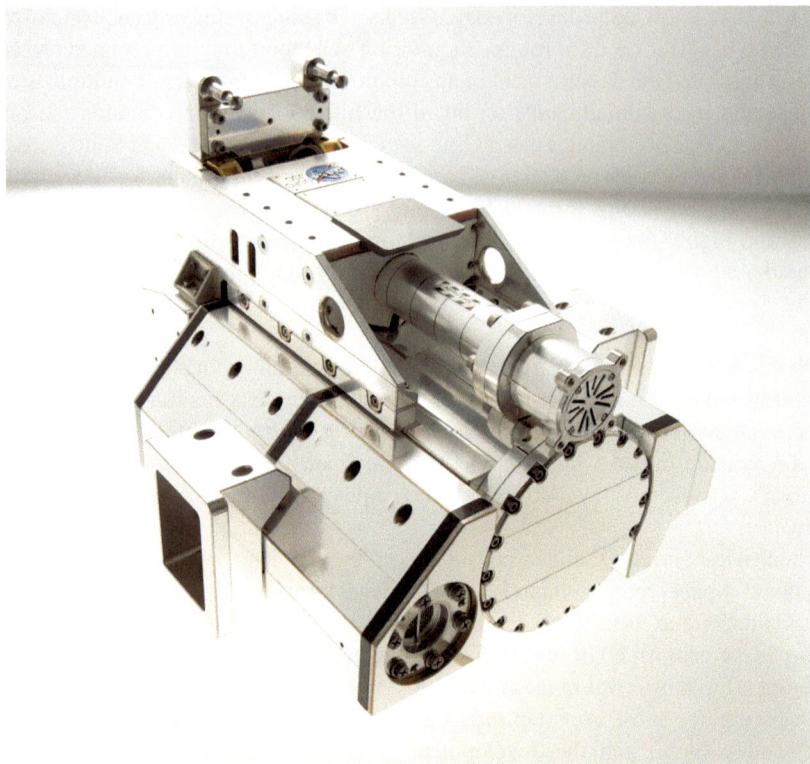

Fig. 5.23 The IRELL instrument for the International Space Station. Photo courtesy of GSFC/Chris Dunn.

5.6.4   Future Applications

By the time of the first missions back to the Moon in the mid-to-late 2020s, and the first to Mars in the mid-to-late 2030s, the lessons from the robotic planetary science missions and the new Orion spacecraft designed to fly humans into deep space will have defined the necessary technologies and flight qualified them. In the general area of remote sensing and the applicable technology in instruments and sensors, this sub-area is fairly well defined for the first decade; especially in relation to what has been discussed in this section; i.e. laser applications and the use of sensors such as spectrometers (both of which will have had a half century of experience). The science of lasers was developed in 1960 and research into its multitude of applications has produced hundreds of instruments. The future will only bring more products that will exploit the latest in technology to interface the sensors to more advanced software. The commercial applications and number of laser derived products are permeating society.

As NASA seeks to return humans to the Moon and establish our presence on Mars, the drive will be toward micro- and nanotechnologies. Already, hand-held and chip-sized ion mobility spectrometer detectors are in development. The lead times for functional, flight-qualified, miniaturized versions of these devices are still long. In the meantime, it is possible that further reductions in the size of mass spectrometry instruments (particularly their in-vacuum systems) might make mass spectrometry a viable technology for future human space missions.

The mass spectrometry instruments currently used by space applications have specifications and performance characteristics that are amazing, even by today's standards. But because these instruments are custom designed, their development costs are very high and a considerable time is required to certify them for flight.

Over the past couple of decades, two approaches have successfully introduced mass spectrometry instruments to space missions. Until recently, their use for the analysis of planetary atmospheres was focused on continual evolution of proven designs rather than the adoption of radically new technologies. This was logical, as the operating characteristics of mass spectrometry mean it is well suited for the vacuum of space. However, the use of mass spectrometry for monitoring cabin atmospheres has undergone a major switch from the use of archival sampling and subsequent analyses to technologies for in-situ analyses.

New technologies for analyzing trace contaminants in the unique environment of a spacecraft have been developed and tested. Early attempts to design volatile organic analyzers for trace contaminants were based upon proven planetary mass spectrometry instruments. Competing technologies based on Direct Sampling Ion Trap Mass Spectrometry (DSITMS), Gas Chromatography/Mass Spectrometry (GC/MS), and Gas Chromatography/Ion Mobility Spectrometry (GC/IMS) were evaluated. GC/IMS technology was chosen and a volatile organic analyzer based on this technique was developed for use on board the International Space Station.

As NASA looks forward to flying robotic probes and human missions in the next several decades, the requirements for analyses during both types of mission will become ever more similar. Hence, development efforts have already started with the objective of serving both robotic planetary probes and human missions. Micro-machined instruments have opened the door for consideration of ion traps, quadrupoles, sector instruments, and many other types of mass analyzers. New approaches such as IMS technology will continue to be developed and assessed for future missions. What is more, NASA has initiated programs in conjunction with academic and commercial sectors to develop technologies that will help in reducing the development costs. The goal of these various programs is to provide NASA with the equipment that is necessary for space missions and to transfer the resulting technologies to the commercial sector for terrestrial applications. It is likely that this approach will result in technologies which revolutionize how air quality monitoring is performed both in space and here on Earth.[23]

---

[23]For example, the Multi-Gas Monitor, which is the first application of laser spectroscopy technology to simultaneous measurement of multiple gases in a confined situation will not only be useful in a spacecraft cabin but also on Earth.

6

Considerations for Cave and Lava Tube Selection

6.1 LANDING SITE AND CAVE LOCATION

The approximate locations of caves or lava tubes will have been determined by mission planners while picking the landing site(s) using inputs from the science community and the flight planning and operations teams, on the basis of satellite reconnaissance and lessons from the robotic missions.

In addition to offering a reasonable possibility of there being a suitable cave and hopefully also water ice, the chosen target will have to meet many science, engineering, and operational considerations.

The site selection criteria are different for the Moon and Mars, but still share some fundamental criteria as follows:

- A latitude and longitude that best satisfies the overall mission and science objectives, including trajectory constraints.
- A location with potential water ice and hopefully biology. (A location that offers a good prospect of water ice might not be compatible with trajectory constraints focused on aborts and return constraints.)
- A location with confirmed caves or lava tubes and will hopefully provide adequate accessibility and protection.
- A site that meets the minimum requirements for radiation protection. (It is assumed this criterion satisfies those for micrometeoroids and temperature extremes.)
- A location that is suitable for landings by both the primary vehicle and the necessary support resource payloads.
- In the case of Mars, consideration for a location to minimize severe dust storms and reduced light levels.
- A skylight entrance to a cave or lava tube is not desirable unless it allows easy access. The first missions would be unlikely to include the option of repelling.

© Springer Nature Switzerland AG 2019
M. von Ehrenfried, *From Cave Man to Cave Martian*,
Springer Praxis Books, https://doi.org/10.1007/978-3-030-05408-3_6

Fig. 6.1   Mobile Home. Two pressurized rovers heading out. Art by Pat Rawlings ©SAIC.

After landing, hopefully close to their pre-positioned resources, the first crew will access those resources, particularly the rover or the equivalent of the latest version of the mobile Multi-Mission Space Exploration Vehicle (MMSEV) that will likely be essential to reaching the selected cave or lava tube area. One lunar vehicle was to support two astronauts for 14 days (half of the diurnal cycle). A proposed Alternate MMSEV would support a crew of four, but the variants for the Moon and Mars would be slightly different to match their mission objectives.

Fig. 6.1 is a 2007 concept for a crew collecting their pre-positioned resources and travelling to the selected site. It is presumed that the transportation vehicle is capable of serving as a temporary habitat for the crew while they are looking for the cave, and then until an inflatable habitat can be positioned inside the cave and pressurized – a phase of the mission that could last many days. The habitat would require a pre-positioned source of breathable gas sufficient to maintain pressure throughout the mission time; this could be a great many heavy pressure bottles. It is also envisioned that other resources are either packaged with the vehicle, or are separately pre-positioned and retrieved by the crew. These packages could include robots, science equipment, and tools.

Once the astronauts have acquired their vehicle and resources and are ready to move out, they must locate the cave or lava tube and carry out their evaluation of its suitability. They will first make observations of the cave from the outside. It is envisioned that one or more robots will be positioned to provide sufficient visual data to determine the usability and safety of the cave. This might take some time, because the crew will want to acquire as much data as possible and even have it evaluated by Mission Control on Earth. Depending on the communications time delays, the crew may begin exploring the outside for possible access to the cave.

Fig. 6.2 depicts a crew exploring a likely cave. Notice that there are a number of potential problems in this concept from 30 years ago. The three astronauts are evaluating what appears to be a cave that is sufficiently large but there is a huge rubble field and no obvious entrance. The floor must be cleared, but the rocks are clearly far too big to be

Fig. 6.2 Astronauts checkout a potential cave. Artistic rendering courtesy of John R.Lowery, 1988.

moved by the crew unaided, even under 1/6 gravity (the Moon) or 3/8 gravity (Mars). Consequently, unless upon closer examination the crew finds a reasonably accessible entrance, a robot would be needed to do the heavy labor of removing the larger rocks. A pathway would be needed not only for the crew, but for the robot or vehicle to maneuver the deflated habitat into the cave, and sufficient rubble cleared to accommodate the size of the habitat once it had been inflated (and expanded by follow-on crews). On the plus side, the roof appears suitably thick and stable. Perhaps a "bot" would be sent in to get 3D laser and video data, and Mission Control would make a recommendation on how to proceed.

Judging by the size of the astronaut in the upper left of the artwork, the cave entrance seems to be about the right size, perhaps 20 ft or so high, and the width is more than adequate. The entrance looks to be horizontal, which is good. After some rocks have been cleared, a crewman will be able to get a better look inside. Is the floor of the cave strewn with rubble? Or is the rubble only in the vicinity of the entrance? Is it likely that the floor inside is relatively flat? Is it stable? What does the ceiling look like? Could a robot clear enough space inside to insert the inflatable habitat? How long would that take? All of these questions (and more) would influence the evaluation process.

A similar evaluation would occur for a Martian cave. The gravity would be about twice that of the Moon and might affect the evaluation, but the questions would be the same. There would be additional assessment for dust storms, as well as the hope of finding water ice inside the cave. And there would also be a greater chance of encountering biology – a topic which presents its own challenges and opportunities.

Fig. 6.3 Hidden Assets. The Cave has been entered. Art by Pat Rawlings ©SAIC.

When it comes to evaluating a vertical cave, the hazards and risks are much greater than a horizontal one. The skylight could be huge, and too dangerous to approach without tethers that are anchored to the exploration vehicle. A form of smart "Mars Bot" would be required to penetrate the cave and transmit the data back to the astronauts and to Mission Control for an assessment. Even if the cave seemed acceptable, in this situation the mission would need the infrastructure to lower the habitat and all of the associated equipment down into the cavity. This may be too great a challenge for the first mission.

In Fig. 6.3 notice that that a pair of astronauts accompanied by a small robot have entered almost horizontally into what looks to be a lunar lava tube through a collapsed skylight. They have cleared a small entrance (to the left) and initiated exploration. The entrance to the cave and the size of the cave itself are certainly large enough for a habitat, but the rubble field seems too large and too rough for them to clear sufficient room without robust robotic assistance.

Horizontal access will be a major factor in assessing the suitability of a cave. Without technologies and techniques to provide simple ingress and egress at all stages from initial

Fig. 6.4 The Pit. Lowering astronaut and robot into a pit. Art by Pat Rawlings ©SAIC.

reconnaissance for site characterization to the emplacement of the infrastructure, establishing a habitat on the Moon or Mars at the bottom of a deep pit might not be very economical in energy terms, as it will require repeated transfer of appreciable masses between the surface and the floor of the cavity.

6.2   THE OPTIMUM HABITABLE CAVE

Bear in mind that NASA has not considered a mission that establishes a habitat in a cave. The current NASA funded contractor habitat designs are primarily for use on the surface or in deep space. Nevertheless, contractors have designs that seem adaptable. Consider that the proposed mission is for the first missions back to the Moon and on to Mars and that they would not require all the equipment currently planned for the larger mission that includes a long stay on the surface of almost a year and a half. A precursor mission that involves a cave or a lava tube would be considerably less expensive (tens, if not a hundred billion dollars less) and could be accomplished much earlier (probably a decade).

Let us now consider the optimum requirements for a cave or a lava tube for a precursor mission.

6.2.1   Sized for the Habitat

The cave would need to protect the crew from intense radiation and a number of other hazards. The first crew to land would be relatively small; probably three or four astronauts, since two crewmembers would remain in orbit in the deep space vehicle for the rendezvous and return home. What is the optimum cave? Well, it depends on the size of the inflatable/expandable habitat needed for the first brief surface mission. As the longest time a crew could spend on Mars during a Short-Stay mission and still catch the trajectory home is in the range of 30–60 days, the habitat must be sized to support that number of crew for that length of time. The following are size comparisons for perspective.

The 1988 JSC Habitat

This 30-year-old design concept was a sphere with a diameter of 16 m (52 ft) and a volume of 2145 m^3 (75,750 ft^3) pressurized at 1 atmosphere. It would be made of Kevlar-29, have a mass of 2200 kg (4850 lbs) and four levels with facilities for eight to twelve astronauts. The structure was to be covered over with 3 m (9.8 ft) of regolith in sand bags. The designers were clearly aiming at a surface habitat. It would likely be too large to be inserted into a cave. It was assumed that the crew was going to do all the work to gather regolith, shovel it into sand bags, and place the bags over the roof to achieve sufficient protection from radiation. (It would be easier to just use a cave or lava tube.)

1998 TransHab

The TransHab was designed to replace the aluminum habitation module intended for the International Space Station. When fully inflated, TransHab would expand up from a diameter of 4.3 m (14 ft) to 8.2 m (27 ft); much greater than the other modules. It was to have a volume of 339.8 m^3 (12,000 ft^3) and a mass of 13,154 kg (29,000 lbs).

In 1999 the National Space Society issued a policy statement recommending that NASA continue R&D of inflatable technologies while ceasing development of a TransHab module for the ISS. Finally, in 2000, rejecting objections from the White House, House Resolution 1654 was signed into law banning NASA from further work on TransHab. However, it included an option to lease an inflatable habitat module from private industry. After NASA had discontinued the project, Bigelow Aerospace bought the patent for this module and is pursuing a variety of configurations, one of which is on the ISS. However, a TransHab-sized facility is rather on the small size for a cave habitat.

NASA Langley Research Center did continue with the R&D, and contracted with ILC Dover in Delaware to design a concept demonstrator for a lunar habitat (see the previous Fig. 5.5 and Fig. 6.5 here). However, it is even smaller than the initial TransHab design.

ISS Harmony

ISS Node 2, Harmony, has a diameter of 4.4 m (14 ft), a length of 7.2 m (24 ft), and a mass of approximately 14,288 kg (31,500 lbs). It was the second of three such modules incorporated into the US Orbital Segment (USOS) of the station. The design had much in common with the Multi-Purpose Logistics Module that was developed for NASA by the

Fig. 6.5 ILC X-Hab Lunar Habitat at Langley Research Center. Photo courtesy of NASA/ LaRC/ILC.

Italian Space Agency, as well as the Columbus Science Module manufactured for the European Space Agency. It contains racks of equipment and crew sleep stations. Such a module would be too small to serve as a cave habitat.

ISS BEAM

The Bigelow Expandable Activity Module (BEAM) experimental space station module was produced by Bigelow Aerospace under contract to NASA. In 2016 it was mated to the International Space Station, then inflated to a diameter of 3.2 m (10.5 ft) and a length of 4 m (13 ft) with a volume of 16 m^3 (565 ft^3). It has a mass of 1400 kg (3086 lbs). Like the other station modules, it is too small for a cave habitat.[1]

B330

The B330 is an inflatable space habitat design concept which is being privately developed by Bigelow Aerospace. The design evolved from NASA's TransHab habitat concept. Its exterior is 6.7 m (22 ft) wide and 13.7 m (45 ft) long, with a pressurized volume of 330 m^3 (11,653 ft^3); hence its designation. Its mass varies between 20,000 kg (44,100 lbs) and 23,000 kg (47,000 lbs) depending on how it is configured.[2] The B330 weighs 14 times as much as the BEAM demonstrator configuration and its volume is 20 times as great.

At first glance, the B330 would seem to be approximately the right size for a precursor mission. The sizes of potential caves on the Moon and Mars aren't yet known, especially their entrances. They are expected to be very large (due to the low gravity), but the key

---

[1]You can watch the expansion of the BEAM in May of 2016 on the following 7 hour video but you can fast forward through it at your leisure. Go to: https://www.youtube.com/watch?v=22GiUvJGj7M.
[2]For a lot of photos of the Bigelow inflatables go to:
https://www.flickr.com/photos/airlinereporter/sets/72157644040818837

factor is the requirement to be able to ease a collapsed, inflatable habitat into the cave. And once inside, there must be sufficient access space around it. The habitat will be inflated once in position. It would appear that the minimum height for the cave opening should be of the order of 75 per cent of the inflated diameter; probably more, given that one would prefer some roof and peripheral clearance. To be on the safe side, the cave opening ought to be a little larger than the diameter of the fully inflated habitat. By this rough estimation, a 20–25 ft opening is in the right ball park. Many caves have a small opening and then a large interior. There may be advantages to a much bigger cave, because it would simplify access. From a protection standpoint, the only disadvantage of a very large cave would be for the Mars dust storm case, where the cave entrance might be too large to adequately shield the interior from the dust and wind.

Although the volume of the B330 is appropriate for such a habitat, its shape (diameter and length) may have to be tailored. And even if the gravity is 1/6 g or 3/8 g, some means will have to be devised to move the deflated 20 ton package into the cave. Perhaps it will have its own transporter, or perhaps the mission's exploration vehicle will undertake this operation, making it the counterpart of a tractor-trailer combination. One key factor in mission planning will therefore be the means by which the habitat is to be transported into a cave.

The optimum size of habitat for a crew of four spending 30–60 days on Mars has yet to be determined, but once that is done it will provide the mass numbers required by mission planners.

6.2.2 Habitat and Equipment Placement

Ideally, the cave would not only have room to accommodate the habitat but also all of the associated equipment and resources; for example, pressurized air bottles, processing, cooling, and communications equipment, cables, robot storage, and surface vehicle. The primary external power source (e.g. the Kilopower system) would have to be located a safe distance away, or outside the cave with the cables running to the habitat (see the previous Fig. 5.20).

The cave floor should be relatively flat or curved slightly for the diameter of the habitat; a lava tube floor may well be smoother than others. There should be adequate clearance around the inflated habitat, including up to the ceiling. If the floor of the cave was fairly rough, perhaps the robots could clear it to ensure the habitat will not suffer abrasion. The habitat would likely be anchored to the cave walls and floor, and possibly the ceiling. If the habitat is retained on its transporter then the condition of the cave floor is less of a factor.

6.2.3 Environment and Safety

Although the Moon has no atmosphere and Mars has a very tenuous one, the crew must sample whatever atmosphere there is in the cave at different levels and at different times. The storms on Mars stir up the fine dust. Some of them can be intense. Once every three local years on average, a storm will grow to become almost global in extent. Owing to the low density of the air, even the most intense dust storm will not tip over or rip apart major mechanical equipment. The crew must guard against contamination of the habitat while cycling the airlock. These precautions will include cleaning the dust from space suits. It is presumed that the habitat will have its own environment and safety equipment, separate from those precautions taken outside related to the cave or during EVAs.

A crew will be at particular risk while riding in the exploration vehicle across rough ground in search of entrances to caves, and while entering caves or lava tubes of unknown stability. The risk will be greater if this exploration involves entering spaces through skylights (Fig. 6.3). Vertical descents would likely not be attempted by the early missions because specialized equipment would be needed.

Apollo astronauts took several trips and falls that looked bad to viewers on Earth, but in the low gravity there were no injuries. If one of the astronauts had been incapacitated, there were few options. Crews on Mars would be given first aid training and supplies, but it would be a long time before an injured astronaut could be returned to Earth (about 10 months, depending on the alignment of the planets at that time). An injured astronaut will receive treatment from colleagues, assisted by medics on Earth via telepresence. In the case of a severe injury, the surface mission might be aborted with the ascent vehicle rendezvousing with the deep space vehicle in orbit, where there will be better medical facilities. For sure, recuperation will be more comfortable in microgravity.

6.3   THE SEARCH FOR WATER ICE

6.3.1   On the Moon

The payload of the Chandrayaan-1 spacecraft that India placed into lunar orbit in 2008 included the NASA-JPL Moon Mineralogy Mapper (M3) instrument which identified three specific signatures that definitively proved the presence of water ice on the surface of the Moon. The instrument was uniquely capable in this role. It not only collected data diagnostic of the reflective properties of ice, it was also able to measure the way in which the molecules absorb infrared light, enabling it to differentiate between liquid water or vapor and solid ice. Most of this ice lies in craters near the lunar poles where, due to the very small tilt of the Moon's rotation axis relative to the plane of Earth's solar orbit, there are permanent shadows in which the temperatures never rise above 116 Kelvin (minus 157°C, minus 250°F).

Dana Hurley Crider of the Catholic University of America in North Carolina and Richard R. Vondrak of NASA's Goddard Space Flight Center carried out a mathematical study of how a "cold trap" would evolve in response to the process called "space weathering" by which the ionized plasma that is delivered by the "solar wind" becomes mixed into the regolith. The results, published 2001–2003, concluded that the regolith would attain a steady state of around 4,000 parts per million of water. Of course, that water had not been present in the plasma; it was created by hydrogen in the solar wind combining with oxygen atoms drawn from oxides in the regolith.

With sufficient ice sitting at or close to the surface, water may well be readily accessible as a resource for future expeditions; potentially even easier to harvest than any water that may be present at depth.

The landing sites for future human missions to the Moon might be different to those for robotic missions. There is one site that seems to be high on the list for a human landing. Shackleton crater lies at 79.13°S, 140.60°E. The peaks along the crater's rim are exposed

Fig. 6.6 First Response. A lunar crewman has fallen and broken his right femur. The responding crew has arrived in a hopper to take him back to base. Art by Pat Rawlings ©SAIC.

to almost continual sunlight for about half the year. The interior is perpetually in shadow, and therefore acts as a perennial cold trap. This combination of resources (solar power and water ice) makes Shackleton a likely candidate for establishing an early base.[3]

Note: I have to wonder how the mission planners expect the crew to get down into the crater to acquire samples. They will land in the well-lit area, not in a place of perpetual darkness. Perhaps a robot will be sent down into the crater to collect samples.

6.3.2   On Mars

The gamma ray spectrometer of the Mars Odyssey orbiter detected the presence of subsurface water ice across large parts of that planet. When Phoenix landed in one such an area it observed water ice beneath a few centimeters of soil. In 2017, the HiRISE camera of Mars Reconnaissance Orbiter found outcropping sheets of water ice in a number of scarps at latitudes in the range 55–58°. These sightings back up the inference from Mars Odyssey that there are extensive sheets of water ice at shallow depth; possibly only a few meters. The layering of ice and soil will provide clues to the climate history of the planet, and the frozen water will be able to be exploited by future robotic or human explorers.

---

[3]For a 1:20 video of Shackleton crater watch:
https://www.youtube.com/watch?v=j2NAZODSdwM
For a two minute video on how data is taken over the south pole watch:
https://www.youtube.com/watch?v=qYW4rTrAA5I

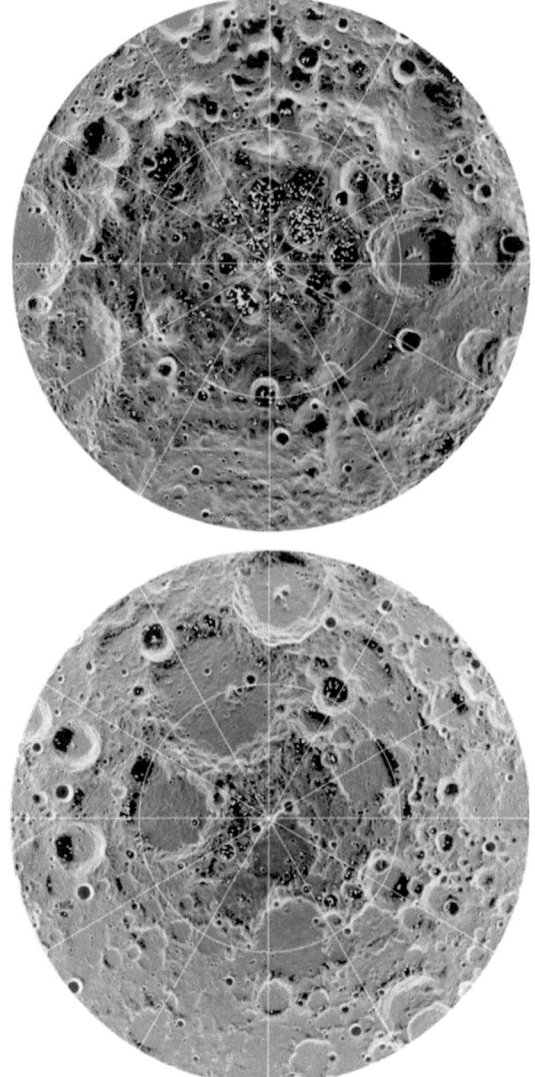

Fig. 6.7 Water ice in the Moon's polar regions. The south pole is at the top and the north pole is at the bottom. The gray scale indicates surface temperature (darker representing colder areas and lighter shades indicating warmer zones). Blue indicates ice, which is concentrated at the darkest and coldest locations, in the shadows of craters. This is the first time scientists have directly observed definitive evidence of water ice on the Moon's surface. Photo courtesy of NASA/JPL-Caltech.

6.3.3   In the Caves and Lava Tubes

In addition to requiring the target landing site to offer a reasonable prospect of a cave or lava tube suitable for a habitat, it would be highly desirable to have ready access to water

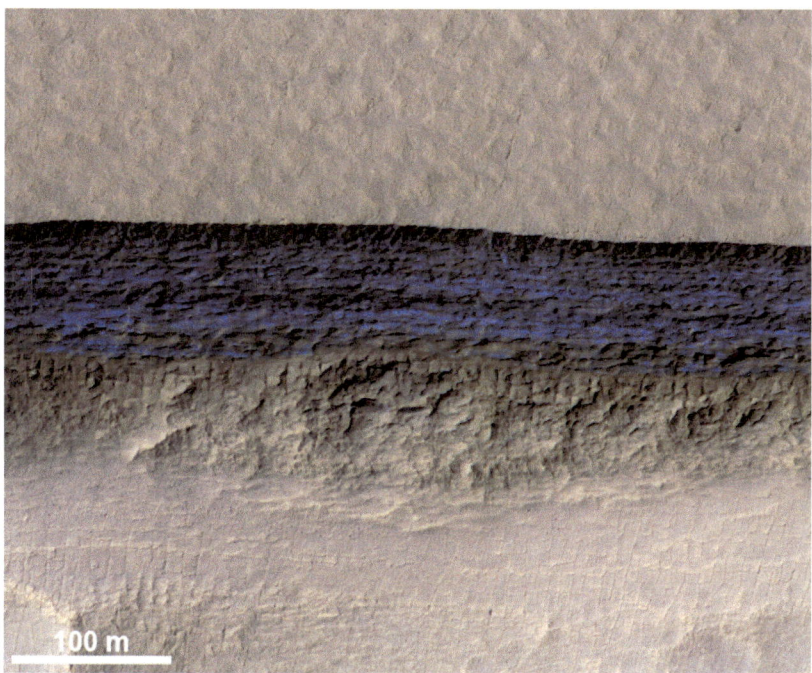

Fig. 6.8 The bright blue in this color enhanced image of a steep scarp (slope) by the HiRISE camera of Mars Reconnaissance Orbiter (MRO) is believed to show a cross-section of subsurface water ice. The scene, located at 56.6°S, 114.1°E, is about 500 m (1640 ft) across. The scarp drops about 128 m (500 ft) from the adjacent surface. The ice sheets extend from just beneath the surface down to a depth of at least 100 m (328 ft). Photo courtesy of NASA/JPL-Caltech/UA/USGS.

ice. A cave with water ice might be more likely on Mars than on the Moon. The mission planners will weigh these requirements against the other science, engineering, and trajectory requirements.

As highly insulated environments, caves that penetrate deep underground will trap and hold cold air to form isolated microclimates. If conditions are favorable, ice could either condense or accumulate at rates that exceed sublimation and thus become trapped. Studies of Mars have shown the presence of water ice at shallow depths across much the planet, particularly in the areas where skylights have been observed.

If the mission planners are confident of finding a cave that contains water ice, they would want to include the equipment for processing that ice in situ; if not by the initial mission then certainly by later missions. Once the presence of ice has been confirmed, astronauts will seek biology. A positive identification will invoke sampling and contamination protocols, strongly influence subsequent activities, and possibly trigger procedures for returning a sample to Earth (see Chapter 7 and Appendix 5).

6.4   THE SEARCH FOR LIFE

6.4.1   On the Moon

It is believed that life does not presently exist on the Moon, but recent research suggests there was a window of time during which it could have supported life. One astrobiologist has theorized that conditions on the surface of the Moon could have supported "simple lifeforms" immediately after its formation, 4 billion years ago, and during the volcanic activity which occurred 3.5 billion years ago. Both periods would have seen large amounts of volatile gases, particularly water vapor, emerging from the interior that could have been at least temporarily supportive of life, presuming that the vapor formed liquid water on the surface.

Although the crew of the first mission back on the Moon won't primarily be seeking life, they will certainly keep this possibility in mind as they investigate potential water ice resources.

6.4.2   On Mars

A variety of robotic orbiters, landers and rovers have dramatically increased our knowledge about Mars, paving the way for future human explorers. As this book was being written, the Curiosity rover was investigating Gale Crater. The results indicated that conditions on the planet would have been suitable for life early in its history, although no evidence has been found to confirm that life was present. The search for evidence of habitability, taphonomy (how organisms decay and become fossilized), and organic molecules on the planet is now a primary NASA and ESA objective.[4]

Scheduled for launch in 2020 as part of NASA's Mars Exploration Program, the Mars 2020 rover is to analyze samples for biosignatures from past microbial life. The NASA-JPL Science Definition Team initially suggested that the rover collect and package as many as 31 interesting samples of surface soil and rock cores for a later mission to bring back on Earth for definitive analysis. Then, in 2015, they expanded the concept by deciding to collect even more samples and leave the packages in small piles or "caches" across the surface for later retrieval. Perhaps by the time of the precursor human mission described in this book, the sample caches would have been retrieved and analyzed on Earth.

Although it would clearly be desirable to have a detailed analysis of Martian samples before a human mission lands, the design of the robotic Sample Return Mission has been debated for many years and has still to be settled. Perhaps one solution, described in my earlier book *Exploring the Martian Moons: A Human Mission to Deimos and Phobos*, would be for these caches to be retrieved by a rover that transfers them to a lander that includes an ascent stage. Once in orbit, that robotic vehicle would rendezvous with a spacecraft whose crew were flying an Apollo 8-style precursor mission. The samples could be investigated either by the astronauts using on board instruments or stored for analysis by state-of-the-art laboratories after returning to Earth.

---

[4]For a 3 minute video on the most likely places to seek life on Mars, go to:
https://www.youtube.com/watch?v=wlc6uujLtng
For a 7 minute video on searching for life on Mars, go to:
https://www.youtube.com/watch?v=LHMQyQ_YwL8

Fig. 6.9 Mars Sample Return. Photo courtesy of NASA/JPL-Caltech.

The Science Definition Team for the Mars 2020 mission down-selected the list of sites to three candidates: Jezero Crater, the Northeastern region of Syrtis Major Planum, and the Columbia Hills in Gusev Crater.

- Jezero, a crater at 18.855°N, 77.519°E, in the Syrtis Major quadrangle, is about 49.0 km (30 mi) in diameter. To put that into perspective, the entire City of Austin, Texas, could fit into it comfortably. Thought to have once been flooded with water, the crater contains a fan-delta deposit that is rich in clays. That occurred when valley networks were forming elsewhere on Mars.
- Syrtis Major Planum is a "dark spot" at the boundary between the northern lowlands and southern highlands, just west of the Isidis impact basin in the Syrtis Major quadrangle.
- During its exploration of Gusev Crater, the Mars Exploration Rover Spirit drove approximately 3 km (1.9 mi) from its landing position to a range of low hills which were named the Columbia Hills in memory of the crew of the Space Shuttle Columbia disaster.

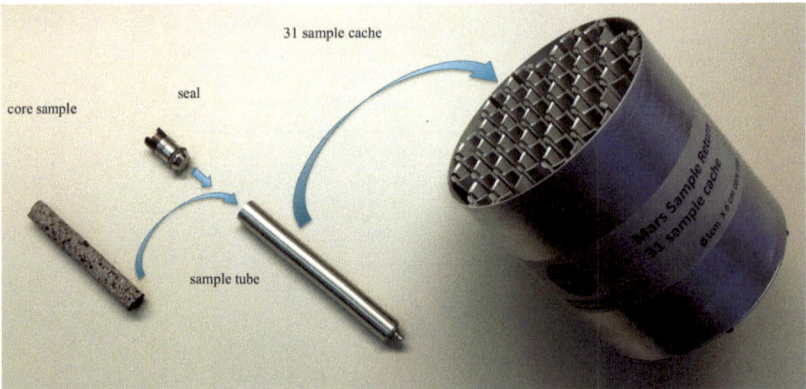

Fig. 6.10 Hardware to cache Martian rock core samples. Photo courtesy of NASA/
JPL-Caltech.

In November 2018 Jezero was chosen, dependent upon extensive analyses and veri-
fication testing of a new capability called Terrain Relative Navigation (TRN) to enable the
"Sky Crane" descent system to avoid hazardous areas. A final report will be presented to
an independent review board and NASA Headquarters in the fall of 2019.

The Exobiology on Mars joint venture by ESA and Russia intends to send the ExoMars
rover equipped with a drill to obtain samples from a depth of 2 m (7 ft) for an investigation of possible morphological and chemical evidence of past life on the planet. The
candidate landing sites had to lie within 30° of the equator for sufficient solar power:

- Mawrth Vallis is located in the Oxia Palus quadrangle at 22.3°N, 343.5°E, with an
  elevation approximately 2 km (1.25 mi) below the Martian mean. Situated between
  the southern highlands and northern lowlands, the valley is a channel formed by
  massive flooding which occurred in Mars' ancient past.
- Oxia Planum is a plain more than 3 km (1.9 mi) below the Martian mean.

The final selection is expected to be made about a year prior to launch.[5]

6.5   PROTECTION AGAINST CONTAMINATION

6.5.1   Background

Long before Neil Armstrong and Buzz Aldrin left boot imprints in the lunar dust, NASA
initiated precautions to protect Earth and its biosphere from extraterrestrial life forms
returned by inbound spacecraft. It devised procedures to prevent "back contamination."
The early Apollo crews returning from the Moon were placed in quarantine in accordance
with a finding by the Interagency Committee on Back-Contamination (ICBC), formed in

---

[5]For a 4 minute video on the ExoMars Rover, go to:
https://www.youtube.com/watch?v=ZSAzWBz6Rny

1966, that most terrestrial disease agents show symptoms within 21 days of infecting the host. NASA is also required to preserve the scientific value of other celestial bodies. It must protect such native life as its missions find, not least because biological contamination from Earth could make it difficult to determine whether any life that is detected was present beforehand.

In recent decades, robots have served as our emissaries to a large variety of other solar system bodies. The Curiosity rover of the Mars Science Laboratory (MSL) that was launched November 2011 was to investigate whether Mars was ever hospitable to microbial life. Planetary protection requirements called for the entire MSL flight system to have no more than 500,000 bacterial spores at the time of launch. This was accomplished mainly through the careful maintenance of clean room protocols, periodic cleansing of vehicle surfaces using alcohol wipes, and dry heat treatment of some spacecraft parts.

Whatever spacecraft or equipment is sent to the Moon or on to Mars will be subjected to similar protocols in manufacturing and launching the payloads. But what are the contamination requirements for returning Martian samples to Earth? And what protocols will apply when we land humans on that planet?

The following section explains the very involved international requirements that influence NASA's protection rules for missions to the Moon and especially to Mars, where the probability of life is greater. These guidelines and the NASA specific protocols will have a significant influence on how we plan and execute a mission, both scientifically and operationally, regardless of whether a habitat is installed on the open surface or (as advocated here) sheltered inside a cave or lava tube.

The Committee on Space Research (COSPAR), which was established by the International Council for Science (ICSU) in 1958, established the first planetary protection guidelines for robotic missions in 2002. Although not legally binding, this Planetary Protection Policy is, very briefly, to avoid organic-constituent and biological contamination in space exploration and to provide accepted guidelines which are compliant with the 1967 Outer Space Treaty. In 2011 the World Space Council revised the COSPAR Planetary Protection Policy to include Principles and Guidelines for Human Missions to Mars.

In May 2012, the Planetary Protection Subcommittee of the NASA Advisory Council (NAC)'s Science Committee formulated a recommendation that NASA Procedural Requirements (NPR) be developed for planetary protection on human missions with NASA Policy Directive (NPD) 8020.7 "Biological Contamination Control for Outbound and Inbound Planetary Spacecraft," and the parallel NPR 8020.12 "Planetary Protection Provisions for Robotic Extraterrestrial Missions." This recommendation was endorsed by the full NAC and forwarded to the NASA Administrator in November 2012, who accepted it on March 8, 2013.

There is presently insufficient scientific and technological knowledge to define detailed requirements and specifications to enable NASA to incorporate planetary protection into the development of spacecraft, equipment, and missions for human exploration. A NASA Policy Instruction (NPI) specifies the policy guidelines and describes the approach for obtaining the scientific information and developing the technologies and procedures that will be required for drafting the NPR for human planetary missions. As NASA prepares to send humans beyond low Earth orbit in order to explore the solar system in search of evidence of extraterrestrial life, it is therefore essential that it develop these guidelines.

As a key international partner with its own robotic programs and ambitions for human exploration, ESA upholds the COSPAR Planetary Protection Policy.

6.5.2   Principles and Guidelines for Missions to Mars

The intent of this planetary protection policy is the same regardless of whether a mission to Mars is undertaken robotically or with human explorers. Accordingly, planetary protection goals should not be relaxed in planning a human mission. In fact, they become even more directly relevant. It is recognized, however, that the specific implementation requirements will differ.

The general principles are paraphrased as follows:

- Safeguarding the Earth from potential back contamination is the highest planetary protection priority in Mars exploration.
- Human explorers can contribute to the astrobiological exploration of Mars only if the scope for forward contamination is understood and controlled.
- For a landed mission conducting surface operations, it will not be possible for all human-associated processes and mission operations to be conducted within entirely closed systems.
- Crewmembers exploring Mars, or their support systems, will inevitably be exposed to Martian materials.

In accordance with these principles, specific implementation guidelines for human missions to Mars include:

- Humans will inevitably carry microbial populations that will vary in both type and quantity, therefore it won't be practical to specify all aspects of a permissible microbial population at the time of launch. Once any baseline conditions for launch are established and satisfied, continued monitoring and evaluation of microbes carried by human missions will be required in order to address both forward and backward contamination concerns.
- A quarantine facility both for the entire crew and for individual members of the crew will have to be provided prior to and after the mission, in case there is potential contact with a Martian lifeform.
- A comprehensive planetary protection protocol for human missions should be developed that encompasses both forward and backward contamination issues, and addresses the overall human and robotic aspects of the mission, including sub-surface exploration, sample handling, and returning samples and the crew to Earth.
- Neither robotic systems nor human activities should contaminate "Special Regions" on Mars (as defined by this COSPAR policy).
- Any uncharacterized Martian site should be evaluated by robots prior to permitting crew access. Information may be obtained by either precursor robotic missions or a robotic component on the human mission.
- Any pristine samples or sampling components from any uncharacterized sites or Special Regions on Mars should be treated according to planetary protection Category V (Restricted Earth Return) with the proper handling and testing protocols.

- An on board crewmember should be given primary responsibility for the implementation of such planetary protection issues as might arise during the mission.
- Planetary protection requirements for the early human missions ought to be based on a conservative approach consistent with a lack of knowledge of Martian environments and possible life, as well as the performance of human support systems in those environments.
- Planetary protection requirements for later missions should not be relaxed without scientific review, justification, and consensus.

COSPAR defined areas on Mars as "Special Regions" as being those in which terrestrial life may have the potential to proliferate, if introduced. However, after a review by NASA's Mars Exploration Program Analysis Group (MEPAG), the committee recommended continuing the practice of revisiting the definition and locations of Special Regions every two years. It also noted that, currently, there are no sites on the planet suggestive of harboring indigenous life. (Nevertheless, scientists certainly have water ice in mind because of the possibility of finding life.)

6.5.3 NASA Office of Planetary Protection

NASA's Planetary Protection function is now part of the Headquarters Office of Safety and Mission Assurance (OSMA). The mission of the Office of Planetary Protection (OPP) is to promote the responsible exploration of the solar system by implementing and developing efforts which seek to protect the science, explored environments, and Earth.

The objectives of planetary protection are several-fold and include:

- Preserving our ability to study other environments in their natural states.
- Avoiding the biological contamination of explored environments that may obscure our ability to find life elsewhere, if it exists.
- Ensuring that we take prudent precautions to protect Earth's biosphere in case life does exist elsewhere.

To achieve these goals, the OPP is involved in facets of mission development which include assistance in the construction of sterile spacecraft (or at least ones that present a low biological burden), the development of flight plans that protect planetary bodies, the development of plans designed to protect Earth's biosphere from returned extraterrestrial samples, and the formulation and application of the space policy as it applies to planetary protection. The OPP works in conjunction with solar system mission planners to assure compliance with NASA policy and international agreements. Ultimately, the objective of planetary protection must be to support the scientific study of chemical evolution and the origins of life in the solar system.

The planetary protection requirements for any given mission and solar system target are determined by the policy guidelines of the relevant space agencies and the best available scientific advice (such as from the Space Studies Board of the National Research Council). Any mission is categorized according to the type of encounter that it will involve (e.g., flyby, orbiter, or lander), and the nature of its destination (e.g., a planet, a moon, a comet, or an asteroid). If the target has the potential to provide clues concerning life or prebiotic chemical evolution, then a spacecraft must meet a higher level of cleanliness, and operating restrictions will be imposed to limit forward contamination. A spacecraft for such a mission must undergo more stringent sterilization and abide by tighter operating restrictions.

The robotic missions that are currently on Mars and those planned for the near future are Category IV. This includes certain types of missions (typically those with an entry probe, lander, or rover) to a target that has chemical evolution or origin-of-life interest, or for which the mission presents a significant chance of contamination that could undermine future biological investigations. Subdivisions of Category IV (designated IVa, IVb, or IVc) address lander and rover missions to Mars (with or without life detection experiments), and missions that will land on or access regions of the planet which are of particularly high biological interest. The Curiosity, Spirt and Opportunity rovers were classified as Category IVa, but the ExoMars rover will be Category IVb.

Category V is for when the spacecraft, or a component of the spacecraft, will return to Earth. The main concern for these missions is the protection of Earth's biosphere from contamination by returning extraterrestrial samples (usually soil and rocks). The Mars 2020 rover is classified as Category V because it includes the future option of a follow-on mission returning its samples to Earth. A human mission to Mars (such as advocated in this book) will necessarily be classified as Category V. For all other Category V missions, there is a subcategory defined as "Restricted Earth Return" in which the highest degree of concern is expressed by requiring the absolute prohibition of destructive impact upon return, the need for containment throughout the return phase of all returning hardware that had direct contact with the target body or unsterilized material from the body, and the need for containment of any unsterilized samples that are to be returned to Earth. Also, there must be timely post-mission analyses of the returned unsterilized samples under strict containment, using the most sensitive techniques. If any sign of the existence of a non-terrestrial replicating organism is detected, the returned sample must remain in containment unless it can be treated using an effective sterilization procedure. Thus Category V expands on the Category IV requirements with the addition of continuous monitoring of mission activities and the need for additional research into sterilization procedures and containment techniques.[6]

---

[6]For more on The NASA Headquarters Office of Planetary Protection, go to their website at: https://planetaryprotection.nasa.gov/about

7

A Proposed Mission for NASA's Design Reference Architecture

7.1 BACKGROUND

Here is a brief history of how NASA has approached human exploration of the Moon and Mars. The documents that are cited represent the dedicated efforts of literally hundreds of people from the science, engineering, and mission planning communities over about a decade. The documents truly represent what NASA is doing and why. I merely propose a slight "tweak" to the emphasis of this effort, and only to one of the Design Reference Missions. As NASA has invited public input, this book is one person's suggestion. The precursor mission proposed has been discussed before by the participants, but was always overshadowed by the "grand" mission in which the first humans to reach Mars spend up to a year and a half on the surface, waiting for the correct planetary alignment to set off back to Earth.

In 2009, NASA-JSC produced NASA-SP-2009-566 "Human Exploration of Mars Design Reference Architecture 5.0." In 2015 they released Addendum #2, with some of the key assessments and studies produced since publication of the original document, predominately those conducted from 2009 through 2012. The latest addendum is 598 pages of very detailed analysis of the possible options to travel to just about anywhere in the solar system but particularly to the Moon and to Mars.

In 2012, NASA developed a set of Technology Roadmaps intended to guide the development of space technologies. The 2015 NASA Technology Roadmaps expanded and enhanced the original roadmaps, providing extensive details about anticipated mission capability needs and the associated technology development needs. These Technology Roadmaps are a set of documents that consider a wide range of needed technology candidates and development pathways for the period 2015 to 2035. They focus on applied research and development activities. There is an introduction that includes a discussion of key crosscutting technologies, and a total of 15 distinct Technology Area (TA) roadmaps.

© Springer Nature Switzerland AG 2019
M. von Ehrenfried, *From Cave Man to Cave Martian*,
Springer Praxis Books, https://doi.org/10.1007/978-3-030-05408-3_7

The 2015 NASA Technology Roadmaps were created through the leadership and dedication of professionals from the Office of the Chief Technologist (OCT), the NASA field Centers, and contributors from other government agencies. The work was led by the OCT and the Chair and Co-Chair of each Technology Area. Each roadmap was developed by a team of eight or nine experts who drew on the support of internal and external contributors. Each roadmap contains a list of the roadmap team members and contributors.

In 2017, the NASA Headquarters Office of the Chief Technologist produced the NASA Strategic Technology Investment Plan (STIP). This prioritized all the technologies deemed to be essential to the pursuit of the Agency's missions and achievement of national goals. The plan (drawn up after the development of the Technology Roadmaps) provides guidance for NASA's technology investments. The plan uses an analytical approach, reflecting input from each NASA Mission Directorate and the ongoing challenges in secure systems to categorize technology investments and provide guidance for balanced investments across the Agency's technology portfolio.

The Office of the Chief Technologist conducts an annual analytical assessment of the portfolio, and compares the year's investments against the guidelines in the STIP. The results of this assessment are reviewed by the Mission Directorates and Offices through NASA's Technology Executive Council (NTEC). As appropriate, NTEC may recommend realigning the portfolio to meet the Agency's goals. Then the Mission Directorates and Offices at the Centers define and implement specific content of the technology portfolio through their programs. These programs drive the development of technologies across NASA, innovating at all of its Centers in partnership with other government agencies, industry, and academia. While some of the activities occur within NASA's focused technology development programs, there are also research and technology development activities taking place as part of the Agency's programs focused on specific missions. Its technology research and development activities strive to be responsive and adapt to the changing needs of missions and the Agency's aspirations.

What this book calls for is a "changing need," or more specifically the need to emphasize early missions to both the Moon and Mars that could be flown prior to the "grand" missions that call for huge emplacement of resources and long stays. What is proposed is not missions as alternatives to those already being assessed, but logical precursors that could be carried out in advance of those missions and deliver useful operational and scientific results.

What is proposed here is totally compatible with the STIPs:

- Strategic Goals to Drive Technology Development.
- Guiding Principles for Implementation.
- Enhancing Technology Investments.
- Transformational Technology Investments.

It utilizes many of the technologies that are being developed (see Chapter 5). It is merely a matter of describing a precursor mission that utilizes the technologies. There is no intent to re-invent the technology wheel, as those already planned are more than adequate. Many are already pushing the state-of-the-art.

The Technology Roadmap defines the following Design Reference Missions and Aeronautic Thrusts:

Fig. 7.1 Technology Roadmaps and Areas. Photo courtesy of NASA.

- 8 Human Exploration missions.
- 38 Science Missions.
- 16 Aeronautic Strategic Thrusts (not applicable to this discussion).

The 8 Human Exploration missions include:

- Asteroid Redirect (cancelled 12/2017).
- Asteroid Redirect in a Distant Retrograde Orbit (cancelled 12/2017).
- Crewed Mission to a Near Earth Asteroid (cancelled 12/2017).
- Crewed Mission to the Lunar Surface.
- Crewed Mission to Mars' Moons (Reference *Exploration of the Martian Moons: A Human Mission to Deimos and Phobos* published by Springer-Praxis).
- Crewed Mission to Mars Orbit.
- Crewed Mission to Mars Surface.
- Crewed Mission to Mars Surface (Minimal) (for example a cave or lava tube during a short duration precursor mission, as discussed below).

7.2 THE PROPOSED PRECURSOR CAVE MISSION

This book proposes a precursor mission defined above as a "Crewed Mission to Mars Surface (Minimal)" and also as Design Reference Mission 9a. This type of mission appears in the Mars Design Reference Architecture 5.0, Addendum #2, "Special Studies

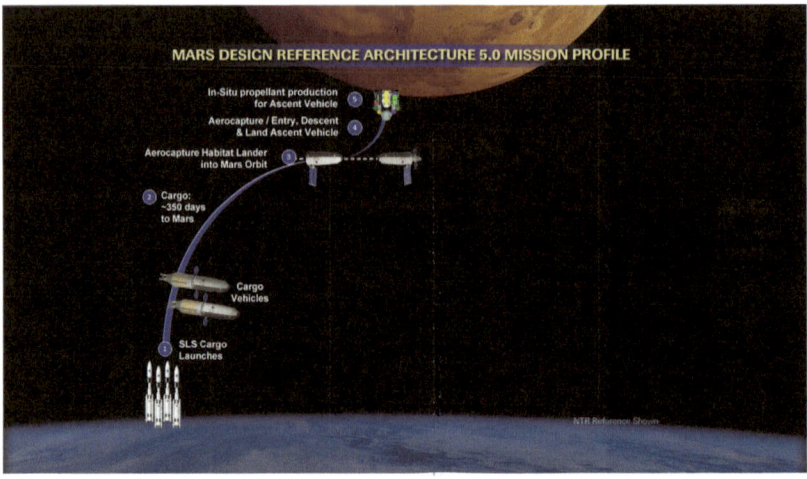

Fig. 7.2 Current Planned Cargo Launch. Photo courtesy of NASA.

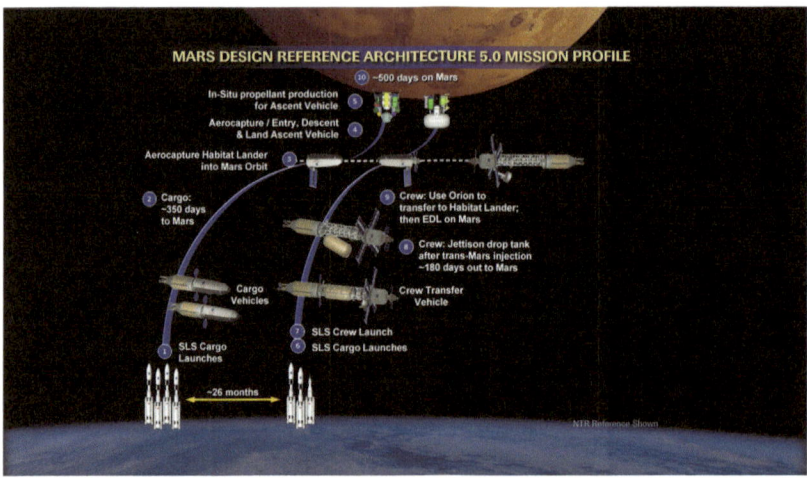

Fig. 7.3 Current Planned Crew Launch. Photo courtesy of NASA.

and Strategic Assessments." However, it was not analyzed as a precursor mission to the surface that utilizes a cave or lava tube for protection. NASA could undertake this study if it wished. It could also be that budgetary and other constraints in the future will make this option even more practical and hence a prudent addition to the mission planning effort.

Eventually, NASA will issue Addendum #3 (or a document which has some other name) and at that time the proposed option could be added. The precursor cave mission for Mars would fall within the category of "Short-Stay" missions, so named because they envisage spending only 30–60 days on the planet. This is also an "Opposition Class" mission. Round-trip times typically range from 560 to 850 days. In contrast, a

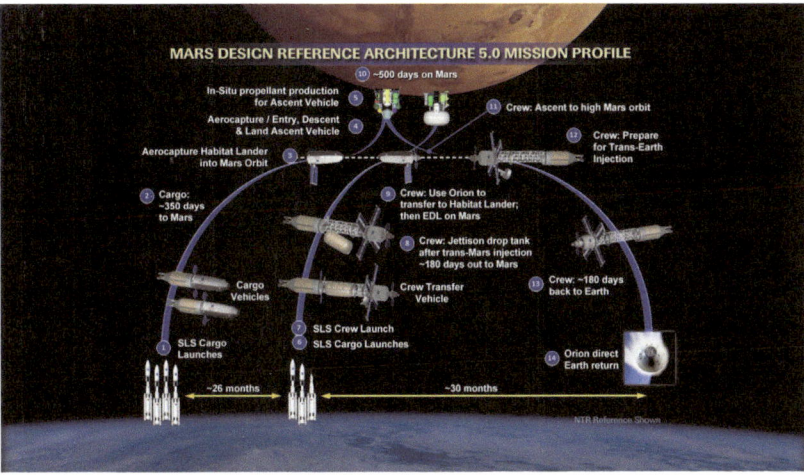

Fig. 7.4   Current Planed Long-Stay and Return to Earth. Photo courtesy of NASA.

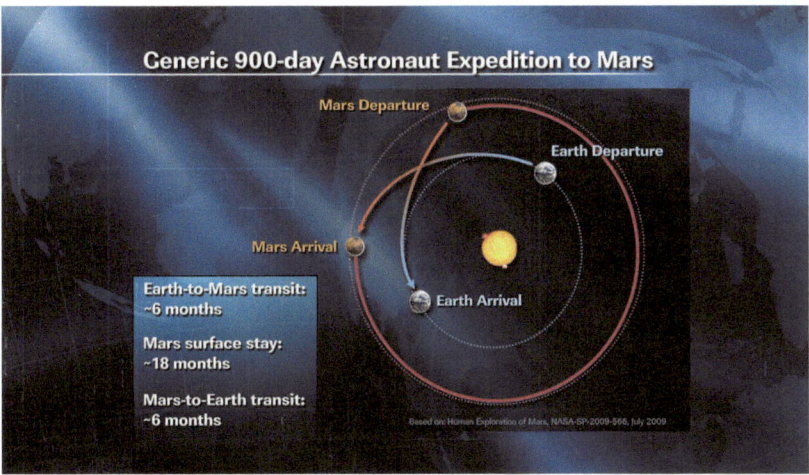

Fig. 7.5   Conjunction Class 900-Day Mission Trajectory. Photo courtesy of NASA.

"Conjunction Class" mission requires the crew to stay on Mars for about 545 days, with a total mission time of around 905 days. The currently envisaged "Long-Stay" mission will not only involve a lot more hardware but also the pre-positioning of logistics on Mars. It will be very expensive. A mission that relies on caves or lava tubes for protection and obviates the need for much of the pre-positioned logistics for the "grand" Conjunction Class mission, ought to be a considerably cheaper option. Such a mission will also increase the overall safety of the crew by not obliging them to spend so long on the surface of Mars before there is a possible trajectory ride home.

However, the Short-Stay class of mission has higher propulsive requirements. They usually have one short transit leg, which may be either on the outbound or the inbound phase, and a long transit that spends part of its time within the orbit of Earth around the Sun, exposed to greater solar heating than for a Conjunction Class mission beyond the orbit of Earth. Once at Mars, rather than waiting for a near optimum return alignment, a crew would initiate the return after a brief stay. However, such a mission may well require a perihelion passage inside the orbit of *Venus* on either the outbound or inbound leg. A trajectory analysis is required to minimize this condition, which will be thermally stressful. This class of mission may also require large total energy propulsive delta-velocity (ΔV) requirements and large variations in ΔV from one flight opportunity to the next. In fact, there are many options available, depending on the actual year and the duration of stay.

Regardless of the stay-time, the same elements will be required:

- SLS.
- Orion/Service Module.
- Exploration Upper Stage.
- Deep Space Habitat.
- Multi-Purpose Logistics Module.
- Tunnel/Airlock.
- Solar Panels.
- In-Space Exploration Vehicle.
- Descent/Ascent Stage.
- Multi-Mission Space Exploration Vehicle.
- The Deep Space Tracking System.
- Mission Control Center and Science Centers.

Given the management directive to analyze a precursor cave-oriented mission, the JSC team (including members from the other Centers and contractors) has the scientific and engineering depth, background, and reference studies to do so. This analysis will hone in on the following types of questions and parameters:

- What is the optimum design for a surface transfer vehicle that can carry a small crew (three, maybe four people), is pressurized, and can serve as a temporary habitat for a duration of several days during collection of pre-positioned resources? (The current wheeled version of the Multi-Mission Space Exploration Vehicle would likely require upgrading to the mission requirements.)
- What is the optimum size and weight for the inflatable habitat?
- Are the airlocks packaged separately?
- Does the inflatable habitat need its own transfer vehicle?
- What are the sizes and weights of all robots and associated equipment?
- What are the tools and instruments for exploration and conversion of a cave or lava tube to accommodate a habitat?
- How many flights would it take to pre-position the necessary equipment?
- How would those packages be configured and launched?
- How accurately can the various payloads be targeted to the landing site?

- How much equipment, if any, could be carried on the descent vehicle?
- Which Short-Stay trajectories would be optimum for the years that are under consideration?

7.3 CONCEPT OF OPERATIONS

NASA created the Human Spaceflight Architecture Team (HAT) to inform the Human Exploration and Operations Mission Directorate (HEOMD) on possible mission architectures and campaigns beyond low Earth orbit. This has adopted a Capability Driven Framework (CDF) approach. No single destination has been considered, but a roadmap of possible destinations was developed that would lead towards human missions to the Moon, asteroids (since cancelled), and Mars.

The Destination Operations Team (DOT) was established within the HAT in October 2012 with a charter to develop destination specific "Point of Departure" Concepts of Operations (ConOps) for HAT Design Reference Missions (DRM) using a systematic analysis approach. The ConOps then identified more detailed products for the HAT Core Team. Using the 2009 NASA Human Exploration of Mars, Design Reference Architecture 5.0 (NASA-SP-2009-566) as a foundation, from April–September 2013 the DOT developed the ConOps for a long duration (500 days or so) human mission to the surface of Mars. By using these ConOps, DOT assessed destination strategies for human exploration in sufficient depth to: (1) capture the range of capabilities required to determine near-term technology investments, and (2) better understand and identify linkages between the Science Mission Directorate (SMD), HEOMD goals and objectives, Strategic Knowledge Gaps (SKG), and required functionality.

An approach was defined to collect and organize information to serve as the foundation for development of a Point of Departure (POD) ConOps for a crewed Mars Long-Stay surface mission. This information gathering approach involved a number of tasks, and was carried out by a number of DOT team members prior to initiating the development of the ConOps.

While many of these concepts apply to the mission proposed by this book, the ConOps focused only on the Long-Stay mission and not the Short-Stay precursor mission. (Also, notice that the following three domains guided their investigations which are, to some extent, applicable to a precursor mission but are quite limited in scope.)

The DOT hosted a series of special briefings for the team by Subject Matter Experts (SME) spanning three domains related to Mars surface operations. These domains included:

- Deep Drilling (on Earth and Mars).
- Biocontainment of Earth and Mars Pathogens.
- Mars Planetary Protection.

A "workbook" was created as a mechanism for the science discipline experts from the team to answer questions involving their discipline, relating to surface operations on Mars. The science disciplines for which detailed information was gathered included geology, geophysics, atmosphere and climate science, and also astrobiology. The workbook questions involved Mars surface operations driven by science objectives and activities; notably drilling and sampling (i.e. location, depth, proximity to landing site), data collection (approach, resources, location), data/sample analysis and return, crew/robotic activities across different mission phases, contamination control requirements, data rate/ frequency, and preliminary information requirements. On completion of the workbook, a teleconference was held for each science discipline. This enabled DOT and science discipline experts to review their sections of the workbook and address outstanding issues. These "interviews" were then transcribed to serve as primary reference material during Mars surface mission ConOps development.

This effort went much further and included:

- The team held an Educational Forum where experts gave briefings on subjects such as suits, EVA, crew medical issues, crew safety, sample handling, planetary protection, etc.
- Key references were identified from other organizations that would be fundamental to understanding the issues.
- A series of key questions were created in order to focus the information gathering on high-level and inter-related issues.
- There was a focus on the "Commuter" strategy which formed part of the "500 day" mission on the surface that included traverses to other sites.
- The "building blocks" approach was employed and then the blocks were combined to create a mission.

This process allowed the ConOps developers to add detail where clarification was needed, and enabled the mission timeline to be readily updated as additional information and understanding was forthcoming. It had the advantage of giving a realistic, representative ConOps which could be traced back to specific objectives and constraints as justification for capability requirements and design of elements. This method also produces a final product that can easily be updated in the future as objectives, constraints, and capabilities change.

This is the process that a team will go through for a precursor mission which utilizes the "Short-Stay" mission focused on the use of caves and lava tubes for protection against the hazards of the Moon and especially Mars.

The report made it clear that "this report represents one feasible crewed Mars surface mission ConOps and should not be viewed as the only feasible ConOps that could meet mission objectives."[1]

7.4 TECHNOLOGIES

The overall goal of establishing a Design Reference Mission for a mission to set up a habitat inside a cave or a lava tube on either the Moon or Mars would be to define the necessary technologies in addition to the mass, velocity and trajectory requirements. A basic scientific understanding of the elements is needed, as well as engineering and operational constraints to determine the viability of potential human habitation and emplacement of the associated infrastructure elements on the target body.

---

[1] The complete report can be read at:
https://www.lpi.usra.edu/lunar/strategies/HAT_Mars-DOT-2013-Final-Report.pdf

The 2015 NASA Technology Roadmaps consist of 16 sections, as follows:

- Introduction, Crosscutting Technologies, and Index.
- TA 1 Launch Propulsion Systems.
- TA 2 In-Space Propulsion Technologies.
- TA 3 Space Power and Energy Storage.
- TA 4 Robotics and Autonomous Systems.
- TA 5 Communications, Navigation, and Orbital Debris Tracking and Characterization Systems.
- TA 6 Human Health, Life Support and Habitation Systems.
- TA 7 Human Exploration Destination Systems.
- TA 8 Science Instruments, Observatories, and Sensor Systems.
- TA 9 Entry, Descent, and Landing Systems.
- TA 10 Nanotechnology.
- TA 11 Modeling, Simulation, Information Technology and Processing.
- TA 12 Materials, Structures, Mechanical Systems, and Manufacturing.
- TA 13 Ground and Launch Systems.
- TA 14 Thermal Management Systems.
- TA 15 Aeronautics.

Note that many of these areas are for the basic elements of the launch, deep space systems, propulsion, reentry and landing systems (back on Earth), and their associated technologies irrespective of the destination. Many relate to making a landing on a planetary body, working there, lifting off and returning to the deep space transit vehicle, then returning to Earth (elements that have been mentioned previously).

Of particular interest for the proposed precursor mission are the following:

- TA 3 Space Power and Energy Storage (specifically the Kilopower plant and habitat/vehicle batteries).
- TA 4 Robotics and Automation Systems (especially those supporting the crews with evaluation and clearing of the cave or lava tube and installing the habitat).
- TA 5 Communications, Navigation (both between vehicles and optical communications back to Earth).
- TA 6 Human Health, Life Support and Habitation Systems (especially the habitat and the support of the crew during the stay).
- TA 7 Human Exploration Destination Systems (for proposed Moon and Mars Short-Stay missions).
- TA 8 Science Instruments, Observatories, and Sensor Systems (including whatever the crews utilize for evaluating a cave, sensing the environment and atmosphere, and also any instruments that are to be left behind in the manner of the ALSEP).
- TA 9 Entry, Descent, and Landing Systems (as it relates to landing on the Moon or Mars; the portion related to the return to Earth is fundamental to all missions).
- TA 10 Nanotechnology (as it relates to the systems of the surface vehicles and robots).
- TA 11 Modeling, Simulation, Information Technology and Processing (as it relates to the proposed mission and systems).

- TA 12 Materials, Structures, Mechanical Systems, and Manufacturing (as it relates to the shielding, habitat, and installation equipment, but not the manufacturing of systems on the surface such as in-situ manufacturing).

Many of the systems associated with these technologies are in various stages of flight readiness and their progress would require monitoring over coming years to determine how they fit into the mission plan and concept of operations. This book could not possibly address all the technologies and their possible application for a mission such as that proposed, but some areas of focus are discussed below.

Many of the habitat studies investigated by the Technology Roadmaps and the Habitat Structures Team identified a lot of evaluation criteria for the construction of habitats. But those studies were surface based and, at the time, the knowledge of existing caves on the Moon and Mars was limited. The focus of these habitat studies was on the Earth-based evaluation type, the in-situ surface type, the deep space transit type, or a combination thereof. Since then, an inflatable habitat has been berthed at the ISS and the future for inflatables is much clearer. Hopefully, satellites and rovers launched in the near future will yield the further knowledge about caves, lava tubes, skylights, and pits on the Moon and Mars that is needed to refine the requirements for insertable, inflatable, and extendable habitats. This in turn will identify technologies for their manufacture, utility and installation.

7.4.1   Examples

Manufacturing Inflatables

The Roadmaps included a Cross-Technology Area. Now it appears that in some cases this cross fertilization is well under way, although the technologies have not yet been fully adapted. One example is the materials, adhesives, and preparation processes that were used for the Hypersonic Inflatable Atmospheric Decelerator (HIAD) might be directly applicable to inflatable, extendable habitats suitable for both surface and subsurface use. Also, there are pockets of expertise across the NASA Centers and their contractors that could be better focused on the concept of a precursor mission that uses insertable, inflatable, and extendable habitats as well as the applicable technology.

Nanotubes

Another example of a Cross-Technology Area is the advance in the development of boron nitride nanotubes that can be processed into materials with potentially useful radiation shielding properties. Consider the potential for a fabric that can stop radiation better than is available today, and how that might reduce the mass of the "stack" of spacecraft required for a deep space mission. This was discussed in Chapter 5, but the list goes on.

Nanotechnology is any technology that relates to features at nanometer scale; that is one billionth of a meter. This includes thin films, fine particles, chemical synthesis, advanced microlithography, and so forth. Nanotechnology is therefore more of a "catch-all" description for activities occurring at the level of atoms and molecules that have applications in the real world. It is now almost everywhere. We must understand what is going on at these levels and take advantage of new space exploration applications. Work in this field has been going on for decades. NASA has related work in the following areas:

- Ultrasensitive Label-Free Electronic Biochips Based on Carbon Nanotube Nano-electrode Arrays.
- Bulk Single-walled Carbon Nanotube Growth. Carbon nanotubes can play a variety of roles in future space systems, including wiring, high-strength lightweight composite materials, thermal protection and cooling systems and electronics/sensors.
- CAD for Miniaturized Electronics and Sensors
- Carbon Nanotube Field Emitters.
- Nanoengineered Heat Sink Materials.
- Human-Implantable Thermoelectric Devices.
- Automatic Program Synthesis for Data Monitors and Classifiers.
- Carbon Nanotubes for Removal of Toxic Gases in Life Support Systems.
- Carbon Nanotube Sensors for Gas Detection.
- Carbon Nanotubes as Vertical Interconnects.
- Nanoelectronics for Logic and Memory.
- Nonvolatile Molecular Memory.
- Large-Scale Fabrication of Carbon Nanotube Probe Tips for Space Imaging and Sensing Applications.
- Nanoelectronics for Space.
- Solid-state Nanopores for Gene Sequencing.
- Research and development to address critical life science questions.
- Optoelectronics and Nanophotonics.
- Plasma Diagnostics.
- Thermal, Radiation and Impact Protective Shields (TRIPS).

Optical Communications

An example of a technology which would enhance a mission to Mars is optical communications. It is a major objective of NASA's Space Technology Mission Directorate (STMD), and its introduction is widely expected to be revolutionary because it addresses the limitations of radio frequency communications. A vastly increased bandwidth will allow a spacecraft (or surface facility) to transmit much more data to Earth.

For example, the Mars Reconnaissance Orbiter uses a 3 m (10 ft) antenna to communicate with Earth. With optical communications, it could use a 20 cm (8 in) aperture telescope instead. Packages for optical communications are smaller, lighter, and less power-hungry than radio frequency units.

The advantages offered by an optical communications are:

- Up to 50 per cent savings in mass. Reduced mass enables decreased spacecraft cost and/or increased science through greater mass for the instruments.
- 65 per cent savings in power. Reduced power enables increased mission life and/or increased science measurements.
- Up to 20 times increase in data rate. Increased data rates enable greater data collection and reduced mission operations complexity.

The Deep Space Optical Communications (DSOC) package aboard NASA's Psyche mission will use photons to transmit more data in a given interval. This mission was selected for flight in early 2017 under NASA's Discovery Program, and is scheduled for

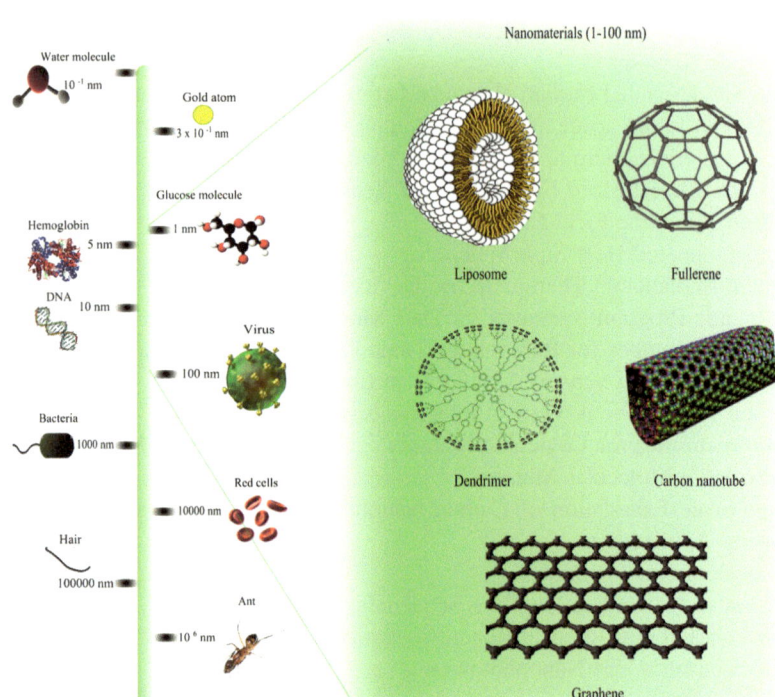

Fig. 7.6 Comparison of nanomaterial sizes. Graphic courtesy of Molecular Diversity Preservation International.

launch in the summer of 2022 to inspect a distinctive metal asteroid (16 Psyche) about three times farther out from the Sun than Earth. It is 200 km (120 mi) in diameter and contains just less than 1 per cent of the overall mass of the asteroid belt. The planned arrival is in 2026.

The goal of DSOC is to increase spacecraft communications performance and efficiency by 10 to 100 times over conventional means, all without increasing the mission burden in terms of mass, volume, power, and/or spectrum. Success with the technologies integrated into a Flight Laser Transceiver (FLT) for deep space missions will advance this method of communications to Technology Readiness Level 6, which means having technology that is a fully functional prototype or a representational model. The laser beacon to DSOC will be sent from JPL's Table Mountain Facility near Wrightwood, in the Angeles National Forest of California. The data from the spacecraft will be received by a telescope at the Mount Palomar Observatory in California.

Robotics

Another example of Cross-Technology is autonomous robots which use artificial intelligence, microelectronics, advanced batteries and new materials to produce cooperative

Fig. 7.7 Psyche arrival. Photo courtesy of NASA/JPL.

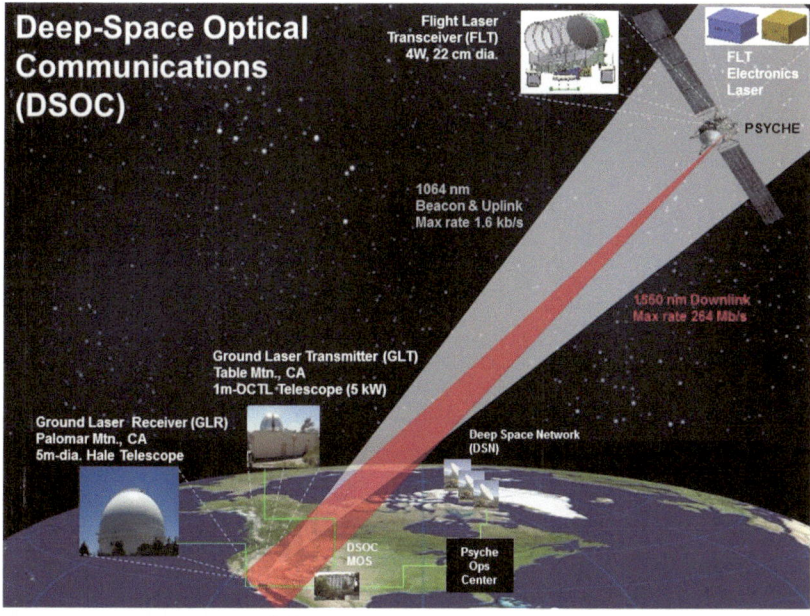

Fig. 7.8 The Deep Space Optical Communications System. Photo courtesy of NASA/JPL/Caltech.

robots. They needn't look like humans for a mission to the Moon or Mars, but just be able to do their jobs. They could have any number of arms and legs; whatever is required to perform tasks better and safer than the crew. Some might walk. Others might fly, for example using thrusters to penetrate caves and lava tubes to perform evaluations. The robotics world needs only to focus on the proposed mission to come up with relevant designs. There is a definite need for a robot to fly into areas such as caves and lava tubes, conduct reconnaissance, and return the data to the crew (and perhaps to Earth for detailed analysis and advice on how to proceed). The Moon is essentially airless and the atmosphere of Mars has a surface pressure of just a few millibars, so traditional propeller drones will not be feasible; instead they will have to maneuver using thrusters. A specialized robot will have to be designed for that purpose. Robots such as a flying LIDAR and 3D multispectral camera could also be helpful for characterizing the volume inside a cave.

These are just a few examples of how the development of technologies feeds into the definition of mission capabilities.

From the work described in Chapter 5 it is clear that progress has been made in nearly all areas, and in some areas the state-of-the-art has been advanced. It is also apparent (in hindsight) that a human mission to Mars could not realistically have been accomplished earlier; it is such a demanding task that it simply had to wait for technological advances to raise the likelihood of success to a reasonable level.

If there is indeed a human mission to Mars in 2033, that will be 64 years after Neil Armstrong placed the first boot print in the lunar dust. Back then, there was optimism in some quarters that we would reach Mars in the 1980's. The fact that it is taking so long to make that next giant leap reflects the enormity of the task. For the most part, the technologies of 2033 will have been hardly imaginable in the relatively primitive 1960's.

Appendix 1

Lunar and Mars Satellites and Rovers

A.1.1 Lunar Satellites and Rovers

Due to page limitations, only missions flown within the last quarter century are listed.

Clementine Orbiter

Officially called the Deep Space Program Science Experiment (DSPSE), the Clementine mission was a joint space project between the DoD's Ballistic Missile Defense Organization (BMDO) and NASA. Launched on January 25, 1994, the objective was to test sensors and spacecraft components under extended exposure to the space environment and to make scientific observations of the Moon and the near-Earth asteroid 1620 Geographos. The project was named after the song "Oh My Darling, Clementine" because after it had finished its mission the spacecraft would be "lost and gone forever."

The lunar observations included imaging at various wavelengths in the visible, ultraviolet and infrared parts of the spectrum, laser ranging altimetry, gravimetry, and charged particle measurements. These observations were for the purposes of obtaining multi-spectral imaging of the entire lunar surface, assessing the surface mineralogy, obtaining altimetry from 60°N to 60°S latitude, and obtaining gravity data for the near side of the Moon. On reaching Geographos it was to determine the size, shape, rotational characteristics, surface properties, and cratering rates of the asteroid.

Clementine carried seven distinct experiments:

- Ultraviolet/Visible Camera.
- Near Infrared Camera.
- Long Wavelength Infrared Camera.

© Springer Nature Switzerland AG 2019
M. von Ehrenfried, *From Cave Man to Cave Martian*,
Springer Praxis Books, https://doi.org/10.1007/978-3-030-05408-3

- High Resolution Camera.
- Two Star Tracker Cameras.
- Laser Altimeter.
- Charged Particle Telescope.

The S-band transponder facilitated communications, tracking, and gravimetry.

The mission had two phases. After two Earth flybys, lunar orbit insertion was achieved approximately a month into flight. Lunar mapping was undertaken over a period of two months using a 5-hour elliptical polar orbit which included an 80-minute mapping phase. After a month of mapping southern latitudes, the orbit was adjusted to conduct another month of mapping northern ones. Over a total of 300 orbits it obtained imaging and altimetry coverage from 60°S to 60°N.

After an Earth to Moon transfer and two more Earth flybys, the spacecraft was to head for Geographos, arriving three months later for a flyby. Unfortunately, on May 7, 1994, after the first Earth transfer orbit, a malfunction caused one of the attitude control thrusters to fire for 11 minutes, using up the supply of propellant and causing the vehicle to spin at about 80 rpm. This meant that the asteroid flyby would not yield useful results. The spacecraft was therefore put into a geocentric orbit that passed through the Van Allen radiation belts in order to test the various components on board in that environment. The mission ended in June 1994 when the power level declined to a point at which the telemetry downlink was no longer intelligible.

Fig. A.1.1 Artist's conception of the Clementine spacecraft. Photo courtesy of the Naval Research Laboratory.

SELENE (KAGUYA) Orbiter

The Japan Aerospace Exploration Agency (JAXA) launched the SELenological and ENgineering Explorer (SELENE), a.k.a. "KAGUYA," on an H-IIA vehicle on September 14, 2007, from the Tanegashima Space Center (TNSC). The name KAGUYA means "radiant night," and is from the Japanese story "The Tale of the Bamboo Cutter" in which Kaguya, a princess who fell to Earth from the Moon, is returned home.

The major objectives of the KAGUYA mission were to obtain data about the origin and evolution of the Moon, and to develop the technology for future lunar exploration. KAGUYA consisted of a main satellite orbiting at about 100 km (62 mi) altitude, plus two small satellites: a Relay Satellite and a Very Long Baseline Interferometer (VRAD) Satellite in polar orbit. On February 1, 2009, KAGUYA lowered its perilune to a 50 km (31 mi) altitude, then lowered it again on April 16, 2009 to 10–30 km (6–19 mi). The mission ended with the spacecraft impacting on June 10, 2009.[1]

JAXA investigated the SELENE Lunar Radar Sounder (LRS) data at locations close to the Marius Hills Hole (MHH), which is a skylight potentially leading to an intact lava tube, and found a distinctive echo pattern exhibiting a precipitous decrease in echo power, subsequently followed by a large second echo peak that may indicate the presence of a lava tube. The search area was further expanded to span 13.00–15.00°N, 301.85–304.01°E around the MHH, obtaining similar echo patterns at several locations. Most of the locations are in regions of underground "mass deficit" suggested by analysis of GRAIL mission gravity data. Some of the observed echo patterns occur along a rille close to the MHH, or on the southwest underground extension of that rille. Intact lava tubes offer a pristine environment for studying the composition of the Moon, and could serve as secure shelters for humans and instruments. Nine images were taken of the MHH at approximately 14.09°N, 303.31°E. The pit is estimated to be 48–57 m (131–197 ft) in diameter, to be about 45 m (148 ft) deep, and to have a roof thickness of 40–60 m (131–197 ft).[2] The pit is believed to have been formed by the partial collapse of a lava tube. See Chapter 3.1 for an overview of the Oceanus Procellarum region of the Moon.

Chang'e 1 Orbiter

Named after a Moon goddess, the robotic lunar-orbiting Chang'e 1 was part of the first phase of the Chinese Lunar Exploration Program. It was launched on October

[1]For a short video of SELENE flying over the Apollo 11 landing site go to:
 https://www.youtube.com/watch?v = sV3GKy8Hr_A
[2]For a video on possible lava tubes at Marius Hills go to:
 https://www.youtube.com/watch?v = fC_gAY48Rrc

24, 2007, from the Yichang Satellite Launch Center, and achieved lunar orbit on November 5, 2007. A year later, a map of the entire lunar surface was published using data collected by Chang'e 1 between November 2007 and July 2008.

The mission was scheduled to last for a year, but it was later extended and the spacecraft was operated until March 1, 2009, when it was taken out of orbit and impacted on the Moon. Data gathered by Chang'e 1 was able to produce the most accurate and highest resolution 3D map yet made of the lunar surface. It was the first lunar probe to conduct passive, multi-channel, microwave remote sensing by using a microwave radiator. Its sister probe, Chang'e 2, was launched on October 1, 2010.

The Chang'e 1 mission had four major goals:

- Obtaining 3D images of the landforms and geological structures of the lunar surface as a reference for planning future soft landings. The orbit of Chang'e 1 was designed to provide full coverage, including areas near the poles that were not covered by previous missions.
- Mapping the abundance and distribution of various chemical elements on the lunar surface for an evaluation of potentially useful resources.
- Probing the features of the lunar soil and assessing its depth, as well as the amount of helium-3 (He-3) present.
- Probing the environment between 40,000 km (24,854 mi) and 400,000 km (248,548 mi) from Earth, studying the solar wind, and the effects of solar activity on the Earth and the Moon.

The lunar probe was composed of five major systems: the satellite system, the launch vehicle system, the launch site system, the monitoring and control system, and the ground application system.

Five goals were accomplished:

- Researching, developing, and launching China's first lunar probe.
- Mastering the basic technology of placing a satellite into lunar orbit.
- Conducting China's first scientific investigation of the Moon.
- Initially forming a lunar probe space engineering system.
- Accumulating experience for subsequent phases of China's Lunar Exploration Program.

Chandrayaan-1 Orbiter/Impactor

Chandrayaan-1 was the Indian Space Research Organization's (ISRO) first lunar probe. The mission included an orbiter and an impactor. It was launched from the Dhawan Space Centre using a PSLV-XL rocket on October 22, 2008.

On November 14, 2008, the Moon Impact Probe separated from the spacecraft and struck the south pole in a controlled manner, making India the fourth country to deliver its flag to the Moon. The probe impacted close to the crater Shackleton, ejecting subsurface soil that could be analyzed for the presence of lunar water ice.

The orbiter carried high resolution remote sensing equipment for visible, near infrared, and soft and hard X-ray frequencies. It was intended to survey the lunar surface over a period of two years in order to map its chemical composition and 3D topography. The polar regions were of particular interest, due to the suspected presence of water ice. There were five ISRO payloads and six that were flown at zero cost for other space agencies which included NASA, ESA, and the Bulgarian Aerospace Agency. One of the mission's greatest achievement was the discovery of the widespread presence of water molecules in the lunar soil.

After almost a year, the orbiter started to suffer a number of technical issues, including losing its star sensors and thermal difficulties. It stopped transmitting on August 28, 2009, and shortly thereafter ISRO officially declared the mission over. Although it operated for 312 days versus the intended two years, the Chandrayaan mission achieved 95 per cent of its planned objectives.

Lunar Reconnaissance Orbiter

NASA-GSFC's Lunar Reconnaissance Orbiter (LRO) was launched on June 18, 2009 by an Atlas V rocket to provide global data about the Moon, including day-and-night temperature maps, a geodetic grid, high resolution color imaging, and the Moon's ultraviolet albedo. There was particular emphasis on the polar regions where it was suspected there might be water ice in permanently shadowed craters and continual solar illumination on elevated terrain.

Although the objectives of the mission were explorative in nature, the payload included instruments with considerable heritage from previous planetary science missions, enabling transition to a science phase under NASA's Science Mission Directorate.

LRO had a complement of six instruments and one technology demonstration:

- Cosmic Ray Telescope for the Effects of Radiation (CRaTER). Its main goal was to characterize the global lunar radiation environment and the biological inplications.
- The Diviner Lunar Radiometer Experiment was to measure lunar surface thermal emission to provide information for future surface operations.
- Lyman-Alpha Mapping Project (LAMP) was to peer into permanently shadowed craters in search of water ice, using ultraviolet light generated by stars as well as the hydrogen atoms that are thinly spread throughout the solar system.
- Lunar Exploration Neutron Detector (LEND) made measurements from which to map possible near-surface water ice deposits.
- Lunar Orbiter Laser Altimeter (LOLA) provided a precise global lunar topographic model and geodetic grid.
- Lunar Reconnaissance Orbiter Camera (LROC) addressed the measurement requirements of landing site certification and polar illumination. The

instrument comprised a pair of narrow-angle push-broom imaging cameras (NAC) and a single wide-angle camera (WAC). The spacecraft flew several times over the historic Apollo lunar landing sites at 50 km (31 mi) altitude and the camera's high resolution revealed the Lunar Module descent stages, along with the Lunar Roving Vehicles and other items left by the astronauts. The mission is still active, and its imagery has revealed a number of pits and skylights which might lead to caves.

- The Miniature Radio Frequency (Mini-RF) radar demonstrated the use of a novel lightweight Synthetic Aperture Radar (SAR) for investigating the lunar surface. Mini-SAR was one of two NASA contributions to India's first lunar mission, Chandrayaan-1, which mapped from a circular polar orbit at 100 km (62 mi). With Mini-RF, LRO mapped the Moon for one year from a 50 km (31 mi) circular polar orbit. Mini-RF is an enhanced version of its cousin on Chandrayaan-1 able to communicate in both S-band (like Chandrayaan-1) and X-band frequencies. Its "zoom" feature enabled it to image details smaller than 30 m (100 ft) across. It was also testing a new communications technology.[3]

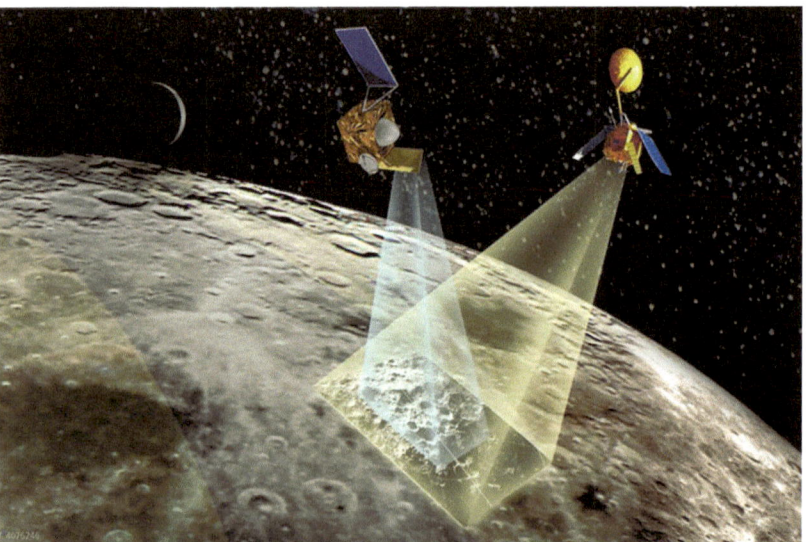

Fig. A.1.2 Lunar Reconnaissance Orbiter (right) and Chandrayaan-1 take strips of data over the lunar poles. Coordinated, bistatic imaging in S-band by the two spacecraft can unambiguously resolve ice deposits. Photo courtesy of JHU/APL & NASA/GSFC.

---

[3]To learn more about Mini-RF, visit: www.nasa.gov/mini-rf

As of this writing, there is evidence that the inertial measurement unit (IMU) on board the Lunar Reconnaissance Orbiter (LRO) spacecraft is nearing its end of life. The IMU has been powered down to reserve its remaining service for highly critical events, such as lunar eclipses and safe mode entries.

The IMU contains a set of three ring laser gyroscopes to measure the rotation of the vehicle in each of three Cartesian axes. The spacecraft, which is currently in its ninth year of operations, was designed for a nominal mission of one year. It was largely single-string, meaning the spacecraft was built with little redundancy. There is only one IMU, so the LRO team must now manage operations without a direct measurement of the rotation rates. In place of the IMU, the team is using successive estimates of spacecraft orientation using two on board star trackers to estimate the rotation rate. The spacecraft is currently being kept in its nadir-only pointing mode, but because it was designed to map the Moon using nadir pointing it is still able to make science measurements. However, reorienting it to point off-nadir was scheduled to resume in late 2018 for special science observations, most notably LROC stereoscopic and oblique-angle imaging.[4]

LCROSS

The Lunar Crater Observation and Sensing Satellite (LCROSS) built by Northrup Grumman was a mission conceived as a low-cost means of determining the nature of the hydrogen which was known to be present in the polar regions of the Moon. The primary objective was therefore to further investigate the presence of water ice in a permanently shadowed crater.

LCROSS was launched along with Lunar Reconnaissance Orbiter (LRO) on June 18, 2009, as part of the shared Lunar Precursor Robotic Program; the first American mission to the Moon for over a decade. Together, these two spacecraft formed the vanguard of NASA's return to the Moon. LCROSS was designed to collect and relay data from the impact and debris plume created by the impact of the launch vehicle's spent Centaur upper stage striking the crater Cabeus, located near the lunar south pole.

The Shepherding Spacecraft separated from the Centaur stage on October 9, 2009, during the approach to the Moon. The Centaur acted as a heavy impactor to produce a debris plume that rose above the lunar surface. Four minutes after the Centaur impacted, the Shepherding Spacecraft passed through this debris plume, collecting and relaying data to Earth before it struck the lunar surface to create a second debris plume. It was successful in confirming water in the crater Cabeus. Although this mission was not looking for caves, it did confirm that there is ice on the Moon.

---

[4]To watch a short video of the Moon taken by the LRO camera go to:
 https://www.youtube.com/watch?v = sjkPeexEdyI

Fig. A.1.3 Artist's rending of the Centaur stage approaching the Moon and the LCROSS spacecraft taking data from the impact of the Centaur, just minutes before it impacted too. Photo courtesy of NASA.

GRAIL Twin Orbiter

The Gravity Recovery and Interior Laboratory (GRAIL) mission that was part of the NASA Discovery Program investigated the interior structure of the Moon by high-quality mapping of the gravitational field. The two small spacecraft GRAIL A (Ebb) and GRAIL B (Flow) were launched on September 10, 2011 on board the most powerful configuration of the Delta II vehicle. GRAIL A separated from the rocket about nine minutes into flight and GRAIL B eight minutes later. On these trajectories, they entered orbit around the Moon 25 hours apart on December 31, 2011 and on January 1, 2012.

Each spacecraft transmitted and received telemetry from the other spacecraft and Earth-based facilities in order to detect changes in the distance between the two vehicles as small as one micron. This enabled the gravitational field of the Moon to be mapped in unprecedented detail and thereby investigate its geological structure.

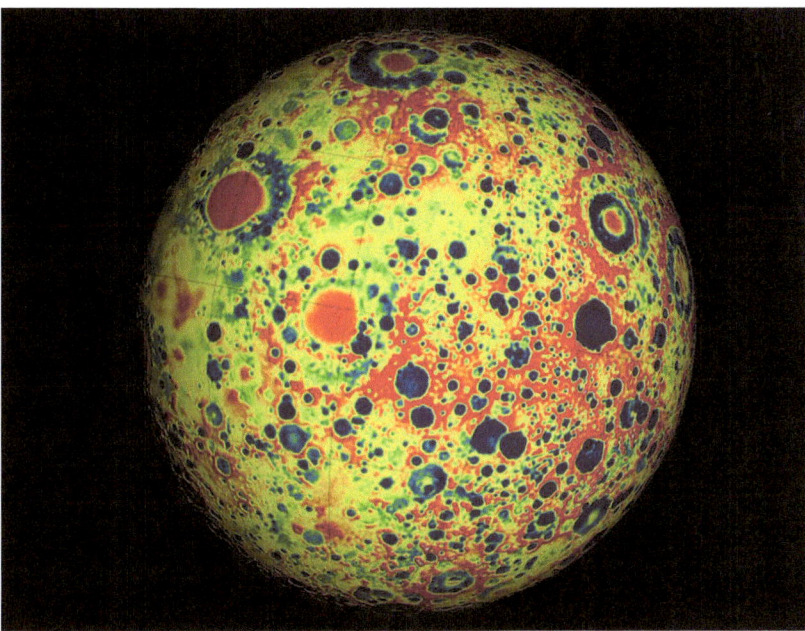

Fig. A.1.4 The Moon's gravity field by GRAIL. This image shows variations in the lunar gravity field. Very precise microwave measurements between two spacecraft, named Ebb and Flow, mapped the lunar gravitational field to a high precision and high spatial resolution. The measurements were three to five orders of magnitude improved over previous data. Red corresponds to mass excesses and blue corresponds to mass deficiencies. The map shows more small-scale detail on the far side of the Moon compared to the near side because the far side has many more small craters. The result of the mission was the highest resolution gravity field map of any celestial body to date. It is revealing the internal structure and composition of the Moon in unprecedented detail. It also shows an abundance of surface detail such as tectonic structures, volcanic landforms, basin rings, crater central peaks and numerous simple, bowl-shaped craters. Data also show the Moon's gravity field is unlike that of any terrestrial planet in our solar system. Photo courtesy of NASA/JPL-Caltech/MIT/GSFC.

The GRAIL mission objectives were:

- Map the structure of the lunar crust and lithosphere.
- Understand the asymmetric thermal evolution of the Moon.
- Determine the subsurface structure of impact basins and the origin of "mascons" (mass concentrations).
- Ascertain the temporal evolution of crustal brecciation (fragmentation) and magmatism.
- Constrain the deep interior structure of the Moon.
- Place limits on the size of the inner core.

The data collection phase of the mission lasted from March 7, 2012 to May 29, 2012; a total of 88 days. A second phase of data collection at a lower altitude was begun on August 31, 2012, and was followed by 12 months of data analysis. On December 5, 2012, NASA released a gravity map of the Moonmade from GRAIL data. This data will provide insight into the evolutionary history of the terrestrial planets and improve the computation of trajectories for future orbiters.

The gravity field of the Moon preserves the record of the impact bombardment that characterized all solid planetary bodies, and reveals evidence for fracturing of the interior extending to the deep crust and possibly into the mantle. The GRAIL mission precisely measured this record for the Moon. It found the bulk density of the Moon's highland crust to be substantially lower than generally assumed. This low bulk crustal density agrees well with data provided by the final Apollo lunar missions in the early 1970s, indicating that local samples returned by astronauts are indicative of global processes. This data indicated that the average thickness of the Moon's crust is between 34 and 43 km (21 and 27 mi), which is about 10 to 20 km (6 to 12 mi) thinner than had been believed. The bulk composition of the Moon is therefore similar to that of Earth, supporting models in which the Moon coalesced from material ejected by a giant impact very early in Earth's history.

At the end of their science phase and a mission extension, the two spacecraft were powered down and decommissioned over a 5-day period. They impacted on the lunar surface on December 17, 2012.

A.1.2 Mars Satellites and Rovers

The Viking missions of 1975 represented the zenith of Mars exploration up to that point. Here is a list of U.S.-led missions to that planet in the last twenty years.

Mars Global Surveyor Orbiter

This was an American robotic spacecraft developed by NASA's Jet Propulsion Laboratory and launched November 7, 1996. It was a global mapping mission to examine the entire planet, from the ionosphere down through the atmosphere to the surface. As part of the larger Mars Exploration Program, it also served as a monitoring relay for sister orbiters during aerobraking and it assisted planners in identifying potential landing sites for rovers and landers and then relayed surface telemetry.

It completed its primary mission in January 2001 and was in its third mission extension phase when, on November 2, 2006, the spacecraft failed to respond to commands. A faint signal detected three days later showed that it had entered safe mode. Efforts to reestablish contact with the vehicle to resolve the problem failed, and NASA declared the mission to be over in January 2007.

Fig. A.1.5 Mars Global Surveyor. Photo courtesy of NASA/JPL-Caltech/Corby Waste.

Mars Odyssey Orbiter

This spacecraft was launched on April 7, 2001, and entered orbit around Mars on October 24 of that year. Since achieving its 4-year primary mission, Odyssey has been granted several mission extensions. At the time of writing, it is still returning science data. Its mission was to use spectrometers and a thermal imager to detect evidence of past or present water and ice, as well as to study the planet's geology and radiation environment. It examined what Mars is made of, detected water and shallow buried ice, and studied the radiation environment in space. It also acted as a relay for communications back to Earth between the Mars Exploration Rovers, the

Curiosity rover of the Mars Science Laboratory mission, and, prior to that, the Phoenix lander. The mission was named as a tribute to Arthur C. Clarke's thought provoking novel *2001: A Space Odyssey*.

The three primary Odyssey instruments were:

- Thermal Emission Imaging System (THEMIS).
- Gamma Ray Spectrometer (GRS).
- Mars Radiation Environment Experiment (MARIE).

The THEMIS instrument assisted in selecting a landing site for the Curiosity rover. Several days before that landing in August 2012, the orbit of Odyssey was adjusted to ensure it would be able to capture signals transmitted during the first few minutes on the Martian surface. Odyssey also serves as a relay for UHF radio signals from Curiosity. It passes over Curiosity at the same local time twice every day, simplifying the scheduling of contact with Earth. The GRS instrument was a collaboration by the University of Arizona's Lunar and Planetary Laboratory, the Los Alamos National Laboratory, and Russia's Space Research Institute, with the latter providing the High Energy Neutron Detector (HEND) that was to search for hydrogen as an indicator of water ice at shallow depths.

Fig. A.1.6 An artist's concept of 2001 Mars Odyssey spacecraft over Syrtis Major Planum. Photo courtesy of NASA/JPL.

Mars Exploration Twin Rovers "Spirit" and "Opportunity"

The next mission to Mars sent a pair of robotic geologists to locations on opposite sides of the planet. Spirit was launched on June 10, 2003 and landed on January 4, 2004 inside a Connecticut-sized crater named Gusev. Its sister Opportunity was launched on July 7, 2003 and landed on January 25, 2004 inside a backyard-sized crater informally named "Eagle." Originally designed to operate for three months, the rovers far exceeded both their expected lifetime and distance travelled. Spirit became bogged down in fine sand in 2009 and continued to function in situ until 2010. The mission was official terminated in 2011. Opportunity lasted longer and travelled farther, until a global dust storm in 2018 denied it solar power and put it into safe mode. As of early December 2018 it had not recovered, even though the dust storm had abated. Both of these rovers found evidence of long-ago Martian environments where water was active and conditions may have been suitable for life.

The scientific objectives of the Mars Exploration Rover mission were to:

- Search for and characterize a variety of rocks and soils that hold clues to past water activity. In particular, the samples sought included those that had minerals deposited by water-related processes such as precipitation, evaporation, sedimentary cementation, or hydrothermal activity.
- Determine the distribution and composition of minerals, rocks, and soils surrounding the landing sites.
- Determine what geological processes had shaped the local terrain and influenced its chemistry. Such processes could include water or wind erosion, sedimentation, hydrothermal mechanisms, volcanism, and cratering.
- Perform calibration and validation of surface observations made by the instruments on Mars Reconnaissance Orbiter. This "ground truth" would

Fig. A.1.7 An artist's rendering of one of the rovers. Photo courtesy of NASA/JPL/ Cornell University/Maas Digital LLC.

help to assess the accuracy and effectiveness of various instruments that were surveying Martian geology from orbit.

- Seek iron-containing minerals, and identify and quantify relative amounts of specific mineral types that contain water or were formed in water, such as iron-bearing carbonates.
- Characterize the mineralogy and textures of rocks and soils to determine the processes that created them.
- Search for geological clues to the environmental conditions that existed when liquid water was present on the surface.
- Assess whether those environments were conducive to life.

Mars Reconnaissance Orbiter

This mission was launched on August 12, 2005 with the most powerful telescopic camera ever sent to another planet, plus five other scientific instruments. It is still operating, monitoring the present water cycle in the atmosphere of Mars and the associated deposition and sublimation of water ice on the surface, while probing the subsurface to see how deep the water ice reservoir detected by Mars Odyssey extends. At the same time, it is searching for surface features and minerals (such as carbonates and sulfates) that indicate the extended presence of liquid water on the surface earlier in the planet's history.

The Mars Reconnaissance Orbiter instruments are:

- Shallow Subsurface Radar (SHARAD).
- Compact Reconnaissance Imaging Spectrometer (CRISM).
- Mars Color Imager (MARCI).
- High Resolution Imaging Science Experiment (HiRISE) .
- Context Camera (CTX).
- Mars Climate Sounder (MCS).

Phoenix Lander

The Phoenix mission to Mars was carried out under the Mars Scout Program. It was launched on August 4, 2007, using a Delta 7925 vehicle. The lander set down on the frozen terrain near the north pole of Mars on May 25, 2008. The mission was led by the Lunar and Planetary Laboratory at the University of Arizona, with project management at the Jet Propulsion Laboratory and project development by Lockheed Martin in Denver, Colorado.

Mars Reconnaissance Orbiter's High Resolution Imaging Science Experiment (HiRISE) camera photographed Phoenix suspended from its parachute during its descent through the Martian atmosphere. This marked the first time ever that one spacecraft had photographed another in the act of landing on a planet.

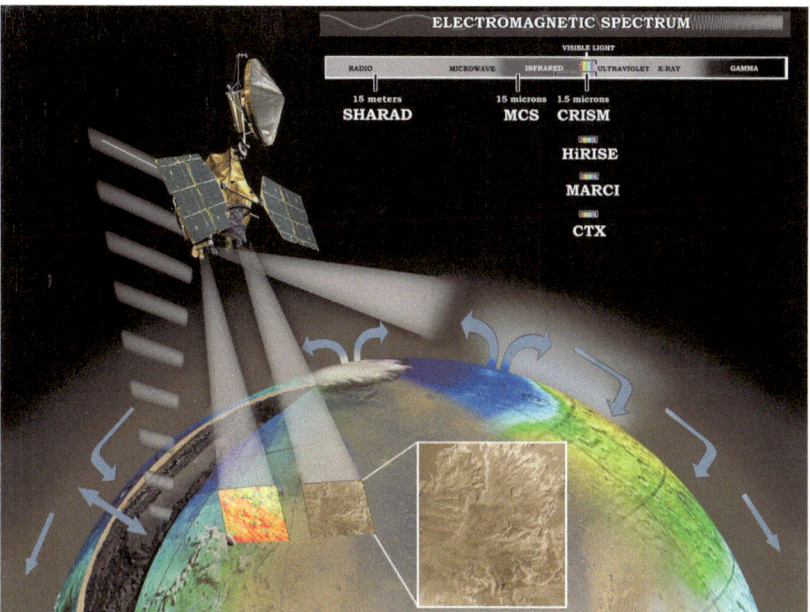

Fig. A.1.8 An artist's concept of the Martian water cycle. To the far left, the MRO radar antenna beams down and "sees" into the first few hundred feet (up to a kilometer) of Mars' crust. Just to the right of that, the next beam highlights the data received from the imaging spectrometer, which identifies minerals on the surface. The next beam represents the high-resolution camera, which can "zoom in" on local targets to provide the highest-resolution orbital images yet of features such as craters and gullies and rocks. The beam that shines almost horizontally is that of the Mars Climate Sounder. This instrument is critical to analyzing the current climate of Mars, since it observes the temperature, humidity, and dust content of the atmosphere, and their seasonal and year-to-year variations. Meanwhile, the Mars Color Imager observes ice clouds, dust clouds and hazes, and the ozone distribution to produce daily global maps in multiple colors that monitor daily weather and seasonal changes. The electromagnetic spectrum is represented on the top right and individual instruments are placed where their capability lies. Photo courtesy of NASA/JPL/Corby Waste.

Phoenix was NASA's sixth successful landing out of seven attempts, and the first successful landing in a Martian polar region (the mission which failed was an attempt to land near the south pole in 1999). The lander completed its mission in August 2008, and made a final brief communication with Earth on November 2 as available solar power dropped with the local winter. The mission was declared to be over on November 10, 2008, after the lander failed to respond to transmissions. The program was deemed to be a success because it achieved its planned science experiments and observations.

Phoenix was equipped with improved versions of the University of Arizona's panoramic cameras and volatiles-analysis instrument from the ill-fated Mars Polar Lander of 1999, as well as experiments created for the cancelled Mars Surveyor 2001

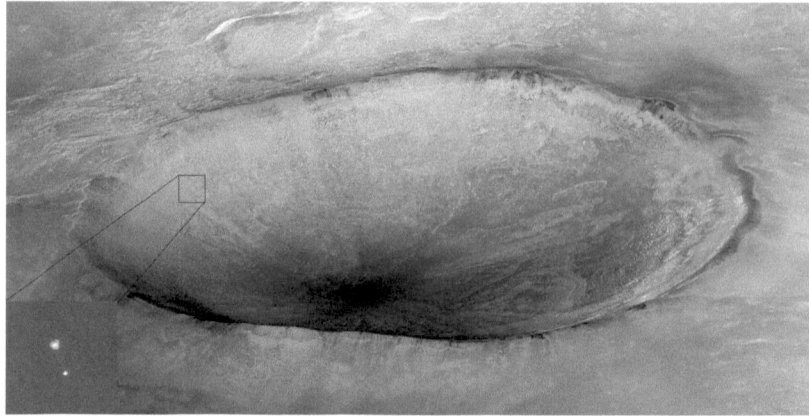

Fig. A.1.9 Phoenix descending by parachute. The MRO HiRISE camera acquired this image of Phoenix on its parachute descending to the Martian surface. Shown here is a 10 km (6 mile) diameter crater informally called "Heimdall," and an improved full-resolution image of the parachute and lander. Although it appears that Phoenix is descending into the crater, it is actually about 20 km (12 mi) in front of the crater. At the time of this observation, MRO was at an altitude of 310 km (192 mi), was travelling at a ground velocity of 3.4 km per second (2.1 mi per second), and was 760 km (472 mi) from the Phoenix lander. Photo courtesy of NASA/JPL-Caltech/UA.

Lander – including a JPL trench-digging robotic arm, a set of wet chemistry laboratories, and optical and atomic force microscopes. The science payload also included a descent imager and a suite of meteorological instruments. The lander spent five months digging trenches with its robotic arm, in the process confirming the inference from orbital sensors that there was water ice at shallow depth at this location.

One of the unexpected findings of the Phoenix mission was that almost 0.5 per cent of the soil contained perchlorate (ClO_4) and thus might not be as life-friendly as previously believed. Laboratory research published in July 2017 demonstrated that when irradiated with a simulated Martian ultraviolet flux, perchlorates would become bacteriocidal. Two other compounds of the Martian surface – iron oxides and hydrogen peroxide – act in synergy with irradiated perchlorates to produce a 10.8-fold increase in cell death compared to cells which are exposed to ultraviolet radiation for 60 seconds. It was also realized that abraded silicates such as quartz and basalt give rise to the formation of toxic reactive oxygen species. Overall, the mission left the issue of the presence of organic compounds open, because heating the samples that contained the perchlorates would have broken down any organics that were present.

The fact that perchlorates have also been detected at the Curiosity rover site (nearer the equator) and in the Martian meteorite EETA79001 is suggestive of a "global distribution of these salts." Only highly refractory and/or well-protected organics are likely to be preserved in the frozen subsurface. An instrument that is planned for the 2020 ExoMars rover to attempt to detect and measure subsurface organics will therefore employ a method that will not be affected by the presence of perchlorates.

Mars Science Laboratory Rover "Curiosity"

On November 26, 2011, the Mars Science Laboratory was launched by Atlas V to deliver JPL's Curiosity rover to Gale Crater on Mars. On August 6, 2012, a new system called the "Sky Crane" hovered just above the surface and gently lowered the rover to the ground. After spending two years studying the floor of the crater, Curiosity reached the base of Mount Sharp in September 2014. Since then, it has been working its way up slope, examining exposed layering to determine how the environment changed over time.

Curiosity had instruments to address eight primary scientific objectives:

Biological

- Determine the nature and inventory of organic carbon compounds.
- Investigate the chemical building blocks of life (carbon, hydrogen, nitrogen, oxygen, phosphorus, and sulfur).
- Identify features that may represent the effects of biological processes (biosignatures).

Geological and geochemical

- Investigate the chemical, isotopic, and mineralogical composition of the Martian surface and near-surface geological materials.
- Interpret the processes that have formed and modified rocks and soils.

Planetary process

- Assess the long-timescale (i.e. 4-billion-year) evolution processes of the Martian atmosphere.
- Determine present state, distribution, and cycling of water and carbon dioxide.

Surfaceradiation

- Characterize the broad spectrum of surface radiation, including cosmic radiation, solar particle events, and secondary neutrons. In addition, the radiation exposure in the interior of the spacecraft was measured during the cruise through interplanetary space. This data will assist in planning future human missions.

About one year into the surface mission, and having found that ancient Mars could have been hospitable to microbial life, the research expanded to developing predictive models for processes that may have preserved organic compounds and biomolecules; a branch of paleontology called taphonomy.

Fig. A.1.10 Curiosity landed in Gale Crater, which is approximately the size of Connecticut and Rhode Island combined. A green dot shows where the rover landed, well within its targeted landing ellipse, outlined in blue. This oblique view of Gale, with Mount Sharp in the center, is derived from a combination of elevation and imaging data from three Mars orbiters. The view is looking toward the southeast. Mount Sharp rises about 5.5 km (3.4 mi) above the floor of Gale Crater. For perspective, that is almost as high as Mount McKinley (Denali) in Alaska. The image combines elevation data from the High Resolution Stereo Camera on the European Space Agency's Mars Express orbiter, image data from the Context Camera on NASA's Mars Reconnaissance Orbiter, and color information from Viking Orbiter imagery. There is no vertical exaggeration in the image. Photo courtesy of NASA/JPL-Caltech/ESA/DLR/FU Berlin/MSSS.

MAVEN Orbiter

The Mars Atmosphere and Volatile EvolutioN (MAVEN) mission was developed by NASA as an orbiter to investigate the Martian atmosphere. The goals included determining how the planet's atmosphere and water, presumed to have once been substantial, were lost over time. It was launched aboard an Atlas V launch vehicle on November 18, 2013, and the Centaur upper stage inserted it into a heliocentric transfer orbit. On September 22, 2014, the spacecraft entered a near-polar, highly elliptical orbit around Mars with an altitude ranging between 6200 km (3900 mi) and 150 km (93 mi).

On November 5, 2015, NASA announced that data from MAVEN indicated the deterioration of Mars' atmosphere increases significantly during solar storms. The leakage of atmosphere to space likely played a key role in Mars' gradual shift from the original carbon dioxide-dominated atmosphere that kept Mars relatively warm and allowed the planet to support liquid surface water, to the cold and arid surface we see today. This shift took place between about 4.2 and 3.7 billion years ago.

The goal of the mission is to determine the history of the loss of atmospheric gases to space, and answer questions regarding the climate evolution of Mars. By measuring the rate with which the atmosphere is currently escaping to space and gathering enough information about the relevant processes, scientists will be able to infer how the atmosphere evolved over time.

The MAVEN mission has four primary scientific objectives:

- Determine the role that loss of volatiles to space from the atmosphere has played through time.
- Determine the current state of the upper atmosphere and ionosphere, and their interactions with the solar wind.
- Determine the current rates of escape of neutral gases and ions to space and the processes that control them.
- Determine the ratios of stable isotopes in the atmosphere.

The Sample Analysis at Mars (SAM) instrument suite on the Curiosity rover made observations of the atmosphere of Mars from Gale Crater to provide data to assist in the interpretation of MAVEN's measurements of the upper atmosphere. The MAVEN data will also provide additional scientific context to test models for current methane formation on Mars.[5]

The instruments for the MAVEN mission include:

Built by the University of California, Berkeley Space Sciences Lab:

- Solar Wind Electron Analyzer (SWEA) measures solar wind and ionosphere electrons.
- Solar Wind Ion Analyzer (SWIA) measures solar wind and magnetosheath ion density and velocity.
- SupraThermal And Thermal Ion Composition (STATIC) measures thermal ions to moderate-energy escaping ions.
- Solar Energetic Particle (SEP) determines the impact of solar particles on the upper atmosphere.

Built by the University of Colorado Laboratory for Atmospheric and Space Physics:

- Imaging Ultraviolet Spectrometer (IUVS) measures global characteristics of the upper atmosphere and ionosphere.
- Langmuir Probe and Waves (LPW) determines ionosphere properties and wave heating of escaping ions and solar extreme ultraviolet (EUV) input to the atmosphere.

Built byGoddard Space Flight Center

- Magnetometer (MAG) measures interplanetary solar wind and ionosphere magnetic fields.
- Neutral Gas and Ion Mass Spectrometer (NGIMS) measures the composition and isotopes of neutral gases and ions.

---

[5]For a short video of Maven go to: https://www.youtube.com/watch?v = 4LQAu_m-BwA

Fig. A.1.11 An artist's rendering of MAVEN. Photo courtesy NASA/GSFC.

InSight Lander

On May 5, 2018, the InSight spacecraft was launched aboard an Atlas V from LC 3-East at Vandenberg AFB. This was the first American interplanetary mission to launch from California. Touchdown in the Elysium Planitia volcanic province of Mars occurred as planned on November 26 of that year. The two MarCo cubesats relayed communications for entry, descent, and landing. Communications during surface activities will be via Mars Odyssey and Mars Reconnaissance Orbiter. Its payload included a robotic arm that was to lower a drill to the surface to emplace sensors to measure the rate at which heat is leaking out of the crust, to investigate the interior structure. The arm was also to emplace a seismometer on the surface. Overall, the mission was to study the early geological evolution of the planet, to gain new understanding of the solar system's terrestrial planets: Mercury, Venus, Earth, Mars, and also our own Moon. The project's cost and risk were reduced by reusing technology from the Mars Phoenix lander, which successfully landed on Mars in 2008.

The science payload consisted of three main instruments:

- The Seismic Experiment for Interior Structure (SEIS) would take precise measurements of quakes and other internal activity on the planet to better understand its history and structure, and to investigate how the crust and mantle responds to meteorite impacts. These observations would provide clues to the interior structure of Mars. The seismometer would also detect sources including atmospheric waves and gravimetric signals (tidal forces) from the planet's larger moon Phobos, up to high frequency seismic waves of 50 Hz. The SEIS instrument was supported by a suite of meteorological tools which would characterize atmospheric disturbances that might affect the experiment.

- The Heat Flow and Physical Properties Package (HP3) that was provided by the German Aerospace Center (DLR) was a self-penetrating heat flow probe. Referred to as a "self-hammering nail" and nicknamed "the mole," it was designed to burrow as deep as 5 m (16 ft) below the surface while trailing a tether bearing attached heat sensors to measure how efficiently heat leaks out of the crust, and thus reveal information about the planet's interior and how it evolved over time. The tether had temperature sensors at 10 cm (3.9 in) intervals to measure a vertical temperature profile. The tractor mole was provided by the Polish company Astronika.
- The Rotation and Interior Structure Experiment (RISE) which was led by the Jet Propulsion Laboratory (JPL) would use the lander's X-band radio to provide precise measurements of planet's rotational modes in order to better understand its interior. This X-band radio tracking with an accuracy of less than 2 cm, will build upon similar measurements using the Viking and Mars Pathfinder landers. The previous data allowed the core size to be estimated, but with more data from InSight the nutation amplitude will be able to be determined. Then, once the spin axis direction, precession, and nutation amplitudes are understood, it ought to be possible to calculate the size and density of the Martian core and mantle. This should increase our understanding of the creation of the terrestrial planets of our solar system and rocky "exoplanets" around other stars.

The weather at the landing site was to be monitored by the Temperature and Winds for InSight (TWINS) provided by Spain's Centro de Astrobiología. Laser RetroReflector for InSight (LaRRI), supplied by the Italian Space Agency, was a corner-cube retroreflector mounted on the top deck of the lander to enable passive laser range-finding by orbiters even after the lander's mission was complete. It is to serve as a node in a proposed Mars geophysical network. (There was a similar device on the failed Schiaparelli landersent by ESA in 2016.)

The Instrument Deployment Arm (IDA) was a 2.4 m (8 ft) manipulator to hoist the SEIS and HP3 instruments from the top deck and then lower them onto the surface. The Instrument Deployment Camera (IDC) was a color camera based on the Navcam used by the Mars Exploration Rover and Mars Science Laboratory missions. It was on the Instrument Deployment Arm and was to give stereoscopic views of the terrain surrounding the landing site and show the instruments both on the lander's deck and after they had been placed on the surface. It featured a 45° field of view and used a 1024 × 1024 pixel CCD detector. The IDC sensor was originally black and white, but was replaced with a color sensor after a program was enacted that tested using a standard Hazcam within development deadlines and budgets. The Instrument Context Camera (ICC), a color camera based upon the Hazcam of the MER and MSL missions, was mounted beneath the lander's deck and its 120° panoramic field of view was to provide a complementary view of the instrument deployment area. As with the IDC, it used a 1024 × 1024 pixel CCD detector.[6]

---

[6]For photos and briefing on the Insight mission go to:
 https://www.flickr.com/photos/nasahqphoto/sets/72157676018862708

Fig. A.1.12 NASA's InSight Lander. The solar arrays are deployed in this 2015 test at Lockheed Martin Space Systems, Denver. This configuration is how the spacecraft will look on the surface of Mars. It will study the deep interior of Mars to improve our understanding of the early history of rocky planets. Photo courtesy of NASA/ JPL-Caltech/Lockheed Martin.

Mars 2020

The Mars 2020 rover is part of NASA's Mars Exploration Program, as were the Opportunity and Curiosity rovers, Mars Odyssey, Mars Reconnaissance Orbiterand MAVEN orbiters. The Mars 2020 rover will explore a site deemed likely to have been habitable in the planet's early history. It will seek signs of past life, and set aside a cache of the most compelling rock cores and soil samples for a future rover to retrieve as part of a sample return mission.

The following are the Mars 2020 instruments and experiments:

- Mastcam-Z, an advanced camera system with panoramic and stereoscopic imaging capability, including a zoom. The instrument also will determine the mineralogy of the Martian surface and assist in rover operations. The Principal Investigator is James Bell of Arizona State University in Tempe.

---

[6] For photos and briefing on the Insight mission go to:
  https://www.flickr.com/photos/nasahqphoto/sets/72157676018862708

Fig. A.1.13 InSight's instruments and equipment. Photo courtesy of NASA/ JPL-Caltech/Lockheed Martin.

- SuperCam, an instrument that can provide imaging, chemical composition analysis, and mineralogy. It will also be able to detect organic compounds in rocks and regolith from a distance. The Principal Investigator is Roger Wiens of the Los Alamos National Laboratory. This instrument includes a major contribution from the Centre National d'Etudes Spatiales, Institut de Recherche en Astrophysique et Plane'tologie (CNES/IRAP) in France.
- Planetary Instrument for X-ray Lithochemistry (PIXL), an X-ray fluorescence spectrometer that will also contain an imager with high resolution to determine the fine-scale elemental composition of surface materials. It will provide capabilities that permit more detailed detection and analysis of chemical elements than ever before on a planetary mission. The Principal Investigator is Abigail Allwood of the NASA Jet Propulsion Laboratory in Pasadena, California.
- Scanning Habitable Environments with Raman and Luminescence for Organics and Chemicals (SHERLOC), a spectrometer that will provide fine-scale imaging, and use an ultraviolet laser to determine fine-scale mineralogy and detect organic compounds. It will be the first ultraviolet Raman spectrometer delivered to the surface of Mars and it will provide complementary measurements with other instruments in the payload. The Principal Investigator is Luther Beegle of JPL.
- The Mars Oxygen ISRU Experiment (MOXIE), an exploration technology investigation that will produce oxygen from carbon dioxide in the Martian atmosphere. This demonstrates technologies for future human exploration. The Principal Investigator is Michael Hecht of Massachusetts Institute of Technology in Cambridge, Massachusetts.

- Mars Environmental Dynamics Analyzer (MEDA), a set of sensors which will measure the temperature, wind speed and direction, pressure, relative humidity, and size and shape of dust particles. The Principal Investigator is Jose Rodriguez-Manfredi of Centro de Astrobiologia, Instituto Nacional de Tecnica Aeroespacial, Spain.
- Radar Imager for Mars' Subsurface Exploration (RIMFAX), a ground-penetrating radar to obtain centimeter-scale resolution of the structure of the subsurface at points along the rover's route. The Principal Investigator is Svein-Erik Hamran of the Norwegian Defence Research Establishment, Norway.

The Mars 2020 mission is scheduled for launch in July or August 2020. It was briefly discussed in Chapter 6.4.2, and an annotated diagram was provided earlier as Fig. 5.9.

-o-O-o-

In addition, the European Space Agency has flown several orbiters and plans to send a rover.

Mars Express Orbiter

Mars Express was launched on June 2, 2003 from Baikonur in Kazakhstan by a Soyuz launch vehicle. It entered capture orbit around Mars on Christmas Day of that year, and by early May 2004 it had "aerobraked" into the desired near-polar orbit ranging between 300 and 10,110 km (186 and 6282 mi). It was equipped with a high-resolution color imager, an infrared spectrometer for mineralogical mapping, two spectrometers to measure the composition of the atmosphere on a local scale to study circulation patterns, and an instrument to investigate how the solar wind interacted with the atmosphere. In addition, it had a long-wavelength radar to penetrate the ground to a depth of several kilometers in search of water ice and possible liquid aquifers. It also served as a telecommunications relay for other orbiters, landers and rovers.

ExoMars

This joint program by ESA and the Russian space agency Roscosmos has passed through many design changes and schedule slips before it was decided to use two heavy-lift Proton launchers to dispatch four spacecraft: two stationary landers, an orbiter, and a rover. The mission was briefly discussed in Chapter 6.4.2.

The Trace Gas Orbiter (TGO) was launched on March 14, 2016, and achieved orbit around Mars on October 19 of that year. Its main scientific objective was to analyze atmospheric gases, but it would also serve as a Mars telecommunications relay. During periapsis passes in its lengthy aerobraking campaign it was able to confirm that its instruments were working satisfactorily. Once in its circular orbit, it was to scan the atmosphere to assemble profiles from the surface up to 160 km (100 mi) to characterize both spatial and temporal variations of methane and other trace gases and localize their sources. If methane occurs in the presence of ethane or propane this will strongly imply a biological origin, but if it is accompanied by gases such as sulfur dioxide this will suggest it is a by-product of geology. Either way, extant life or active geology, the result will be significant. Its data helped in selection of the landing site for the future ExoMars rover mission.

On its initial approach to Mars, the TGO spacecraft released the Schiaparelli Entry, Descent and Landing Demonstrator Module (EDM), but this crashed while attempting to land.

The second mission, scheduled for launch in July 2020, will use an 1800 kg (3968 lbs) Russian landing system developed using experience gained from the 2016 Schiaparelli EDM lander. The Russian company Lavochkin will fabricate most of the hardware for the landing system and ESA will supply items such as the guidance, radar, and navigation systems. The current intention is to use two parachutes – one will open while the module is still moving at supersonic speed and the other one after the probe is subsonic. The heat shield will eventually fall away from the entry capsule to allow the ExoMars rover, riding its retro-rocket-equipped platform, to achieve a soft landing. The lander will then deploy ramps to enable the 207 kg (456 lbs) ExoMars rover to drive down on to the surface.

Fig. A.1.14 ExoMars Rover with the lander in the background. Photo courtesy of ESA.

The solar powered rover is to spend seven months seeking evidence of past life on Mars. It will relay data to Earth via the TGO satellite. The "Pasteur" analytical laboratory is a suite of instruments with which to seek evidence of biomolecules and biosignatures: theMars Organic Molecule Analyzer (MOMA), MicrOmega-IR, and Raman Laser Spectrometer (RLS). Among its other instruments will be a 2 m (6.6 ft) drill to pull up core samples for analysis on board.

Appendix 2

Analog Projects

A.2.1 Introduction

The word "analog" suggests a form of inference based on the assumption that if two things are similar in some respects, they will be alike in other respects. The analogs in this chapter have both systems and human applications. For example, there are situations on Earth that produce effects on the body that are similar to those experienced in space, both physical and mental/emotional. These studies help prepare crews for long duration missions. NASA is associated with at least 22 analog missions throughout the world. However, these are mostly focused on its Human Research Program, and while many may be applicable to a precursor mission to the Moon and Mars they do not specifically focus on a mission that uses caves or lava tubes as protection from the hazards of such environments.[7, 8] A summary of those analogs was included as Appendix 3 of my earlier *Exploring the Martian Moons: A Human Mission to Deimos and Phobos*.

Scientists, astronauts, and mission planners are undertaking projects which specifically address research that is applicable to precursor missions to the Moon and Mars that utilize caves and/or lava tubes. They come from the U.S., ESA, and other countries involved in space exploration. They conduct these projects for the same reasons that NASA conducts the research mentioned above, such as:

- Not all experiments can be carried out in space. There is not enough time, money, equipment, and manpower.
- Countermeasures can be tested in analogs prior to trying them in space. Those that do not work in analogs will not be flown in space.

---

[7] To read about the analogs NASA is conducting, go to the following sites:
 For NASA's website on analogs, go to: https://www.nasa.gov/analogs
[8] There are 22 analog videos at this site. https://www.nasa.gov/analogs/video

© Springer Nature Switzerland AG 2019
M. von Ehrenfried, *From Cave Man to Cave Martian*,
Springer Praxis Books, https://doi.org/10.1007/978-3-030-05408-3

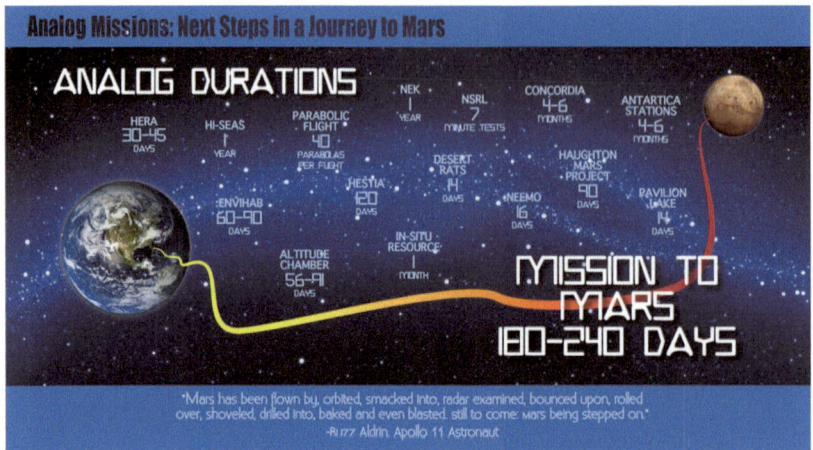

Fig. A.2.1 NASA Analog Mission Calendar. Photo courtesy of NASA.

- Ground-based analog studies can be completed more quickly and less expensively.

Observe that the caves and lava tubes listed in Chapter 2 were not necessarily separate analog projects, rather research of interest to speleologists, geologists, or biologists. In addition to geology and astrobiology research by some of NASA's organizations and universities, Chapter 4 mentioned work such as the Caves of Mars Project and PERISCOPE. Some sites were occasionally visited by analog project scientists whose main interest was geology, notably the Nördlinger Ries crater in Germany, the Italian Dolomites, and the volcanoes of Lanzarote in the Canary Islands.

A.2.2 PANGAEA

The European Space Agency (ESA) runs the PANGAEA course to provide its astronauts with introductory and practical knowledge of Earth and insights into planetary geology. The acronym reflects the ancient supercontinent which once unified the entire land surface, and stands for Planetary ANalogue Geological and Astrobiological Exercise for Astronauts.

The course is to prepare astronauts to become effective partners with planetary scientists and engineers in designing the next exploration missions. It also gives astronauts a solid basis in the geology of the solar system from leading European scientists. It is the first step in preparing astronauts to become planetary explorers, allowing them to communicate effectively with science advisors on Earth, using geologically correct language to increase rapid and fruitful decision-making while selecting scientifically relevant places to take samples.

Through PANGAEA, Europe is developing operational concepts for surface missions in which astronauts and robots work together in situ, and with scientists and engineers on Earth, to apply the best field geology and planetary observation techniques.

The project's goals are:

- To attain basic knowledge on geological processes and environments of Earth, Moon, Mars, and asteroids.
- Develop observational and decisional skills to identify geological features, carry out efficient sampling, and report concisely and correctly to ground control.
- Recognize and describe environments that could host extra-terrestrial life.

The course is split into three parts over the period of a year in these locations:

- The planetary geology introduction course is in Bressanone, Italy.
- The geological field training is in Lanzarote, Spain.
- A astrobiology and microbiology part of the course is conducted in the field as appropriate.

The following institutions provide the PANGAEA course with scientific and logistic support:

- Center of Studies and Activities for Space (CISAS) at the University of Padova, Italy.
- Cabildo Insular of Lanzarote, National Park of Timanfaya.
- Instituto de Geociencias (IGEO), Centro Mixto del Consejo Superior de Investigaciones Científicas (CSIC) y de la Universidad Complutense de Madrid (UCM).
- Geoparque Lanzarote.

In 2018, PANGAEA concentrated on interpreting orbital imagery, planetary protection, and lunar geology and impact cratering. It was primarily designed for astronauts and featured ESA astronaut Thomas Reiter taking part with Eurocom and ESA spacewalk instructor Hervé Stevenin.

In addition to astronauts, PANGAEA will also prepare mission developers to work with European planetary geology field scientists in developing operational concepts for exploration and the design of sample-return missions. The team of astronauts, spacewalk experts, engineers and scientists will exploit the course to address aspects of taking geological samples, discussing issues from operational field tools to mobility constraints, location selection, avoiding contamination of samples, support tools, and the documentation of such work.

Three field trips were planned for 2018 starting with the Ries impact crater in Germany that was also visited by the Apollo 14 astronauts in 1970. After the last session of the course in Lanzarote, Canary Islands, an extension campaign tested a series of technologies and protocols to increase awareness and prepare astronauts and operation engineers working in environments which resemble other planets in certain respects.[9, 10]

---

[9]For a 3 minute video about the team's general training go to:
https://www.youtube.com/watch?v=_9npxsMzKIg
[10]For a 3 minute video on the PANGAEA team in Lanzarote, Canary Islands go to: https://www.esa.int/esatv/Videos/2018/08/Pangaea-X_test_campaign_gearing_up_to_explore_other_planets

The trainees participating in PANGAEA 2018 were:

- Thomas Reiter, ESA astronaut.
- Sergei Kud-Sverchkov, Roscosmos cosmonaut.
- Aidan Cowley, ESA's science advisor.
- ESAastronaut Matthias Maurer
- Trainer Hervé Stevenin.

A.2.3 CAVES

Another ESA astronaut training course in which international astronauts train in a space-analog cave environment is CAVES (Cooperative Adventure for Valuing and Exercising human behavior and performance Skills). By means of a realistic scientific and exploration mission within a multicultural, ISS-representative team, the European Astronaut Center prepares astronauts to carry out safe and efficient long duration spaceflight operations.

Each training cycle lasts for approximately two weeks. In the first week, the so-called "cavenauts" are provided with the necessary scientific knowledge and technical skills to operate effectively and safely in the underground environment, then they visit some simple caves in order to get acquainted with the conditions they will find on their expedition. Afterwards, they spend six uninterrupted days exploring a complex cave system. The primary purpose of the mission is to foster skills in communication, decision-making, problem-solving, leadership, and team dynamics by way of scientific exploration conducted in a space-like environment.

CAVES takes place in several caves in the Italian island of Sardinia, part of a Karst System in the Supramonte region of the Gennargentu National Park. Rocks most commonly found in the area are limestone and dolomite.

The CAVES training targets the following objectives for participants:

- Working together effectively in a challenging environment.
- Adapting to a lack of comfort and privacy.
- Exploring a cave system.
- Conducting scientific and technological research.
- Managing logistical problems and coping with limited resources.
- Facing the psychological effects of the mission.
- Handling critical situations.
- Being aware of safety requirements at all times.
- Training the participants in leadership skills.

The cave environment is an exceptional space analog because it permits the trainers to recreate, on Earth, most of the stress elements and numerous specific characteristics which are encountered in long duration spaceflight. Such peculiar elements include:

- Unknown/unfamiliar environment. All that the crew knows of the cave is what previous expeditions have documented.

- Permanent darkness. The need for artificial illumination.
- Lack of time parameters. This is a direct consequence of the condition of permanent darkness in the cave.
- Alteration of circadian rhythm and sleep disturbance. The lack of time references and limited facilities affect sleep quality and cycles length.
- Sensory alteration/deprivation. Not only are caves lightless, they offer almost no auditory or olfactory stimuli.
- Limited privacy. Small, confined spaces do not permit much room for privacy or to allow team members to move apart.
- Social and cultural aspects/crew size. Each team is representative of the membership of current ISS crews (ESA, NASA, Roscosmos, CSA, JAXA and CNSA) and so are the team dynamics that emerge during the mission.
- Limited resources/hygiene. The transportation logistics inside the cave is extremely complex, only limited supplies can be carried inside.
- Isolation/limited communication with outside world. The crew can only rely on a telephone line up to just beyond the base camp or radio devices working only in specific locations.
- Large degree of autonomy. A direct result of the isolation condition, the crew must operate autonomously with minimal inputs from outside.
- Real physical danger. Even with all reasonable safety measures, a cave is still a dangerous environment, presenting risks of falling, slipping, rocks tumbling down, being blocked by landslides and floods. It is necessary to pay constant attention to these aspects.
- Limited mission abort/rescue capabilities. Given the complexity of the environment, both evacuation and rescue operations will require several hours (or even days) to plan and execute.

Another analog aspect is astronaut progression; i.e. how they move inside the cave. The speleological techniques employed require safety principles resembling those of EVA, such as the need to be attached to a safe item: the wall of the cave in CAVES or the ISS or other vehicle in space. The mission performance of the crew during the exploration of a cave has several elements in common with ISS operations, including timed activities, daily planning calls to the support team, standardized procedures, and requirements for data collection.

In the same way that astronauts in space spend part of their time doing science, "cavenauts" are tasked to perform a real crew mission involving several different experiments and activities in their cave.

In addition to the primary objective of team training, a CAVES course is able to explore and document previously unknown areas of cave systems and conduct real scientific and technological research.[11]

---

[11]The following is a 16 minute video on ESA CAVE training. While this is for 2012, it is an excellent overview of the training. https://www.youtube.com/watch?v = lIEPMUFteJM

A.2.4 PSTAR/BRAILLE/FELDSPAR

The Planetary Science and Technology from Analog Research (PSTAR) program is sponsored by the NASA Ames Research Center and addresses the requirement for integrated interdisciplinary field experiments as an integral part of preparation for planned human and robotic missions. It also solicits proposals for studies that are focused on exploring the Earth's extreme environments in order to develop a sound technical and scientific basis for astrobiological research on other bodies in the solar system. The focus of this program element is on providing high-fidelity scientific investigations, scientific input, and science operations constraints in the context of planetary field campaigns.

PSTAR particularly emphasizes analog missions. For example, Biologic and Resource Analog Investigations in Low Light Environment (BRAILLE) engaged in astrobiology mission simulation activities at Lava Beds National Monument in a high-visibility field campaign in the summer of 2018, with a second scheduled for 2019.

Lava tubes are a terrestrial analog for caves that have already been detected by satellites orbiting the Moon and Mars. Although lava tubes located at Lava Beds National Monument are not a perfect analog in terms of the arid surface of Mars, they are a good analog environment for early Mars activities. BRAILLE will also be the first project of its kind to combine the use of rover technology and remote science operations in an extreme environment in which there is limited light and resources.

By conducting field research inside lava tubes, BRAILLE 's findings can help NASA scientists and engineers on future missions to caves on the Moon and Mars to search for microbial life and biosignatures of interest while mitigating the risks associated with planetary exploration. See Chapter 4.3.5Kansas State University (KSU).

Fig. A.2.2 The CaveR rover at the Lava Beds National Monument. Photo courtesy of Kansas State University.

The issue of whether life, specifically microbial life, exists on other planets is one of the most important questions in astrobiological research. Caves and lava tubes are sheltered from harmful surface conditions that may well have destroyed evidence of life. If microbes were to be preserved, they would be in minerals in the cave walls, and will provide scientists with a geological record of the types of microbial communities that are present in lava tubes. If we can characterize the presence of microbial life and biosignatures in terrestrial lava tubes, and refine procedures for conducting remote science operations using a rover, then we can justify support for future expeditions of extraterrestrial caves.

PSTAR also funds analog research in the volcanic regions of Iceland. These are excellent Mars analogs owing to their desiccation, low nutrient availability, and temperature extremes, in addition to the advantages of geological youth and isolation from sources of anthropogenic contamination. Many of the analog sites are also sufficiently remote and rugged to offer the same type of instrumentation and sampling constraints typically faced by robotic exploration. Such a study is Field Exploration and Life Detection Sampling for Planetary Analogue Research (FELDSPAR). It is an ongoing Mars analog study which is undertaken at recent basaltic eruption sites and basaltic sandy plains in Iceland, and carries out field operations which preview a Mars sample return mission, including prior remote sensing and employment of nested sampling grids to characterize the variability within, and degree of correlation between, physicochemical data and a variety of types of biomarkers at different spatial scales.

In-situ exploration of planetary environments allows biochemical analysis of sub-centimeter scale samples. However, landing sites are selected a priori based on measurable meter-to-kilometer-scale geological characteristics. Therefore the optimization of life detection mission science return requires both understanding the expected distribution of biomarkers across sample sites at different scales and efficiently using early-stage in-situ geochemical instruments to justify later-stage biological or chemical analysis.

See also The Mars Institute's work in the Lofthellir Cave Complex in Iceland discussed in Chapter 4.4.

A.2.5 Austrian Analogs

The Austrian Space Forum (ÖWF) is a national network of aerospace specialists and space enthusiasts. It serves as a communications platform between the space sector and the public and is embedded in a global network of specialists from the space industry, research, and politics. Its highly active membership contributes to space endeavors, in many cases in cooperation with other space organizations on an international basis. The spectrum of activities ranges from simple classroom presentations to visitor space exhibitions, from expert reports for the Ministry of Transport to space technology transfer activities for terrestrial applications.

Over recent decades the ÖWF has been active in various Mars analogs, such as:

- "AustroMars," a Mars analog simulation at the Mars Desert Research Station in Utah. For the first time the crew, the support team, and the experiments, as well as the major part of the hardware originated from Austria.
- PolAres, an interdisciplinary program in cooperation with international partners to develop strategies for human-robotic interaction procedures that emphasize planetary protection, in preparation for a future human-robotic Mars surface expedition. This program included the following elements:

  - LIFE fluorescence laser.
  - Development of a simulation involving Mars rovers Phileas and Magma.
  - Aouda.X space suit simulator (Aouda is the name of the Indian Princess in Jules Vernes' novel *Around the World in 80 Days*).

- MARS 2013, a four-week Mars analog field test in the northern Sahara. Nineteen experiments were conducted under simulated Martian surface exploration conditions by a field crew in Morocco, who worked with a supervisory team in Innsbruck, Austria.

AMADEE (named after Austrian Amadeus Mozart) is the current flagship research program of the ÖWF and is expected to span 2018 to 2028. It represents the framework for the development of hardware, workflows and science of future human-robotic planetary surface exploration. In terms of the implementation of the Vienna Statement for Planetary Analog Research (VSAPR), AMADEE is a catalyst and test platform for the search for life in advance of a mission. It applies a heuristic approach to mission-driven innovation which integrates technological innovation, advances in scientific understanding, and operational experience. The intention is that every 2 or 3 years, a new field mission will build upon the results of its predecessors.

Combining strategic partnerships, cutting-edge technologies, and state-of-the-art science, as well as a strong outreach and education role, AMADEE is open to international collaboration. It is built upon extensive knowledge and operational expertise from its predecessor PolAres, including eleven major international Mars field campaigns, the development of space suit simulators, stratospheric balloon flights, and the use of rovers. This developed the infrastructure and training and certification system for a Mission Support Center, a group of Analog Astronauts, and a multi-mission Science Data Archive.[12]

The participating astronauts were:

- Stefan Dobrovolny, MD, Austria.
- Carmen Köhler, PhD, Germany.
- Kartik Kumar, MSc, The Netherlands.
- João Lousada, MSc, Portugal.
- Iñigo Muñoz Elorza, MSc, Spain.

---

[12]For a short video on the Oman camp go to:
  http://www.planetary.org/blogs/jason-davis/2018/20180215-amadee-18.html

The ÖWF has conducted Mars analog simulations at:

- Dharfor, Oman.
- Rio Tinto, Spain.
- Dachstein, Austria ice caves (participated with ten nations).
- Erfoud, Morocco northern Sahara.
- Kaunertal Glacier in Austria.
- Hanksville, Utah.

The simulations are supported by a Remote Science Support team, a Mission Control Center, and a Flight Control Team.

Space Suits

Aouda.X is a novel space suit simulator developed by the Austrian Space Forum. It is not a real suit, but simulates the functions and feel of a space suit that would possibly be used by the astronauts on a planetary surface. This lets us learn here on Earth how difficult it is to operate and perform science on a distant planetary body. With a helmet featuring a head-up display (HUD), full air circulation with carbon dioxide control, medical monitoring, joints and gloves which emulate the stiffness of a real suit, and radio equipment, Aouda.X simulates extravehicular activity on a planetary surface with an advanced human-machine interface, a set of sensors, and purpose-designed software to assist the simulated astronaut. It is designed to interact with other field components such as a rover and a variety of instruments.

Fig. A.2.3 An astronaut in the Aouda.X simulated space suit in the Dachstein ice caves. Photo courtesy of ÖWF.

Fig. A.2.4 An Aouda.X suited astronaut riding a Eurobot rover near a lunar lander at Rio Tinto site. Photo courtesy of ÖWF.

The second generation of Mars space suit simulator built on the field-proven Aouda.X suits, the Serenity suit was developed in participation with international and national high-tech companies and universities. Development started in 2018 and the commissioning will take place in 2019, then it will be introduced for the AMADEE-20 mission.

The main departure from the Aouda.X space suit for the Serenity suit was the implementation of a rear-entry concept, which means instead of donning several parts of the suit (e.g. shoes, trousers, hard-upper torso) the analog astronaut will enter the integrated suit through the rear, with the suit itself docked at a suitport on the outside of the habitat. In this way, sand and dust stays outside the habitat and the donning time can be reduced by about 50 per cent, time which becomes available to do more experiments on a simulated Mars excursion. This concept correlates to the most recent standards of Mars analog research, and the team is coordinating with the University of North Dakota, which is conducting research into Mars analog habitat and space suit designs. See Chapter 4.3.7.

A.2.6 Potrillo/Goddard Instruments

The Goddard Planetary Geology, Geophysics and Geochemistry Laboratory is conducting research and carrying out instrument development that is relevant to investigating the structure, dynamics, and evolution of the terrestrial planets. It

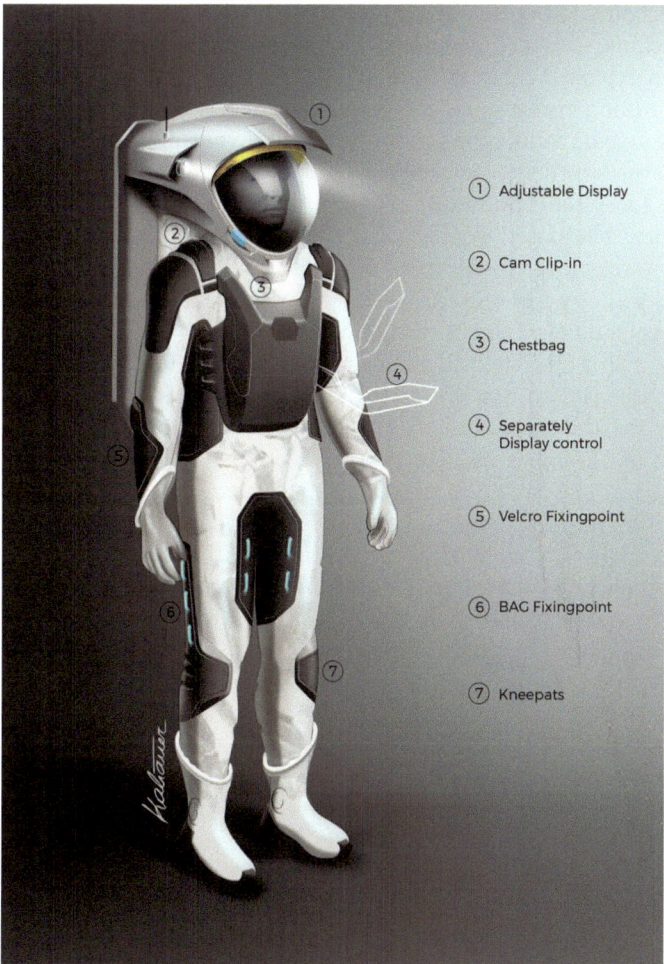

Fig. A.2.5 The next-generation Serenity space suit simulator. Photo courtesy of Bernhard Kaliauer Design Studio BKDS.

combines in-situ and remote sensing data, analog field work, laboratory sample analysis, and geochemical and geophysical modeling in order to understand:

- The timing and nature of volcanic, cratering, and other physical and chemical processes that modify planetary surfaces.
- The orbital and tidal processes that affect the surfaces, internal structure and evolution of planetary and smaller bodies.
- The comparative evolution of planetary and smaller bodies and how that determines the location of habitable environments in the solar system.
- How human exploration of solar system bodies can be optimized in terms of science return and safety.

To support this effort, the NASA Goddard Instrument Field Team (GIFT) is testing and refining the chemical-analyzing and land-surveying tools which will assist future human explorers on the Moon and Mars. This effort offers a direct benefit to astronauts working in caves and lava tubes. The field work portion of the instrument development activity is undertaken in the Potrillo volcanic field, located on the Rio Grande Rift in southern New Mexico. The volcanic field lies 35 km (22 mi) southwest of Las Cruces and occupies more than 1000 km^2 (400 mi^2) near the U.S.-Mexico border. The desert geology was shaped over millions of years by volcanic and tectonic forces that closely resemble those of the Moon, which is why Potrillo has provided geology training for astronauts dating back to the Apollo era.

When astronauts land again on the Moon or Mars their limited resources will provide only a short window of time each day to explore their new surroundings. Instruments designed to quickly reveal the terrain's chemistry and form will help them understand the environments around them, and how they change over time. Many of the technologies used to develop new instruments build upon ones that have already equipped robotic orbiters and rovers. NASA's Curiosity rover, for instance, studies the composition of soil with the help of spectrometers, or tools that identify what rocks are made of by measuring how their chemical elements interact with electromagnetic radiation. On Earth too, instruments that reveal the chemical makeup of objects are widely used in subjects including art restoration, archaeology, and geology. Though the technologies powering these tools already exist, NASA's objective is to make the instruments small and efficient enough to help robots (and future astronauts) to analyze on the spot, the composition of the surface of planets, moons and asteroids. This will facilitate quick adjustments to exploration plans and permit astronauts to make well-informed decisions about which samples they can return to Earth on a spacecraft that will have a limited capacity.

The GIFT team is analyzing a number of new and smaller instruments. One example is the portable X-ray fluorescence spectrometer. It emits X-rays when placed up against a point on the surface of a rock. The incoming X-ray kicks an electron out of close orbit around the nucleus of an atom, triggering an electron from an outer orbit to replace it, thereby returning the atom to equilibrium. This creates more X-ray energy, or fluorescence, which is picked up by a detector in the instrument. Such an analysis can identify the chemical elements which objects are made of in a matter of minutes (in recent tests). This spectrometer resembles a retail-price scanner. It is powered by a rechargeable lithium-ion battery, has a large-area silicon drift detector that senses and measures X-rays, and an anode tube that emits X-rays. With a mass of just 2 kg (4 lbs), it could readily be hand-carried by an astronaut in the reduced gravities of the Moon or Mars. What the team has learned from analog experiments, particularly with the assistance of ISS astronaut Barry Eugene "Butch" Wilmore, is that speed and ease-of-use are vital for space tools. Astronauts doing EVAs have limited oxygen and other resources, so device features such as instrument size, number of buttons, and the data display are crucial.

Another seemingly benign consideration in instrument design is that thorough field testing is significant. One example is whether or not hand-held instruments should have buttons. A device designed like a grocery-store price scanner makes for

Fig. A.2.6 Geologist Kelsey Young demonstrates the portable X-ray fluorescence spectrometer. From left to right: geologist Liz Rampe, Kelsey Young, and (to the far right almost out of the image) ISS astronaut Barry "Butch" Wilmore. This photo was taken at the Valentine Cave at Lava Beds National Monument, Tulelake, California. Photo courtesy of NASA/GSFC.

a comfortable hand-held spectrometer shape on Earth, but squeezing a trigger in space is not so straightforward. Space suits are pressurized, and it takes more effort for astronauts to move against the extra force of their space suits, making small movements strenuous.[13] Before selecting tools for use on space missions it will be essential to understand their ergonomics.

---

[13]To watch a short video of how this instrument is used in the field, go to: https://www.youtube.com/watch?v = Zb-ka1ny0aA&feature = youtu.be

Appendix 3

Scientists in Planetary Cave and Lava Tube Research

The following people have contributed to this book in one form or another. Some of them I have contacted seeking information, others have referenced papers and reports germane to the topics addressed in the book. One is even deceased but his work lives on. Unfortunately, not everyone could be included here owing to page limitations. They are in alphabetical order.

Dr. Penny Boston is the Director of NASA's Astrobiology Institute (NAI), at the NASA Ames Research Center at Moffett Field, California. She is leading the

Fig. A.3.1 Dr. Penelope Boston. Photo courtesy of the NASA Ames Research Center.

© Springer Nature Switzerland AG 2019
M. von Ehrenfried, *From Cave Man to Cave Martian*,
Springer Praxis Books, https://doi.org/10.1007/978-3-030-05408-3

scientific activities of the Institute's member teams and all operational aspects of the organization. Penny leads the NAI in fulfilling its mission to perform, support, and catalyze interdisciplinary astrobiology research; train the next generation of astrobiologists; provide scientific and technical leadership for astrobiology space mission investigations; and develop new information technology approaches for collaborations among widely distributed investigators.

Prior to joining NASA, Penny founded and then directed the Cave and Karst Studies Program at New Mexico Tech at Socorro, New Mexico, where she is a professor, and chaired the Earth and environmental sciences department there. In addition, from 2002 to 2016 she was Associate Director of the National Cave and Karst Research Institute, the congressionally mandated institute in Carlsbad, New Mexico.

Dr. Boston is a fellow of the NASA Institute for Advanced Concepts, a past President of the Association of Mars Explorers, the recipient of the NSS Science Award in 2010, and the 2013 recipient of the David P. McKay Memorial Life on Mars Award.

Dr. Philip R. Christensen is a Regents Professor and the Ed and Helen Korrick Professor in the Department of Geological Science at Arizona State University in Phoenix. His research interests focus on the composition, physical properties and processes, and morphology of planetary surfaces, with an emphasis on Earth and Mars.

A major element of Philip's research has been the design and development of infrared remote sensing instruments for space. He was the Principal Investigator for the Mars Odyssey Thermal Emission Imaging System (THEMIS) instrument and the Thermal Emission System (TES) for Mars Global Surveyor. He was also a Co-Investigator on the Mars Exploration Rover missions, being responsible for building and operating their Mini-TES instruments.

Fig. A.3.2 Dr. Philip R. Christensen. Photo courtesy of NASA/Arizona State University.

Dr. Christensen was selected to provide the OSIRIS-REx Thermal Emission Spectrometer (OTES) for the OSIRIS-REx mission that was launched in 2016 to rendezvous with asteroid 101955 Bennu in December 2018 to map its surface as a prelude to collecting a sample for return to Earth.

Dr. Dean Eppler was awarded a Bachelor of Science degree in geology by St. Lawrence University in 1974, a Master of Science in geology from the University of New Mexico in 1976, and a Doctorate of Philosophy in geology from Arizona State University in 1984. He was a Senior Scientist with the Science Applications

Fig. A.3.3 Dr. Dean Eppler. Photo courtesy of Digital Space.

International Corporation from 1986 to 2009, which included 20 years of support to NASA at the Johnson Space Center (JSC). During that time, he was a Lead Suit Test Subject for advanced planetary space suit development and geological field testing from 1996 to 2005; the International Space Station (ISS) Payloads Office Program Lead on development of a high-fidelity window to conduct research on the ISS from 1994 to 2005; the Program Originator and Lead Scientist on the ISS Window Observational Research Facility (WORF) from 1998 to 2003; and the Lead for Science Operations and Logistics Concept Development for Advanced Planetary Exploration Programs, including two years in the lunar surface systems for the Constellation Program.

In 2009, he transitioned to NASA to work in the Astromaterials Research and Exploration Science (ARES) Directorate, doing science operations development for lunar missions, including working up science operations concepts for Desert Research and Technology Studies (RATS) and developing and implementing the geological training curriculum for the 2009 Astronaut Class. Spanning his career, Dean has published at least 30 scientific publications and been awarded the Army Commendation Medal, the Antarctic Service Medal, and the NASA Exceptional Public Service Medal.

Dr. Glen Cushing is with the USGS Astrogeology Science Center in Flagstaff, Arizona. He has analyzed Mars Reconnaissance Orbiter data of caves on Mars. In

Fig. A.3.4 Dr. Glen Cushing. Photo courtesy of the USGS/LinkedIn.

particular he has looked for atypical pit craters (nearly cylindrically shaped pits, deeper than they are wide), some of which may offer access to caves. He has also proposed using lava tubes and caves as protection against the hazards that humans would face on the surface. Since this is precisely the subject of this book, I have benefited considerably from some of his papers.

Dr. Saugata Datta is involved in a NASA study to explore microbial life in lava caves using a four-wheeled rover named CaveR to assess accessibility. This work is part of the NASA PSTARanalog mission, BRAILLE. He teaches a number of courses in hydrogeology, low temperature geochemistry, and water resources. His principal expertise is trace element and oxyanion migration and contamination in local environments, particularly in ground waters, urban air particulates, subway microenvironments, and unproductive soil environments using hydrological and geochemical tools including synchrotron spectroscopy. He was recently active in research in geochemical implications of carbon dioxide sequestration.

Fig. A.3.5 Dr. Saugata Datta. Photo courtesy of Kansas State University.

Dr. Pablo de León is currently a professor in extravehicular activities (EVA) and space suit design at the Department of Space Studies of University of North Dakota in Grand Forks. He is also preparing a new course on Human Spaceflight for UND, and is project manager of a NASA-funded program on planetary space suit design.

He worked with the Space Shuttle as payload manager and general designer of the Project PADE (G-761) science experiments package which flew on STS-108 to the ISS in 2001. He was the chief designer and fabrication manager for several underwater-simulation, EVA-analog, pressure suit systems. He has also flown as a payload specialist on more than 80 weightless parabolas in the NASA KC-135 aircraft to perform four Zero-G fluid dynamics experiments.

Fig. A.3.6 Dr. Pablo de León. Photo courtesy of the University of North Dakota.

Pablo has published a number of books and reports about space, with a special focus on human spaceflight. One of these books, *108 Minutes in Space*, describes the first flight of a human in space. He is editor-in-chief of the *Latin-American Journal of Space Science and Technology*. He has authored dozens of papers on space engineering and life support systems, presented at international congresses. He also belongs to several professional aerospace engineering societies. He was selected Regent of the United Societies in Space (USIS) in 2000. It was during that year that he co-founded the Latin American Space Association.

Dr. Sarah Fagents is a researcher at the Hawaii Institute of Geophysics and Planetology (HIGP) of the University of Hawaii at Manoa. Her research is on planetary volcanism, volcanic fluid dynamics, the geology of icy satellites, and volcano-ice interactions on Mars (in particular the distribution of small volcanic cones believed to have been created by the explosive vaporization by lava flows of ice deposits in the Martian soil). She is the Co-Investigator on the Mastcam-Z; a multispectral, stereoscopic imaging instrument selected by NASA for the Mars 2020 rover mission.

Fig. A.3.7 Dr. Fagents on Raupehu Volcano in the Tongariro National Park, New Zealand. Photo courtesy of University of Hawaii.

The late Dr. Ronald Greeley was a Regents Professor in the School of Earth and Space Exploration at Arizona State University, the Director of the NASA-ASU Regional Planetary Image Facility (RPIF), and Principal Investigator of the Planetary Aeolian Laboratory which is at the NASA Ames Research Center. He became involved with lunar and planetary sciences in the 1960's and prior to his death in 2011 was studying planetary surface processes and geological histories. He authored many papers relating to lava tubes and caves.

Fig. A.3.8 Dr. Ronald Greeley. Photo courtesy of Arizona State University/Wikimedia.

Dr. Laszlo Keszthelyi (a.k.a. Kestay) is a planetary volcanologist working for the U.S. Geological Survey's Astrogeology Science Center in Flagstaff, Arizona. He has studied volcanism across the solar system with remote sensing, numerical modeling, and field studies. He is involved in assessing natural resources across the solar system and the hazards of meteorite impacts. He is the former Director of the Astrogeology Science Center, which aides exploration of the solar system with support for space missions from concept through to fruition. He is member of the MRO HiRISE team and also an imaging system for the ExoMars Trace Gas Orbiter.

Dr. Evelynn Mitchell is a professor of Environmental Science at St. Mary's University in San Antonio, Texas. She has worked with students in studying carbon dioxide levels in local caves using ground penetrating radar to identify voids, and using electrical resistivity to study subsurface locations. She taught physics at the University of the Incarnate Word before pursuing her Ph.D. She also gained experience in engineering and applied physics while working as a manufacturing engineer at Sony Semiconductor from 2000 to 2003. Mitchell focused on studying the Edwards Aquifer in her dissertation research, in which she applied hydrogeology and geophysical skills to determine specific storage values by studying the compression of earthquake waves. She has also carried out research using seismic and resistivity applications to investigate the subsurface geology and water table in the Texas Hill Country and in Jalisco, Mexico.

Fig. A.3.9 Dr. Lazlo Kestay in the field. Photo courtesy of the USGS.

Fig. A.3.10 Dr. Evelynn Mitchell. Photo courtesy of the St. Mary's University.

Dr. Pascal Lee is chairman of the Mars Institute and a planetary scientist at the SETI Institute. He is also director of the NASA Haughton-Mars Project (HMP) at NASA-Ames. The HMP is an international interdisciplinary field research project centered on the scientific study of the Haughton impact structure and surrounding terrain on Devon Island, High Arctic, the largest uninhabited island on Earth. This serves as a terrestrial analog for similar features on the Moon and Mars.

From 2009 to 2011, Pascal led the "Northwest Passage Drive Expedition," a record-setting 750 km (466 mi) traverse from the North American mainland to Devon Island by way of the Northwest Passage in a customized Humvee (known as the Okarian) that simulated a pressurized rover for future human exploration. It was sponsored by the Mars Institute, the SETI Institute, and NASA. The journey took three seasons to complete, and yielded lessons learned for a future road trip on Mars. In November 2018, Dr. Lee conducted research in Iceland's Lofthellir Lava Cave Complex as an analog for assessing the use of lava tubes on the Moon and Mars as protection against the hazards of the environment. (See Chapter 4.4.)

In other work, Dr. Lee identified candidate lava tube skylights on the floor of the Philolaus Crater near the north pole of the Moon.[14]

Fig. A.3.11 Planetary scientist Pascal Lee crawling along the roof of an ice-filled section of the Lofthellir Lava Cave, Iceland. Photo courtesy of Pascal Lee.

---

[14]A video of Dr. Lee's research of possible skylights in Philolaus Crater can be seen at: https://www.youtube.com/watch?v = doMXNGc_N-I

Dr. Peter J. Mouginis-Mark is the Vice Chancellor for Research and Graduate Education at University of Hawaii at Manoa, the Director of the Hawaii Institute of Geophysics and Planetology at the School of Ocean and Earth Science and Technology (SOEST), and the Director of the Pacific Regional Planetary Data Center. He previously served as an interim Associate Dean for Research in the College of Engineering and as Chief Scientist of the Pacific Disaster Center. As the Director of the Hawaii Space Grant Consortium he oversaw interdisciplinary education, research, and public service programs related to space science, Earth sciences, remote sensing, human exploration and development of space, and use of small satellites. His current activity is finishing a 1:200,000 scale geological map of the Hrad Vallis area of Mars, as well as a review paper on the geology of Olympus Mons, the largest volcano on that planet. He previously made a study of Tooting Crater in the Amazonis Planitia Region of Mars.

Fig. A.3.12 Dr. Peter J. Mouginis-Mark. Photo courtesy of the University of Hawaii.

Dr. Diana E. Northup is an Associate Professor of Biology at the University of New Mexico. She has been studying things that live in caves since 1984 and as a member of SLIME (Subsurface Life In Mineral Environments) is researching how microbes help to produce the colorful ferromanganese deposits that coat the walls of caves, and how microbes participate in the precipitation of calcium carbonate formations called "pool fingers." Her other research includes the investigation of the potential for White Nose Syndrome to infect New Mexico bats, the microbial diversity of lava caves, and microbes that masquerade as minerals.

Dr. Aaron Parness gained two Bachelor of Science degrees in 2004 from the Massachusetts Institute of Technology, one in Creative Writing and the other in Mechanical Engineering, and then a Master of Science and Doctor of Philosophy from Stanford University, both in Mechanical Engineering. His research focuses on the attachment interfaces between robotic systems and their surroundings, in

Fig. A.3.13 Dr. Diana Northrup. Photo courtesy of the New Mexico University.

particular robots that grip and climb. As an expert in novel methods of prototype manufacturing, Aaron has experience in microfabrication, polymer prototyping, and traditional machining and has also experience of mechanical part design and mechatronic systems.

Dr. Parness is the Principal Investigator on several research projects at NASA-JPL on extreme terrain mobility. He led the gripper development of the Asteroid Redirection Mission. He also works in the formulation of mission proposals for the Discovery and New Frontiers programs and leads innovation and technology initiatives for the Office of the Chief Technologist and the Space Technology Program. His current research interests include anchoring systems for Near Earth Objects, which are so small that they offer almost no gravitational attraction, the development of climbing robots for natural terrain including cliff faces and cave ceilings, micro ground vehicles (< 100 grams) with dynamic mobility, and new methods of robotic manipulation which do not rely on traditional friction-based force-closure.

Dr. William "Red" Whittaker is currently the Fredkin Research Professor at the Robotics Institute and the Director of the Field Robotics Center and Chief Scientist

Fig. A.3.14 Dr. Aaron Parness. Photo courtesy of JPL.

Fig. A.3.15 Roboticist "Red" Whittaker. Photo courtesy of Carnegie Mellon University.

of the Robotics Engineering Consortium, both of which are located at Carnegie Mellon University. Over the years, he and his team have developed at least 60 robots.

He led Tartan Racing to its first-place victory in the DARPA Grand Challenge (2007) Urban Challenge and brought Carnegie Mellon University the $2 million prize. Previously, Whittaker competed for the DARPA Grand Challenge, placing second and third place simultaneously in the Grand Challenge Races. He also led the Carnegie Mellon University team in the competition for the Google Lunar X Prize. As the Chairman and CSO of Astrobotic Technology, Whittaker will play an instrumental role in future lunar robot development. He has advised 26 Ph.D. students, holds 16 patents, and authored over 200 publications. His ambition is to drive nanobiology technology to fulfillment and create nanorobotic agents which can carry out a variety of tasks on Earth and beyond.

Judson "Jut" Wynne gained his Ph.D. in Biological Sciences from Northern Arizona University (NAU) in 2014 with a focus on community ecology and cave biology. In 2003, the same institution awarded him an M.Sc. in Environmental Science and Policy. He has written about wildlife habitat modeling, cave biology, NASA education programs, and cave detection methods for Earth, the Moon, and

Fig. A.3.16 Dr. "Jut" Wynne. Photo courtesy of Northern Arizona University.

Mars. He is also a guest writer for *Scientific American*, a Fulbright Scholar, holds fellowships of The Explorers Club and the Royal Geographical Society, and is an elected member of Sigma Xi.

Dr. George Veni is the Executive Director of the National Cave and Karst Research Institute (NCKRI). He is an internationally recognized hydrogeologist specializing in caves and karst terrains Although much of his work has been in Texas, he has participated in karst research throughout the U.S.A. and in several other countries. Over the years he has published over 200 papers and six books on hydrogeology, biology, and environmental management.

Fig. A.3.17 Dr. George Veni in the Parks Ranch Gypsum Cave near Carlsbad, New Mexico. Photo courtesy of National Cave and Karst Research Institute.

He is a long-time caver, a life member of the National Speleological Society (NSS), and has led many caving projects. Since 2002, he has also served on the Union Internationale de Spéléologie (UIS) Bureau, first as an Adjunct Secretary and then eight years as the Vice President of Administration. Since 2017, he has been UIS President.

Appendix 4

Space Policy Directives

(I have placed some sentences in bold either for clarity or for emphasis)

A.4.1 President Trump's Space Policy Directive 1

On Monday, Dec. 11, 2017, President Donald J. Trump signed Space Policy Directive 1 titled Reinvigorating America's Human Space Exploration Program.

It called for Human Expansion Across the Solar System. The signing was at the White House before Representatives of Congress and Vice President Mike Pence, President of the National Space Council. Among other dignitaries on hand for the signing, were NASA astronauts Sen. Harrison "Jack" Schmitt, Buzz Aldrin, Peggy Whitson andChristina Koch. Policy Directive 1 represents a change in national space policy that provides for a U.S.-led, integrated program with private sector partners for a human return to the Moon, followed by missions to Mars and beyond.

The policy calls for the NASA administrator to "**lead an innovative and sustainable program of exploration with commercial and international partners to enable human expansion across the solar system and to bring back to Earth new knowledge and opportunities**." The effort will more effectively organize government, private industry, and international efforts toward returning humans on the Moon, and will lay the foundation that will eventually enable human exploration of Mars.

"The directive I am signing today will refocus America's space program on human exploration and discovery," said President Trump. "It marks a first step in returning American astronauts to the Moon for the first time since 1972, for long-term exploration and use. **This time, we will not only plant our flag and leave our footprints; we will establish a foundation for an eventual mission to Mars, and perhaps someday, to many worlds beyond.**"

The policy grew from a unanimous recommendation by the new National Space Council, chaired by Vice President Mike Pence, after its first meeting October 5, 2017. **In addition to the direction to plan for human return to theMoon, the policy also**

© Springer Nature Switzerland AG 2019
M. von Ehrenfried, *From Cave Man to Cave Martian*,
Springer Praxis Books, https://doi.org/10.1007/978-3-030-05408-3

ends NASA's existing effort to send humans to an asteroid. The president revived the National Space Council in July to advise and help implement his space policy with exploration as a national priority. Work toward the new directive will be reflected in NASA's Fiscal Year 2019 budget request next year.

"NASA looks forward to supporting the president's directive strategically **aligning our work to return humans to theMoon, travel to Mars and opening the deeper solar system beyond**," said acting NASA Administrator Robert Lightfoot. "This work represents a national effort on many fronts, with America leading the way. We will engage the best and brightest across government and private industry and our partners across the world to reach new milestones in human achievement. Our workforce is committed to this effort, and even now we are developing a flexible deep space infrastructure to support a steady cadence of increasingly complex missions that strengthens American leadership in the boundless frontier of space. The next generation will dream even bigger and reach higher as we launch challenging new missions, and make new discoveries and technological breakthroughs on this dynamic path."

Republican and Democrats on the Senate space subcommittee insisted that the goal for NASA's human spaceflight program is Mars, not the Moon, at a hearing yesterday. **President Trump formally restored theMoonto NASA's plans in December and NASA's FY2019 budget request reflects that change**. The Senators raised no objections as long as it does not distract from what they consider the primary goal of landing humans on Mars in the 2030s.

A.4.2 President Trump's Space Policy Directive 2

On May 24, 2018, President Donald J. Trump issued Space Policy Directive 2 titled: Streamlining Regulations on Commercial Use of Space.

Section 1. Policy.

It is the policy of the executive branch to be prudent and responsible when spending taxpayer funds, and to recognize how government actions, including Federal regulations, affect private resources. It is therefore important that regulations adopted and enforced by the executive branch promote economic growth; minimize uncertainty for taxpayers, investors, and private industry; protect national security, public-safety, and foreign policy interests; and encourage American leadership in space commerce.

Section 2. Launch and Reentry Licensing.

(a) No later than February 1, 2019, the **Secretary of Transportation** shall review regulations adopted by the Department of Transportation that provide for and govern licensing of commercial spaceflight launch and reentry for consistency with the policy set forth in section 1 of this memorandum and shall rescind or revise those regulations, or publish for notice and comment proposed rules rescinding or revising those regulations, as appropriate and consistent with applicable law.

(b) Consistent with the policy set forth in section 1 of this memorandum, the Secretary of Transportation shall consider the following:

(i) requiring a single license for all types of commercial spaceflight launch and reentry operations.

(ii) replacing prescriptive requirements in the commercial spaceflight launch and reentry licensing process with performance-based criteria.

(c) In carrying out the review required by subsection (a) of this section, the Secretary of Transportation shall coordinate with the members of the National Space Council.

(d) The Secretary of Defense, the Secretary of Transportation, and the Administrator of the National Aeronautics and Space Administration shall coordinate to examine all existing U.S. Government requirements, standards, and policies associated with commercial spaceflight launch and reentry operations from Federal launch ranges and, as appropriate and consistent with applicable law, to minimize those requirements, except those necessary to protect public safety and national security, that would conflict with the efforts of the Secretary of Transportation in implementing the Secretary's responsibilities under this section.

Section 3. Commercial Remote Sensing.

(a) Within 90 days of the date of this memorandum, the **Secretary of Commerce** shall review the regulations adopted by the Department of Commerce under Title II of the Land Remote Sensing Policy Act of 1992 (51 U.S.C. 60101 et seq.) for consistency with the policy set forth in section 1 of this memorandum and shall rescind or revise those regulations, or publish for notice and comment proposed rules rescinding or revising those regulations, as appropriate and consistent with applicable law.

(b) In carrying out the review required by subsection (a) of this section, the Secretary of Commerce shall coordinate with the Secretary of State, the Secretary of Defense, the Administrator of the National Aeronautics and Space Administration, and, as appropriate, the Chairman of the Federal Communications Commission.

(c) Within 120 days of the date of the completion of the review required by subsection (a) of this section, the Secretary of Commerce, in coordination with the Secretary of State and the Secretary of Defense, shall transmit to the Director of the Office of Management and Budget a legislative proposal to encourage expansion of the licensing of commercial remote sensing activities. That proposal shall be consistent with the policy set forth in section 1 of this memorandum.

Section 4. Reorganization of the Department of Commerce.

(a) To the extent permitted by law, the Secretary of Commerce shall consolidate in the Office of the Secretary of Commerce the responsibilities of the Department of Commerce with respect to the Department's regulation of commercial spaceflight activities.
(b) Within 30 days of the date of this memorandum, the Secretary of Commerce shall transmit to the Director of the Office of Management and Budget a legislative proposal to create within the Department of Commerce an entity with primary responsibility for administering the Department's regulation of commercial spaceflight activities.

Section 5. Radio Frequency Spectrum.

(a) The Secretary of Commerce, in coordination with the Director of the Office of Science and Technology Policy, shall work with the Federal Communications Commission to ensure that Federal Government activities related to radio frequency spectrum are, to the extent permitted by law, consistent with the policy set forth in section 1 of this memorandum.
(b) Within 120 days of the date of this memorandum, the Secretary of Commerce and the Director of the Office of Science and Technology Policy, in consultation with the Chairman of the Federal Communications Commission, and in coordination with the members of the National Space Council, shall provide to the President, through the Executive Secretary of the National Space Council, a report on improving the global competitiveness of the United States space sector through radio frequency spectrum policies, regulation, and United States activities at the International Telecommunication Union and other multilateral forums.

Section 6. Review of Export Licensing Regulations.

The Executive Secretary of the National Space Council, in coordination with the members of the National Space Council, shall:

(a) initiate a review of export licensing regulations affecting commercial spaceflight activity;

(b) develop recommendations to revise such regulations consistent with the policy set forth in section 1 of this memorandum and with applicable law; and

(c) submit such recommendations to the President, through the Vice President, no later than 180 days from the date of this memorandum.

Section 7. General Provisions.

(a) Nothing in this memorandum shall be construed to impair or otherwise affect:

  (i) the authority granted by law to an executive department or agency, or the head thereof; or

  (ii) the functions of the Director of the Office of Management and Budget relating to budgetary, administrative, or legislative proposals.

(b) This memorandum shall be implemented consistent with applicable law and subject to the availability of appropriations.

(c) This memorandum is not intended to, and does not, create any right or benefit, substantive or procedural, enforceable at law or in equity by any party against the United States, its departments, agencies, or entities, its officers, employees, or agents, or any other person.

(d) The Secretary of Transportation is authorized and directed to publish this memorandum in the Federal Register.

A.4.3 President Trump's Space Policy Directive 3

On June 18, 2018 President Donald J. Trump issued Space Policy Directive 3 titled: National Space Traffic Management Policy.

Section 1. Policy.

For decades, the United States has effectively reaped the benefits of operating in space to enhance our national security, civil, and commercial sectors. Our society now depends on space technologies and space-based capabilities for communications, navigation, weather forecasting, and much more. Given the significance of space activities, the United States considers the continued unfettered access to and freedom to operate in space of vital interest to advance the security, economic prosperity, and scientific knowledge of the Nation.

Today, space is becoming increasingly congested and contested, and that trend presents challenges for the safety, stability, and sustainability of U.S. space operations. Already, the **Department of Defense** (DoD) tracks over 20,000 objects in space, and that number will increase dramatically as new, more capable sensors come online and are able to detect smaller objects. DoD publishes a catalog of space objects and makes notifications of potential conjunctions (that is, two or more objects coming together at the same or nearly the same point in time and space). As the number of space objects increases, however, this limited traffic management activity and architecture will become inadequate. At the same time, the contested nature of space is increasing the demand for DoD focus on protecting and defending U.S. space assets and interests.

The future space operating environment will also be shaped by a significant increase in the volume and diversity of commercial activity in space. Emerging commercial ventures such as satellite servicing, debris removal, in-space manufacturing, and tourism, as well as new technologies enabling small satellites and very large constellations of satellites, are increasingly outpacing efforts to develop and implement government policies and processes to address these new activities.

To maintain U.S. leadership in space, we must develop a new approach to space traffic management (STM) that addresses current and future operational risks. This new approach must set priorities for space situational awareness (SSA) and STM innovation in science and technology (S&T), incorporate national security considerations, encourage growth of the U.S. commercial space sector, establish an updated STM architecture, and promote space safety standards and best practices across the international community.

The United States recognizes that spaceflight safety is a global challenge and will continue to encourage safe and responsible behavior in space while emphasizing the need for international transparency and STM data sharing. Through this national policy for STM and other national space strategies and policies, the United States will enhance safety and ensure continued leadership, preeminence, and freedom of action in space.

Section 2. Definitions.

For the purposes of this memorandum, the following definitions shall apply:

(a) Space Situational Awareness shall mean the knowledge and characterization of space objects and their operational environment to support safe, stable, and sustainable space activities.
(b) Space Traffic Management shall mean the planning, coordination, and on-orbit synchronization of activities to enhance the safety, stability, and sustainability of operations in the space environment.
(c) Orbital debris, or space debris, shall mean any human-made space object orbiting Earth that no longer serves any useful purpose.

Section 3. Principles.

The United States recognizes, and encourages other nations to recognize, the following principles:

(a) Safety, stability, and operational sustainability are foundational to space activities, including commercial, civil, and national security activities. It is a shared interest and responsibility of all spacefaring nations to create the conditions for a safe, stable, and operationally sustainable space environment.
(b) Timely and actionable SSA data and STM services are essential to space activities. Consistent with national security constraints, basic U.S. Government-derived SSA data and basic STM services should be available free of direct user fees.
(c) Orbital debris presents a growing threat to space operations. Debris mitigation guidelines, standards, and policies should be revised periodically, enforced domestically, and adopted internationally to mitigate the operational effects of orbital debris.
(d) A STM framework consisting of best practices, technical guidelines, safety standards, behavioral norms, pre-launch risk assessments, and on-orbit collision avoidance services is essential to preserve the space operational environment.

Section 4. Goals.

Consistent with the principles listed in section 3 of this memorandum, the United States should continue to lead the world in creating the conditions for a safe, stable, and operationally sustainable space environment. Toward this end, executive departments and agencies (agencies) shall pursue the following goals as required in section 6 of this memorandum:

(a) Advance SSA and STM Science and Technology. The United States should continue to engage in and enable S&T research and development to support the practical applications of SSA and STM. These activities include improving fundamental knowledge of the space environment, such as the characterization of small debris, advancing the S&T of critical SSA inputs such as observational data, algorithms, and models necessary to improve SSA capabilities, and developing new hardware and software to support data processing and observations.
(b) **Mitigate the effect of orbital debris on space activities**. The volume and location of orbital debris are growing threats to space activities. It is in the interest of all to minimize new debris and mitigate effects of existing debris. This fact, along with increasing numbers of active satellites, highlights the need to update existing orbital debris mitigation guidelines and practices to enable more efficient and effective compliance, and establish standards that can be adopted internationally. These trends also highlight the need to establish satellite safety design guidelines and best practices.
(c) Encourage and facilitate U.S. commercial leadership in S&T, SSA, and STM. Fostering continued growth and innovation in the U.S. commercial space sector, which includes S&T, SSA, and STM activities, is in the national interest

of the United States. To achieve this goal, the U.S. Government should streamline processes and reduce regulatory burdens that could inhibit commercial sector growth and innovation, enabling the U.S. commercial sector to continue to lead the world in STM-related technologies, goods, data, and services on the international market.

(d) Provide U.S. Government-supported basic SSA data and basic STM services to the public. The United States should continue to make available basic SSA data and basic STM services (including conjunction and reentry notifications) free of direct user fees while supporting new opportunities for U.S. commercial and non-profit SSA data and STM services.

(e) Improve SSA data interoperability and enable greater SSA data sharing. SSA data must be timely and accurate. It is in the national interest of the United States to improve SSA data interoperability and enable greater SSA data sharing among all space operators, consistent with national security constraints. The United States should seek to lead the world in the development of improved SSA data standards and information sharing.

(f) Develop STM standards and best practices. As the leader in space, the United States supports the development of operational standards and best practices to promote safe and responsible behavior in space. A critical first step in carrying out that goal is to develop U.S.-led minimum safety standards and best practices to coordinate space traffic. U.S. regulatory agencies should, as appropriate, adopt these standards and best practices in domestic regulatory frameworks and use them to inform and help shape international consensus practices and standards.

(g) **Prevent unintentional radio frequency (RF) interference**. Growing orbital congestion is increasing the risk to U.S. space assets from unintentional RF interference. The United States should continue to improve policies, processes, and technologies for spectrum use (including allocations and licensing) to address these challenges and ensure appropriate spectrum use for current and future operations.

(h) **Improve the U.S. domestic space object registry**. Transparency and data sharing are essential to safe, stable, and sustainable space operations. Consistent with national security constraints, the United States should streamline the interagency process to ensure accurate and timely registration submissions to the United Nations (UN), in accordance with our international obligations under the Convention on Registration of Objects Launched into Outer Space.

(i) Develop policies and regulations for future U.S. orbital operations. Increasing congestion in key orbits and maneuver-based missions such as servicing, survey, and assembly will drive the need for policy development for national security, civil, and commercial sector space activities. Consistent with U.S. law and international obligations, the United States should regularly assess existing guidelines for non-government orbital activities, and maintain a timely and responsive regulatory environment for licensing these activities.

Section 5. Guidelines.

In pursuit of the principles and goals of this policy, agencies should observe the following guidelines:

(a) Managing the Integrity of the Space Operating Environment.

    (i) Improving SSA coverage and accuracy. Timely, accurate, and actionable data are essential for effective SSA and STM. The United States should seek to minimize deficiencies in SSA capability, particularly coverage in regions with limited sensor availability and sensitivity in detection of small debris, through SSA data sharing, the purchase of SSA data, or the provision of new sensors.

       New U.S. sensors are expected to reveal a substantially greater volume of debris and improve our understanding of space object size distributions in various regions of space. However, very small debris may not be sufficiently tracked to enable or justify actionable collision avoidance decisions. As a result, close conjunctions and even collisions with unknown objects are possible, and satellite operators often lack sufficient insight to assess their level of risk when making maneuvering decisions. The United States should develop better tracking capabilities, and new means to catalog such debris, and establish a quality threshold for actionable collision avoidance warning to minimize false alarms.

       Through both Government and commercial sector S&T investment, the United States should advance concepts and capabilities to improve SSA in support of debris mitigation and collision avoidance decisions.

    (ii) **Establishing an Open Architecture SSA Data Repository**. Accurate and timely tracking of objects orbiting Earth is essential to preserving the safety of space activities for all. Consistent with section 2274 of title 10, United States Code, a basic level of SSA data in the form of the publicly releasable portion of the DoD catalog is and should continue to be provided free of direct user fees. As additional sources of space tracking data become available, the United States has the opportunity to incorporate civil, commercial, international, and other available data to allow users to enhance and refine this service. To facilitate greater data sharing with satellite operators and enable the commercial development of enhanced space safety services, the United States must develop the standards and protocols for creation of an open architecture data repository. The essential features of this repository would include:

        • Data integrity measures to ensure data accuracy and availability.
        • Data standards to ensure sufficient quality from diverse sources.
        • Measures to safeguard proprietary or sensitive data, including national security information.
        • The inclusion of satellite owner-operator ephemerides to inform orbital location and planned maneuvers.

- Standardized formats to enable development of applications to leverage the data.

To facilitate this enhanced data sharing, and in recognition of the need for DoD to focus on maintaining access to and freedom of action in space, a civil agency should, consistent with applicable law, be responsible for the publicly releasable portion of the DoD catalog and for administering an open architecture data repository. The Department of Commerce should be that civil agency.

(iii) **Mitigating Orbital Debris**. It is in the interest of all space operators to minimize the creation of new orbital debris. Rapid international expansion of space operations and greater diversity of missions have rendered the current U.S. Government Orbital Debris Mitigation Standard Practices (ODMSP) inadequate to control the growth of orbital debris. These standard practices should be updated to address current and future space operating environments.

The United States should develop a new protocol of standard practices to set broader expectations of safe space operations in the 21st century. **This protocol should begin with updated ODMSP, but also incorporate sections to address operating practices for large constellations, rendezvous and proximity operations, small satellites, and other classes of space operations**. These overarching practices will provide an avenue to promote efficient and effective space safety practices with U.S. industry and internationally.

The United States should pursue active debris removal as a necessary long-term approach to ensure the safety of flight operations in key orbital regimes. This effort should not detract from continuing to advance international protocols for debris mitigation associated with current programs.

(b) Operating in a Congested Space Environment.

(i) Minimum Safety Standards and Best Practices. The creation of minimum standards for safe operation and debris mitigation derived in part from the U.S. Government ODMSP, but incorporating other standards and best practices, will best ensure the safe operation of U.S. space activities. These safety guidelines should consider maneuverability, tracking, reliability, and disposal.

The United States should eventually incorporate appropriate standards and best practices into Federal law and regulation through appropriate rulemaking or licensing actions. These guidelines should encompass protocols for all stages of satellite operation from design through end-of-life. Satellite and constellation owners should participate in a pre-launch certification process that should, at a minimum, consider the following factors:

- Coordination of orbit utilization to prevent conjunctions.
- Constellation owner-operators' management of self-conjunctions.
- Owner-operator notification of planned maneuvers and sharing of satellite orbital location data.
- On-orbit tracking aids, including beacons or sensing enhancements, if such systems are needed.
- Encryption of satellite command and control links and data protection measures for ground site operations.
- Appropriate minimum reliability based on type of mission and phase of operations.
- Effect on the national security or foreign policy interests of the United States, or international obligations.
- Self-disposal upon the conclusion of operational lifetime, or owner-operator provision for disposal using active debris removal methods.

(ii) **On-Orbit Collision Avoidance Support Service**. Timely warning of potential collisions is essential to preserving the safety of space activities for all. Basic collision avoidance information services are and should continue to be provided free of direct user fees. The imminent activation of more sensitive tracking sensors is expected to reveal a significantly greater population of the existing orbital debris background as well as provide an improved ability to track currently listed objects. Current and future satellites, including large constellations of satellites, will operate in a debris environment much denser than presently tracked. Preventing on-orbit collisions in this environment requires an information service that shares catalog data, predicts close approaches, and provides actionable warnings to satellite operators. The service should provide data to allow operators to assess proposed maneuvers to reduce risk. To provide on-orbit collision avoidance, the United States should:

- Provide services based on a continuously updated catalog of satellite tracking data.
- Utilize automated processes for collision avoidance.
- Provide actionable and timely conjunction assessments.
- Provide data to operators to enable assessment of maneuver plans.

To ensure safe coordination of space traffic in this future operating environment, and in recognition of the need for DoD to focus on maintaining access to and freedom of action in space, a civil agency should be the focal point for this collision avoidance support service. The Department of Commerce should be that civil agency.

(c) Strategies for Space Traffic Management in a Global Context.

(i) Protocols to Prevent Orbital Conjunctions. As increased satellite operations make lower Earth orbits more congested, the United States should develop a set of standard techniques for mitigating the collision risk of increasingly congested orbits, particularly for large constellations. Appropriate methods, which may include licensing assigned volumes for constellation operation and establishing processes for satellites passing through the volumes, are needed.

The United States should explore strategies that will lead to the establishment of common global best practices, including:

- A common process addressing the volume of space used by a large constellation, particularly in close proximity to an existing constellation.
- A common process by which individual spacecraft may transit volumes used by existing satellites or constellations.
- A set of best practices for the owner-operators of utilized volumes to minimize the long-term effects of constellation operations on the space environment (including the proper disposal of satellites, reliability standards, and effective collision avoidance).

(ii) **Radio Frequency Spectrum and Interference Protection**. Space traffic and RF spectrum use have traditionally been independently managed processes. Increased congestion in key orbital regimes creates a need for improved and increasingly dynamic methods to coordinate activities in both the physical and spectral domains, and may introduce new inter-dependencies. U.S. Government efforts in STM should address the following spectrum management considerations:

- Where appropriate, verify consistency between policy and existing national and international regulations and goals regarding global access to, and operation in, the RF spectrum for space services.
- Investigate the advantages of addressing spectrum in conjunction with the development of STM systems, standards, and best practices.
- Promote flexible spectrum use and investigate emerging technologies for potential use by space systems.
- Ensure spectrum-dependent STM components, such as inter-satellite safety communications and active debris removal systems, can successfully access the required spectrum necessary to their missions.

(iii) Global Engagement. In its role as a major spacefaring nation, the United States should continue to develop and promote a range of norms of behavior, best practices, and standards for safe operations in space to minimize the space debris environment and promote data sharing and coordination of space activities. It is essential that other spacefaring nations also adopt best practices for the common good of all spacefaring states. The United States should encourage the adoption of new norms of behavior and best practices for space operations by the international community through bilateral and multilateral discussions with other spacefaring nations, and through U.S. participation in various organizations such as the Inter-Agency Space Debris Coordination Committee, International Standards Organization, Consultative Committee for Space Data Systems, and UN Committee on the Peaceful Uses of Outer Space.

Section 6. Roles and Responsibilities.

In furtherance of the goals described in section 4 and the guidelines described in section 5 of this memorandum, agencies shall carry out the following roles and responsibilities:

(a) Advance SSA and STM S&T. Members of the National Space Council, or their delegees, shall coordinate, prioritize, and advocate for S&T, SSA, and STM, as appropriate, as it relates to their respective missions. They should seek opportunities to engage with the commercial sector and academia in pursuit of this goal.

(b) Mitigate the Effect of Orbital Debris on Space Activities.

   (i) The Administrator of the National Aeronautics and Space Administration (NASA Administrator), in coordination with the Secretaries of State, Defense, Commerce, and Transportation, and the Director of National Intelligence, and in consultation with the Chairman of the Federal Communications Commission (FCC), shall lead efforts to update the U.S. Orbital Debris Mitigation Standard Practices and establish new guidelines for satellite design and operation, as appropriate and consistent with applicable law.

   (ii) The Secretaries of Commerce and Transportation, in consultation with the Chairman of the FCC, will assess the suitability of incorporating these updated standards and best practices into their respective licensing processes, as appropriate and consistent with applicable law.

(c) Encourage and Facilitate U.S. Commercial Leadership in S&T, SSA, and STM. The Secretary of Commerce, in coordination with the Secretaries of Defense and Transportation, and the NASA Administrator, shall lead efforts to encourage and facilitate continued U.S. commercial leadership in SSA, STM, and related S&T.

(d) Provide U.S. Government-Derived Basic SSA Data and Basic STM Services to the Public.

   (i) The Secretaries of Defense and Commerce, in coordination with the Secretaries of State and Transportation, the NASA Administrator, and the Director of National Intelligence, should cooperatively develop a plan for providing basic SSA data and basic STM services either directly or through a partnership with industry or academia, consistent with the guidelines of sections 5(a)(ii) and 5(b)(ii) of this memorandum.

(ii) The Secretary of Defense shall maintain the authoritative catalog of space objects.

(iii) The Secretaries of Defense and Commerce shall assess whether statutory and regulatory changes are necessary to effect the plan developed under subsection (d)(i) of this section, and shall pursue such changes, along with any other needed changes, as appropriate.

(e) Improve SSA Data Interoperability and Enable Greater SSA Data Sharing.

    (i) The Secretary of Commerce, in coordination with the Secretaries of State, Defense, and Transportation, the NASA Administrator, and the Director of National Intelligence, shall develop standards and protocols for creation of an open architecture data repository to improve SSA data interoperability and enable greater SSA data sharing.

    (ii) The Secretary of Commerce shall develop options, either in-house or through partnerships with industry or academia, assessing both the technical and economic feasibility of establishing such a repository.

    (iii) The Secretary of Defense shall ensure that release of data regarding national security activities to any person or entity with access to the repository is consistent with national security interests.

(f) Develop Space Traffic Standards and Best Practices. The Secretaries of Defense, Commerce, and Transportation, in coordination with the Secretary of State, the NASA Administrator, and the Director of National Intelligence, and in consultation with the Chairman of the FCC, shall develop space traffic standards and best practices, including technical guidelines, minimum safety standards, behavioral norms, and orbital conjunction prevention protocols related to pre-launch risk assessment and on-orbit collision avoidance support services.

(g) Prevent Unintentional Radio Frequency Interference. The Secretaries of Commerce and Transportation, in coordination with the Secretaries of State and Defense, the NASA Administrator, and the Director of National Intelligence, and in consultation with the Chairman of the FCC, shall coordinate to mitigate the risk of harmful interference and promptly address any harmful interference that may occur.

(h) Improve the U.S. Domestic Space Object Registry. The Secretary of State, in coordination with the Secretaries of Defense, Commerce, and Transportation, the NASA Administrator, and the Director of National Intelligence, and in consultation with the Chairman of the FCC, shall lead U.S. Government efforts on international engagement related to international transparency and space object registry on SSA and STM issues.

(i) Develop Policies and Regulations for Future U.S. Orbital Operations. The Secretaries of Defense, Commerce, and Transportation, in coordination with the Secretary of State, the NASA Administrator, and the Director of National Intelligence, shall regularly evaluate emerging trends in space missions to recommend revisions, as appropriate and necessary, to existing SSA and STM policies and regulations.

Section 7. General Provisions.

(a) Nothing in this memorandum shall be construed to impair or otherwise affect the authority granted by law to an executive department or agency, or the head thereof; or the functions of the Director of the Office of Management and Budget relating to budgetary, administrative, or legislative proposals.
(b) This memorandum shall be implemented consistent with applicable law and subject to the availability of appropriations.
(c) This memorandum is not intended to, and does not, create any right or benefit, substantive or procedural, enforceable at law or in equity by any party against the United States, its departments, agencies, or entities, its officers, employees, or agents, or any other person.
(d) The Secretary of Commerce is authorized and directed to publish this memorandum in the Federal Register.

A.4.4 NASA Policy on Planetary Protection Requirements for Human Extraterrestrial Missions

1. Background

In May 2012, the Planetary Protection Subcommittee of the NASA Advisory Council (NAC) Science Committee formulated a recommendation that NASA Procedural Requirements (NPR) be developed for planetary protection on human missions under NASA Policy Directive (NPD) 8020.7, "Biological Contamination Control for Outbound and Inbound Planetary Spacecraft," as a parallel document to NPR 8020.12, "Planetary Protection Provisions for Robotic Extraterrestrial Missions." This recommendation was endorsed by the full NAC and forwarded to the Administrator in November 2012, and was agreed upon by the NASA Administrator in a letter dated March 8, 2013.

There is presently insufficient scientific and technological knowledge to establish detailed requirements and specifications to enable NASA to incorporate planetary protection into the development of crewed spacecraft and missions. Thus, this NASA Policy Instruction (NPI) establishes policy guidelines and describes the approach for obtaining the scientific information and developing the technologies and procedures over the next few years that are needed to draft an NPR for crewed planetary missions.

Space exploration is now conducted by the space agencies of nations around the globe. The International Council for Science, a nongovernmental organization, established the Committee on Space Research (COSPAR) in 1958 as an interdisciplinary scientific body concerned with the progress on an international scale of all kinds of scientific investigations carried out with space vehicles, rockets and balloons.

The Treaty on Principles Governing the Activities of States in the Exploration and Use of Outer Space, including the Moon and Other Celestial Bodies, which established the basic legal framework of international space law, entered into force in 1967. Article IX of this treaty provides in relevant part, that:

"States Parties to the Treaty shall pursue studies of outer space, ..., and conduct exploration of them so as to avoid their harmful contamination ("forward contamination") and also adverse changes in the environment of the Earth resulting from the introduction of extraterrestrial matter ("back contamination") and, where necessary, shall adopt appropriate measures for this purpose."

COSPAR established the first planetary protection guidelines for robotic missions in 2002. While not legally binding, COSPAR's Planetary Protection Policy is:

"...for the reference of spacefaring nations, both as an international standard on procedures to avoid organic-constituent and biological contamination in space exploration, and to provide accepted guidelines in this area to guide compliance with (Article IX of the 1967 Outer Space Treaty) and other relevant international agreements."

In March 2011, amendments to the COSPAR Planetary Protection Policy were approved by the Bureau and Council, World Space Council to include Principles and Guidelines for Human Missions to Mars. (see Attachment A).

As NASA, in collaboration with our international partners, prepares to return humans beyond low Earth orbit to explore the solar system and search for signs of life beyond Earth, it is critical that NASA guidelines be developed for crewed missions. A key NASA international partner, the European Space Agency (ESA) adheres to COSPAR Planetary Protection Policy for both crewed and robotic missions, as expressed in ESA/C(2007)112.

2. History

Even before Neil Armstrong's boot first touched the Moon, NASA has been concerned with the protection of Earth and its inhabitants from extraterrestrial life forms returned from inbound spacecraft. In order to protect against possible disease or other health issues incurred upon Earth's inhabitants, procedures were created to prevent such back contamination. Each of the early Apollo astronauts endured 21 days of quarantine upon their return to Earth, as determined by the Interagency Committee on Back-Contamination based on the fact that most terrestrial disease agents were capable of invading a host and causing evident disease symptoms within 21 days after exposure of the host. In addition to protecting against back-contamination, NASA is also dedicated to the preservation of any native extraterrestrial life forms and maintaining the scientific purity of the celestial bodies to which NASA travels. Contamination by biological material from Earth could make it impossible to determine if life was present before humans visited.

Since the end of the Apollo era, robotic missions have served as humankind's emissary to other solar system bodies, including the Sun, planets and small solar system objects. As an example, launched November 2011, the Mars Science Laboratory's (MSL) Curiosity rover was designed to assess whether Mars ever had a habitable environment, able to support small life forms called microbes. Planetary protection requirements called for the entire MSL flight system to launch with no more than 500,000 bacterial spores. This was accomplished mainly through the careful maintenance of clean room protocols, periodic cleaning of spacecraft surfaces with alcohol wipes, and dry heat treatment of some spacecraft parts.

Space exploration is now conducted by the space agencies of nations around the globe. The International Council for Science, a nongovernmental organization, established the Committee on Space Research (COSPAR) in 1958 as an interdisciplinary scientific body concerned with the progress on an international scale of all kinds of scientific investigations carried out with space vehicles, rockets and balloons.

The Treaty on Principles Governing the Activities of States in the Exploration and Use of Outer Space, including the Moon and Other Celestial Bodies, which established the basic legal framework of international space law, entered into force in 1967. Article IX of this treaty provides in relevant part, that:

"States Parties to the Treaty shall pursue studies of outer space, …, and conduct exploration of them so as to avoid their harmful contamination ["forward contamination"] and also adverse changes in the environment of the Earth resulting from the introduction of extraterrestrial matter ["back contamination"] and, where necessary, shall adopt appropriate measures for this purpose."

COSPAR established the first planetary protection guidelines for robotic missions in 2002. While not legally binding, COSPAR's Planetary Protection Policy is:

"…for the reference of spacefaring nations, both as an international standard on procedures to avoid organic-constituent and biological contamination in space exploration, and to provide accepted guidelines in this area to guide compliance with [Article IX of the 1967 Outer Space Treaty] and other relevant international agreements."

In March 2011, amendments to the COSPAR Planetary Protection Policy were approved by the Bureau and Council, World Space Council to include Principles and Guidelines for Human Missions to Mars (see Attachment A).

As NASA, in collaboration with our international partners, prepares to return humans beyond low Earth orbit to explore the solar system and search for signs of life beyond Earth, it is critical that NASA guidelines be developed for crewed missions. A key NASA international partner, the European Space Agency (ESA) adheres to COSPAR Planetary Protection Policy for both crewed and robotic missions, as expressed in ESA/C(2007)112.

3. Policy Guidance

NASA adheres to the COSPAR guidelines. NPD 8020.7G (Biological Contamination Control for Outbound and Inbound Planetary Spacecraft (expires February 19, 2018), quoting the COSPAR policy statement, requires Agency compliance with

COSPAR policy regarding biological contamination control for outbound and inbound planetary spacecraft, covering all spaceflight missions which may intentionally or unintentionally carry Earth organisms and organic constituents to the planets or other solar system bodies, including spacecraft which are intended to return to Earth and/or its biosphere from extraterrestrial targets of exploration. All missions in which NASA will participate are required to adhere to NPD 8020.7G and to be consistent with the COSPAR policy and guidelines for human missions (Attachment A).

4. Studies

Detailed studies must be conducted in order to obtain information critical to developing planetary protection requirements for human spaceflight missions. NASA will gather community input to determine the topics that should be studied; for example:

- Developing capabilities to comprehensively monitor the microbial communities associated with human systems and evaluate changes over time.
- Developing technologies for minimizing/mitigating contamination release, including but not limited to closed-loop systems; cleaning/re-cleaning capabilities; support systems that minimize contact of humans with the environment of Mars and other solar system destinations.
- Understanding environmental processes on Mars and other solar system destinations that would contribute to transport and sterilization of organisms released by human activity.

5. Path Forward

NASA shall utilize the following roadmap to develop the necessary understanding of the scientific and technological basis to take sufficient steps to ensure planetary protection and then to develop an NPR setting forth requirements for planetary protection and carry out the NPR's mandates.

- Present the required studies report to senior management for approval and commitment of funding, through a Memorandum of Understanding or other documentation.
- Include sufficient funding for approved planetary protection studies as part of the NASA budget development process, leading to approval of funding for these studies no later than Fiscal Year 2016.
- Conduct studies and develop planetary protection requirements.
- Integrate funding for planetary protection requirements into the ongoing budgets of all developing human missions that will come in contact with another celestial body.
- Develop and formalize NPR for Planetary Protection for Crewed Missions.

In response to the Planetary Protection Subcommittee's recommendation, a cross-disciplinary ad hoc team was established that developed this NPI and is responsible for:

- Conducting a literature review to identify completed studies and investigations relevant to the development of verifiable planetary protection requirements for human missions.
- Seeking input from scientific and space operations community through a variety of sources, including, a workshop.
- Oversight of the recommended studies and following through on their completion to the development of specific requirements.
- Developing a draft NPR for planetary protection for human spaceflight that includes these specific requirements for mission development and follow the necessary NASA coordination and approval processes to baseline the NPR.
- Coordinating with relevant mission management teams within NASA, to ensure understanding of the requirements in order to achieve compliance.

The team is led by the Human Exploration and Operations Mission Directorate, with the Planetary Protection Officer serving as a technical advisor. Other participants include representatives from the following organizations: Science Mission Directorate, Space Technology Mission Directorate, Office of the General Counsel, Office of the Chief Scientist, Office of the Chief Medical Officer, and Office of International and Interagency Relations. Other organizations may be added as appropriate.

6. References

Attachment A: COSPAR Policy and Guidelines for Human Missions.
Attachment B: Letter from NAC Planetary Protection Subcommittee Chair
to NAC Science Committee Chair.

Appendix 5

Quotes

A.5.1 Philosophy

"For once you have tasted flight you will walk the Earth with your eyes turned skywards, for there you have been and there you will long to return."

– Leonardo da Vinci

"In spite of the opinions of certain narrow-minded people, who would shut up the human race upon this globe, as within some magic circle which it must never outstep, we shall one day travel to the Moon, the planets, and the stars, with the same facility, rapidity, and certainty as we now make the voyage from Liverpool to New York."

– Jules Verne

"Earth is the cradle of humanity, but one cannot live in the cradle forever."

– Konstantin Tsiolkovsky

"There can be no thought of finishing for 'aiming for the stars.' Both figuratively and literally, it is a task to occupy the generations. And no matter how much progress one makes, there is always the thrill of just beginning."

– Robert H. Goddard

"This is the goal: To make available for life every place where life is possible. To make inhabitable all worlds as yet uninhabitable, and all life purposeful."

– Hermann Oberth

"The earth is not a mere fragment of dead history, stratum upon stratum like the leaves of a book, to be studied by geologists and antiquaries chiefly, but living poetry like the leaves of a tree, which precede flowers and fruit – not a fossil earth, but a living earth; compared with whose great central life all animal and vegetable life is merely parasitic. Its throes will heave our exuviæ from their graves... You may melt your metals and cast

© Springer Nature Switzerland AG 2019
M. von Ehrenfried, *From Cave Man to Cave Martian*,
Springer Praxis Books, https://doi.org/10.1007/978-3-030-05408-3

them into the most beautiful moulds you can; they will never excite me like the forms which this molten earth flows out into."

– Henry David Thoreau

"The greatest gain from space travel consists in the extension of our knowledge. In a hundred years this newly won knowledge will pay huge and unexpected dividends."

– Wernher von Braun

"TheMoonis the first milestone on the road to the stars."

"The inspirational value of the space program is probably of far greater importance to education than any input of dollars....A whole generation is growing up which has been attracted to the hard disciplines of science and engineering by the romance of space."

– Arthur C. Clarke

"We are all . . . children of this universe. Not just Earth, or Mars,or this System, but the whole grand fireworks. And if we areinterested in Mars at all, it is only because we wonder over ourpast and worry terribly about our possible future."– Ray Bradbury in *Mars and the Mind of Man*, 1973

"I wonder what responsibility society at large has for the sponsorship of exploration?"

– James Michener, 1976

"In science it often happens that scientists say, 'You know that's a really good argument; my position is mistaken,' and then they would actually change their minds and you never hear that old view from them again. They really do it. It doesn't happen as often as it should, because scientists are human and change is sometimes painful. But it happens every day. I cannot recall the last time something like that happened in politics or religion."

– Carl Sagan, 1987

"Our only chance of long-term survival is not to remain lurking on planet Earth, but to spread out into space."

– Stephen W. Hawking

"Space exploration is a force of nature unto itself that no other force in society can rival. Not only does that get people interested in sciences and all the related fields, but it transforms the culture into one that values science and technology, and that's the culture that innovates. And in the 21st century, innovations in science and technology are the foundations of tomorrow's economy."

– Neil deGrasse Tyson, 2012

"I believe that developing a flexible deep space infrastructure to support a steady cadence of increasingly complex missions will strengthen American leadership in a quest to go where no person has gone before."

– Dr. Peggy Whitson, ISSastronaut with enough time in space for a round trip to Mars; over 665 days.

A.5.2 Quotes by Apollo Astronauts

"If we die, we want people to accept it. We're in a risky business, and we hope that if anything happens to us it will not delay the program. The conquest of space is worth the risk of life."

– Virgil I. Grissom, Apollo 1 Astronaut

"The world itself looks cleaner and so much more beautiful. Maybe we can make it that way – the way God intended it to be – by giving everybody that new perspective from out in space."

– Roger B Chaffee, Apollo 1 Astronaut

"I think you have to understand the feeling that a pilot has, that a test pilot has, that I look forward a great deal to making the first flight. There's a great deal of pride involved in making a first flight."

– Edward White II, Apollo 1 Astronaut

"You guys want to fix this ship or not?" If so let me see you down on the factory floor with the rest of us."

– Walter Schirra, Apollo 7 Astronaut

"Dedication and commitment to yourself, your family, your country and your life's goals is not to be taken lightly. They will ensure that you remain your own person and be the one who determines the path you will travel. You can be a leader or a follower in any field, anywhere. A leader sets his own pace and direction and attracts others to follow. A follower is merely one of the pack, doing what his "friends" do with little regard for how it will affect his life. Living for today, spending all you make and associating with the wrong crowd can get you into big trouble and strip you of the valuable rewards life has to offer."

– Walt Cunningham, Apollo 7 Astronaut

"When you're finally up at theMoonlooking back on Earth, all those differences and nationalistic traits are pretty well going to blend, and you're going to get a concept that maybe this really is one world and why the hell can't we live together like decent people?"

– Frank Borman, Apollo 8 Astronaut

"Well, Frank, my thoughts are very similar. The vast loneliness up here at theMoonis awe-inspiring, and it makes you realize what you have back there on Earth. The Earth from here is a grand oasis in the big vastness of space."

– James Lovell, Apollo 8 Astronaut

"We came all this way to explore theMoon, and the most important thing is that we discovered the Earth" and the first pictures taken of the Earth from theMoon, inspired environmentalists everywhere."

– Bill Anders, Apollo 8 Astronaut

"From theMoon, the Earth is so small and so fragile, and such a precious little spot in that Universe, that you can block it out with your thumb. Then you realize that on that spot, that little blue and white thing, is everything that means anything to you – all of history and music and poetry and art and death and birth and love, tears, joy, games, all of it right there on that little spot that you can cover with your thumb. And you realize from that perspective that you've changed forever, that there is something new there, that the relationship is no longer what it was."

– Russell Schweickart, Apollo 9 Astronaut

"Time Dreams. They are memories of the soul. They encompass all Time. They exist in the space of a dream."

– Tom Stafford, Apollo 10 Astronaut

"Anyone who sits on top of the largest hydrogen-oxygen fueled system in the world; knowing they're going to light the bottom, and doesn't get a little worried, does not fully understand the situation."

– John Young, Apollo 10 Astronaut

"It suddenly struck me that that tiny pea, pretty and blue, was the Earth.I put up my thumb and shut one eye, and my thumb blotted out the planet Earth.I didn't feel like a giant. I felt very, very small."– Neil Armstrong, Apollo 11 Astronaut

"Whenever I gaze up at theMoon, I feel like I'm on a time machine. I am back to that precious pinpoint of time, standing on the foreboding - yet beautiful - Sea of Tranquility. I could see our shining blue planet Earth poised in the darkness of space."

– Buzz Aldrin, Apollo 11 Astronaut

"We are gliding across the world in total silence, with absolute smoothness; a motion of stately grace which makes me feel godlike as I stand erect in my sideways chariot, cruising the night sky."

– Michael Collins, Apollo 11 Astronaut

"Here Men From Planet Earth First Set Foot Upon TheMoonJuly 1969 A.D. We Came In Peace For All Mankind."

– Plaque left on the Moon by Apollo 11

"Whoopee! Man, that may have been a small one for Neil, but it's a long one for me."

– Pete Conrad, Apollo 12 Astronaut

"Apollo was an impossible dream. To get to theMoonand return safely we worked, prayed, and cheered together as a world. Together we made an impossible dream come true."

– Alan Bean, Apollo 12 Astronaut

"There's nothing between you and oblivion except a pressure suit, and you just can't afford to get out there and get in a big rush and tangle yourself up where nobody can

help you. ... The biggest thing I've learned from the people that have gone in the past, you simply have to take your time, and you can't exhaust yourself."

– Richard Gordon, Apollo 12 Astronaut

"It's the abject smallness of the Earth that gets you."

– Stuart Roosa, Apollo 14 Astronaut

"If somebody had said before the flight, "Are you going to get carried away looking at the Earth from the Moon?" I would have say, "No, no way." But yet when I first looked back at the Earth, standing on the Moon, I cried."

– Alan Shepard, Apollo 14 Astronaut

"One of my most memorable trips was to the volcanically active and very remote region of central Askja, Iceland, in July 1967. Known for its volcanic craters called calderas, this region had a very rocky terrain with black volcanic sand, as well as a large lake and hot springs. It was a misty, surreal place unlike anything I'd ever seen in my travels. And because we were there during the summer it seemed like the Sun never set."

– Edgar Mitchell, Apollo 14 Astronaut

"I spent around ten days exploring the volcanically active regions of Iceland, a place so stark and barren I felt as if I were already on the Moon. We were there in the summertime, and it seemed like the sun never set. You could be out at 3 a.m. and see people strolling the city streets, the stores still open"

– Al Worden, Apollo 15 Astronaut

"As we got farther and farther away it diminished in size. Finally it shrank to the size of a marble, the most beautiful marble you can imagine. That beautiful, warm, living object looked so fragile, so delicate, that if you touched it with a finger it would crumble and fall apart. See this has to change a man."

– James Irvin, Apollo 15 Astronaut.

"As I stand out here in the wonders of the unknown at Hadley, I sort of realize there's a fundamental truth to our nature. Man must explore. And this is exploration at its greatest."

– David Scott, Apollo 15 Astronaut

"I think the future of lunar bases has to be somewhere around the South or North Pole. You have less variation in temperature and more daylight hours."

– Charles Duke, Apollo 16 Astronaut

"I had this very palpable fear that if I saw too much, I couldn't remember. It was just so impressive. And these things kept coming for the next 10 days. They never stopped,"

– Ken Mattingly, Apollo 16 Astronaut

"One-sixth gravity on the surface of the Moon is just delightful. It's not like being in zero gravity, you know. You can drop a pencil in zero gravity and look for it for three days. In one-sixth gravity, you just look down and there it is."

– John Young, Apollo 16 and 17 Astronaut

"The exposure of Apollo Astronauts to the geology of Iceland contributed greatly to the experience of Apollo astronauts as they prepared for lunar exploration and sampling. All the lunar landing crews benefited from examination of the varied rock assemblages found in glacial outwash channels that resemble the complexities of the lunar surface debris layer. Apollo 11,12, 15, and 17 explorations of lunar volcanic terrain also gained insights from exposure to the varieties of newly-formed volcanic rocks and structures found in Iceland."

– Harrison Schmitt, Apollo 17 Astronaut

"As I take man's last step from the surface, back home for some time to come… I'd like to just (say) what I believe history will record. That America's challenge of today has forged man's destiny of tomorrow. As we leave theMoonat Taurus-Littrow, we leave as we came and, God willing, as we shall return, with peace and hope for all mankind. Godspeed the crew of Apollo 17."

– Gene Cernan, Apollo 10 and 17 Astronaut

"It was shown as a symbol to the rest of the world that two great superpowers with different languages, different units of measurement, and certainly different political systems could have a common goal they could work together to achieve. It was really the highpoint of the opening of the Iron Curtain and a great goodwill in the middle of the Cold War."

– Tom Stafford, Apollo Soyuz Test Project (ASTP) Astronaut

"Man, I tell you, this is worth waiting 16 years."

– Deke Slayton, ASTP Astronaut

"Valery and I had a telecast over the Soviet Union, which was a little bit larger than what Russia is now. It was a huge country still is and to get across it, it was nine or 10 time zones, at least. I remember looking down at that terrain and Valery and I were describing it to people. We were describing the mountains, the deserts, the fields, the cities and that was an example of something that was a lot of fun to do."

– Vance Brand, ASTP Astronaut

"I think that crews now in orbit on the International Space Station use the lessons from the Soyuz-Apollo mission for how to fly in space together with other countries like the United States, Russia and European countries."

– Valery Kubasov, Soviet Cosmonaut on the ASTP Mission

"The best part of our joint flight was the occasion when we opened the hatch and I saw the face of Tom Stafford "I said, 'Hello Tom! Hello Deke!' and at this moment we shook hands."

– Alexei Leonov, Soviet commander on the ASTP mission

A.5.3 Quotes from Scientists and Engineers

"The discovery on Mars of fossils or evidence of contemporary Martian life would be epocal, since it would prove that the evolution of nonliving matter toward life is not unique to Earth. This would imply that the universe we inhabit is filled with living things. And, if Mars was once a home for life, must it not be so again? Finding the truth in such matters is worth."

– Robert M. Zubrin and Christopher P. McKay

"Today's robots are great at driving on the surfaces of planets, but they cannot reach or explore the caves that lie below. These caves are important because they could protect explorers fromradiation, meteorites and extreme temperatures on the surface. We will develop technology to guide flying robots into steep, confined underground spaces where traditional robots cannot tread."

– Dr. William "Red" Whittaker, Carnegie Mellon University Professor of Robotics, Astrobotic's Chairman and Chief Science Officer.

"There's never been a more exciting time especially in robotics, automation, and artificial intelligence where the rate of change and innovation has become exponential and is poised to surpass anything we've ever seen before."

– William Studebaker, President & CIO, ROBO Global.

"NASA is counting on robots to setup and care for deep space exploration facilities and equipment pre-deployed ahead of astronauts. Robots are also excellent precursors for conducting science missions ahead of human exploration."

– Sasha Congiu Ellis, NASA's Langley Research Center.

"Extreme space environments are dangerous for humans. And, robots are ideal for dangerous tasks. NASA already has rovers on Mars. This is an effort to advance autonomy of humanoid robots. We will have a better understanding of when and how humanoid robots will help with future deep space exploration missions as we continue our research and development in this field."

– Professor Taskin Padir, Northeastern University.

"A fuselage needs thousands of holes drilled and drilling them manually is not feasible."

– Chris Blanchette, National Account Manager, FANUC Robotics.

"That (secondary encoders) allows better control of the robot by minimizing backlash to achieve a higher level of accuracy. Secondary encoders reduce settling time, because the robot can react more quickly in drilling applications."

– David Masinick, Aerospace Account Manager, KUKA Robotics Corp.

"Robotics is playing a key role aboard the International Space Station and will con- tinue to be critical as we move toward human exploration of deep space," said. What's extraordinary about space technology and our work with projects likeRobonautare the unexpected possibilities space tech spinoffs may have right here on Earth. It's exciting

to see a NASA-developed technology that might one day help people with serious ambulatory needs begin to walk again, or even walk for the first time. That's the sort of return on investment NASA is proud to give back to America and the world."

– Michael Gazarik, Former Director of NASA's Space Technology Program.

"We greatly value our collaboration with NASA. The X1's (robotic exoskeleton) high-performance capabilities will enable IHMCto continue performing cutting-edge research in mobility assistance while expanding into the field of rehabilitation."

– Ken Ford, Director and CEO, Florida Institute for Human and Machine Cognition.

"We've made progress on reducing and shielding against these energetic particles, but we're still working on finding a material that is a good shield and can act as the primary structure of the spacecraft."

– Sheila Thibeault, a materials researcher at NASA Langley.

"Ultimately, the solution toradiationwill have to be a combination of things. Some of the solutions are technology we have already, like hydrogen-rich materials, but some of it will necessarily be cutting edge concepts that we haven't even thought of yet."

– Johathan A. Pellish, NASA Goddard.

"In our quest for space exploration, we should not forget that we take our humanity wherever we go. The things we do here on Earth, we will have to do in the places we go. We should not view being human as a handicap, but a virtue of what we have become and what we can achieve with our limited means."

– Dr. Pablo de León, Director, UND Human Spaceflight Laboratory
Department of Space Studies, John D. Odegard School of Aerospace Sciences
University of North Dakota

"The dawn of the age of 'astrospeleology' is upon us."

– Dr. Penny J. Boston, NASA Astrobiology Institute (NAI).

"Geologists have a saying: rocks remember."

– Neil Armstrong

Appendix 6

Cave and Lava Tube Terminology

Geologists have their own language. Planetary geologists also use this language and have added a few words of their own to the lexicon. Geologist that are cave researchers also added to that lexicon; especially those researching terrain with distinctive hydrology and landforms. These hydrogeologist also have their own lexicon.

For those unfamiliar with the language of caves, the first word to learn is the German word "karst." Karst is terrain with distinctive hydrology and landforms arising from the combination of high rock solubility and well-developed solution channel (secondary) porosity underground. Aqueous dissolution is the principal process at work. It creates the secondary porosity, and may be largely or wholly responsible for a specific landform on the surface. Where it is not quantitatively predominant, it is the essential (trigger) process that allows others to operate; for example, where a doline forms by mechanical collapse of insoluble strata into a solution cavity below.

In hydrogeology, a karst aquifer is one modified from earlier granular and/or fracture aquifer conditions by development of interconnected solutional channels ("caves" when large enough for human entry). Their larger aperture and efficient interconnection usually imply that water will circulate much more rapidly than in an unmodified aquifer. In the study of landforms, the karst geomorphic system is distinguished from the fluvial, glacial, eolian, coastal, and other systems because of the role of dissolution which causes water to circulate underground rather than run off at the surface in river channels.

The geology of the Moon is characterized by impact cratering and volcanism. The geology of Mars is more complex and involves volcanism, tectonism, water, ice, and impacts. While water has been discovered on both the Moon and Mars, the hydrogeologist would have more interest on Mars. The scientist astronaut on either body will be familiar with the following terms.

© Springer Nature Switzerland AG 2019
M. von Ehrenfried, *From Cave Man to Cave Martian*,
Springer Praxis Books, https://doi.org/10.1007/978-3-030-05408-3

Most of these terms originated in English-speaking countries, but a few of the more common terms are given in French, German, and Spanish. These terms are old, and the list has been expanded over many years by geologists from different countries.

abime	(French.) 1. Abyss. 2. Wide, deep shaft, in limestone, the walls of which are vertical or overhanging.
abris sous roche	(French.) See rock shelter.
active cave	1. Cave containing a running stream. 2. Cave in which speleothems are growing. Compare live cave.
aeolianite	See eolian calcarenite.
Aeration	See zone of aeration.
aggressive water	Water having the ability to dissolve rocks. In the context of limestone and dolomite, this term refers especially to water containing dissolved carbon dioxide (carbonic acid) or, rarely, other acids.
aguada	(Spanish for watering place.) In the Yucatan, shallow depression generally covering several hectares used for water supply.
air pocket	An enclosed air space between the water surface and the roof of a cave.
aisle	An elongated high narrow traversable passage in a cave. See also crawl, crawl way; corridor; passage. alternative. Adjective used to designate an intake or resurgence operating only during rainy seasons; in some areas reversible; equivalent to intermittent. Used also as a noun.
alveolization	(From the Latin word "alveolatus," meaning hollowed out.) Pitting of a rock surface produced by wind loaded with sand, by water charged with carbonic acid, or by plant roots.
anastomosis	A network of tubular passages or holes in a cave or in a solution sculptured rock. A complex of many irregular and repeatedly connected passages. Synonym, labryrinth.
anentolite	A helictite in which the eccentricity is ascribed to the action of air currents.
anthodite	A cave formation composed of feathery or radiating-masses of long needlelike crystals of gypsum or aragonite, which radiate outward from a common base. See also cave flower.
apron	A smooth bulging mass of flowstone covering sloping projections from walls of caves or limestone cliffs.
aquifer	A ground-water reservoir. Pervious rock that is completely saturated and will yield water to a well or spring.
aragonite	A mineral composed of calcium carbonate, $CaCO_3$, like calcite but differing in crystal form.
arete, pinnacle karat	A landscape of naked reticulated saw-topped ridges having almost vertical slopes and a relief of as much as 120 m. The

	ridges rise above forest-covered corridors and depressions. Found in New Guinea at elevations of 2000 m and more.
aven	1. (French.) A vertical or highly inclined shaft in limestone, extending upward from a cave passage, generally to the surface; smaller than an abtme. Commonly related to enlarged vertical joints. Compare cenote; natural well; pothole. 2. (British.) A vertical extension from a shaft in a passage or chamber roof that tapers upward rather like a very elongate cone. Compare dome a very elongate cone. Compare dome pit.
backflooding	Temporarily rising water level in a cave caused by a downstream passage being too small to pass an abnormally high discharge. The excavation and reexcavation of some caves is ascribed to the enlargement of a passage at or near the water table by gravity flow alternating with periods of calcite precipitation.
bacon	Thin, elongated, translucent flowstone having parallel colored bands on or projecting from roofs and walls of some caves. See also blanket; curtain; drapery.
balcony	Any projection on the wall of a cave large enough to support one or more persons.
bare karst	See naked karst. base level, karst. See karst base level.
beachrock	A friable to indurated rock consisting of sand grains of various minerals cemented by calcium carbonate; occurs in thin beds dipping seaward at less than 15°. Also known as beach standstone.
bed	A layer in sedimentary rocks; a stratum.
bedding plane	A plane that separates two strata of differing characteristics.
bedding-plane cave	A passage formed along a bedding plane, especially where there is a difference in susceptibility to corrosion in the two beds.
bicarbonate	A salt containing the radical HCO3, such as Ca(HCOs).
biospeleology	The study of subterranean living organisms, particularly in caves in limestone regions.
blade	In a cave, a thin sharp projection jutting out from roof, wall, or floor, of which it is an integral part; generally the remains of a partition or bridge.
blanket	A thick layer of dripstone, not translucent, see also bacon; curtain; drapery.
blind valley	A valley that ends suddenly at the point where its stream disappears underground; some blind valleys have no present-day streams. See also half-blind valley.
blowhole	1. A hole on land near the shore through which air and water are forced by incoming waves. 2. (Australia) A small hole in the surface of the Nullarbor Plain through which air

blows in and out with observable force, sometimes audibly.
blowing cave. A cave out of which or into which a current of air flows intermittently.

blowing well
A well into which and from which air blows with noticeable force owing to changes in barometric pressure or tidal action on the underlying aquifer.

blue hole
1. (Jamaica.) A major emergence where water rises from below without great turbulence. See also boiling spring. 2. (Bahamas.) A drowned solution sinkhole.

bogaz
(Slavic.) A long narrow chasm enlarged by solution of the limestone. See also corridor; struga; rinjon.

boiling spring
(Jamaica.) A large turbulent spring. See also blue hole.

bone-breccia
Cave breccia including much bone.

botryoid
A grapelike deposit of calcium carbonate generally found on walls of caves. Synonyms, clusterite; grape botryoid. A grapelike deposit of calcium carbonate generally found on walls of caves. Synonyms, clusterite; grape formation.

bourne
(British.) Intermittent stream in a normally dry valley in chalk country.

boxwork
Network of thin blades of calcite or gypsum etched out in relief on the limestone walls and ceiling of a cave.

branchwork
A dendritic system of subterranean watercourses having many incoming branches and no risible outgoing ones.

breakdown
See cave breakdown.

bridge
In a cave, a residual rock span across a passage. See also natural bridge.

buried karst
Karst topography buried by younger sediments. See also covered karst; paleokarst.

calanque
(French.) 1. Cove or small bay. 2. A valley excavated in limestone or formed by collapse of the roof of a cave and subsequently submerged by a rise in sea level.

calc-
Prefix meaning limy; containing calcium carbonate.

calcarenite
Limestone or dolomite composed of coral or shell sand or of grains derived from the disintegration and erosion of older limestones. Size of particles ranges from 1/16 to 2 millimeters. calcareous. Containing calcium carbonate.

calcareous tufa
See sinter.

calcification
Replacement of the original hard parts of an animal or plant by calcium carbonate.

calcilutite
Clastic limestone or dolomite in which the grains have an average diameter of less than 1/16 mm; calcareous mudstone.

calcirudite
A fragmental limestone in which the particles are generally larger than 2 mm.

calcite	A mineral composed of calcium carbonate, CaCO3, like aragonite but differing in crystal form; the principal constituent of limestone.
calcite bubble	A hollow sphere formed by the deposition of calcite around a gas bubble; the interior is smooth, and the exterior consists of small jagged crystals.
calcite flottante	(French.) See floe calcite.
calcrete	(South Africa.) See caliche.
caliche	1. (Chile and Peru.) A natural deposit of nitrates and other salts precipitated at the soil surface. 2. (Mexico and Southwestern U.S.) Indurated calcium carbonate and other salts found in the soil at the surface in arid and semiarid regions, generally formed by evaporation of lime-bearing waters drawn to the surface by capillary action. 3. In some by capillary action. 3. In some areas, refers to hardpan resulting from concentration of carbonate in the soil by downward leaching and reprecipitation. See also kankar, kunkar.
canalc	(Italian.) Long drowned valley on the Dalmatian coast. Some canali may be drowned poljes.
canopy	A compound cave formation consisting of flowstone hanging from a sloping wall projection and forming a fringe of shawls or stalactites on the outer edge.
canyon	1. A steep-walled chasm, gorge, or ravine cut by running: water. 2. A chasm that has been formed by a cave stream. 3. A valley formed by collapse of the roof of a long fairly straight cave; a karst valley.
capillary stalagmite	Hollow stalagmite formed by saturated karst water pushed up through capillaries and small cracks in a sinter; covering permeable fluvial deposits on the floor of a cave; first reported from Cuba, where such stalagmites are composed of aragonite.
carbonate	1. A salt or ester of carbonic acid; a compound containing the radical CO2, such as calcium carbonate, CaCO3. 2. A rock consisting mainly of carbonate minerals, such as limestone or dolomite.
causse	(French.) A limestone plateau in the southeastern part of the central massif of France characterized by closed depressions, caves, and avens; a number of such plateaus in and around the basin of the river Tarn constitute Les Grandes Gausses. This region was considered to exemplify karst development intermediate between holokarst and merokarst.
cave	1. A natural underground room or series of rooms and passages large enough to be entered by a man; generally

formed by solution of limestone. 2. A similar artificial opening.

cave blister
A small pimplelike cave formation, roughly oval in shape, generally loose, and having a core of mud.

cave breakdown
1. Enlargement of parts of a cave system by fall of rock masses from walls and ceiling. 2. Heaps of rock that have collapsed from the walls and ceiling of a cave, generally called cave breccia.

cave breccia
Angular fragments of rock forming a fill in a cave, either cemented together by dripstone or in a matrix of cave earth. Sec also solution breccia.

cave coral
A rough, knobby growth of calcite resembling coral in shape, generally small; found on floor, walls, or ceiling of a cave. Synonym, coral formation. See also knobstone. cave earth, cave fill. Insoluble deposits of clay, silt, sand, or gravel flooring or filling a cave passage. In a more restricted sense, cave earth includes only the finer fractions: clay, silt, and fine sand deposits.

cave flower
An elongate curved deposit of gypsum or epsomite on a cave wall in which growth occurs at the attached end. Synonyms, gypsum flower; oulopholite.

cave formation
Secondary mineral deposit formed by the accumulation, dripping, or flowing of water in a cave. See also speleothem.

cave group
A number of caves or cave systems, not interconnected but geographically associated in some relief feature or particular geological outcrop. See also cave series.

cave guano
Accumulations of dung in caves, generally from bats; in some places partially mineralized.

cave ice
Ice formed in a cave by natural freezing of water. Loosely but incorrectly applied to calcium carbonate dripstone and flowstone.

cave-in
The collapse of the ceiling or side walls of a cave or of the land surface into a subterranean passage as a result of undermining or of pressure from above.

cave marble
Banded deposit of calcite or aragonite capable of taking a high polish. See also flowstone; onyx marble.

cave onyx
See onyx marble.

cave pearl
Small concretion of calcite or aragonite formed by concentric precipitation around a nucleus. Synonyms, pisolite, pisolith.

cave series
A group of caves of similar morphology in a particular district. See also cave group.

cave spring
A spring rising in a cave.

cave system
1. An underground network of passages, chambers, or other cavities. 2. The caves in a given area related to each other

	hydrologically, whether continuous or discontinuous from a single opening. See also cave group, cave series.
caver	One who explores caves as a sport. Synonyms, potholer; spelunker.
cavern	A cave, often used poetically or to connote larger-than-average size.
cavernous	Containing numerous1 cavities or caverns.
caving	The sport of exploring caves. Synonyms, potholing; spelurking. 2. A method of mining in which the ore is allowed to cave or fall.
ceiling block	Roughly cubical joint bounded large block, which has fallen from the ceiling of a cave. See also cave breakdown; ceiling slab.
ceiling cavity	Solutional concavity in the ceiling of a cave. The orientation is determined by joints or a bedding plane.
ceiling channel	Sinuous channel developed in the ceiling of a cave, presumably during the phreatic (underground water) phase of cave development.
ceiling meander	A winding upside-down channel in a cave ceiling.
ceiling pocket	See pocket. ceiling slab, roof slab. A thin but extensive piece of rock that has fallen from the ceiling of a cave in roughly horizontal limestone. See also cave breakdown, ceiling block.
ceiling tube	A half tube remaining in the ceiling of a cave.
cenote	(Spanish, after Mayan tzouet or dzonot.) Steep-walled natural well that extends below the water table; generally caused by collapse of a cave roof. Term used only for features in Yucatan. See also natural well.
chalk	Soft poorly indurated (hardened) limestone, generally light in color; commonly composed of the tests of floating microorganisms in a matrix of very finely crystalline calcite.
chamber	1. (America.) The largest order of cavity in a cave or cave system; it has considerable length and breadth but not necessarily great height. 2. (England.) A room in a cave.
chasm	1. A deep, fairly narrow breach in the ground; an abyss; a gorge; a deep canyon. 2. A deep, wide, elongated gap in the floor of a cave.
chert	Light-cream or gray to black rock composed of silica, found occurring as nodules or layers in limestone, or as a replacement of limestone.
chimney	A narrow vertical shaft in the roof of a cave, generally smaller than an aven; a dome pit.
chockstone	A rock wedged between the walls of a cave passage.
choke	Rock debris or cave fill completely blocking a passage.
chute	An inclined channel or trough in a cave.

clay fill	Dry or wet clay that fills a cave passage.
clay filling	According to Bretz (1942), time interval between end of phreatic solution of a cave and beginning of deposition of flowstone.
clint	(England.) slabs or blocks of limestone, parallel to the bedding, forming a pavement. Widened joints, or grikes, isolate individual clints. Synonym, Flachkarren.
closed depression	A general term for any enclosed topographic basin having no external drainage, regardless of origin or size.
clusterite	See botryoid.
cockpit	(Jamaica.) 1. Any closed depression having steep sides. 2. A star- shaped depression having a conical or slightly concave floor. The surrounding hill slopes are steep and convex. Cockpits are the common type of closed depressions in a Kegelkarst.
cockpit karst	Tropical karst topography containing many closed depressions surrounded by conical hills. Divided by French and German geographers into several types depending on shape of hills. See also cone karst; Halbkugelkarst; Kegelkarst; Spitzkegelkarst; tower karst.
collapse breccia	A mass of rock composed of angular to rounded fragments of limestone or dolomite that has formed as the result of the collapse of the roof of a cave, of an underlying cave, or of an overhanging ledge. See also solution breccia.
collapse sink	A closed depression formed by the collapse of the roof of a cave. See also doline.
column	A flowstone formation, generally cylindrical, formed by the union of a stalactite and stalagmite. See also pillar.
conduit	A subterranean stream course filled completely with water and always under hydrostatic pressure. See also siphon.
cone karst	A type of karst topography, common in the tropics, characterized by star-shaped depressions at the base of many steep-sided cone shaped hills. A variety of Kegelkarst. See also cockpit karst.
conical wall niche	See meander niche, constructive waterfall. A large rimstone dam on a surface stream.
conuiite	A hollow, cone-shaped speleothem formed when a conical depression is drilled in cave mud by falling water. Subsequent erosion may remove the mud, isolating the calcite lining of the depression.
corrasion	Mechanical erosion performed by such moving agents as water, ice, and wind, especially when armed with rock fragments. See also corrosion.
corridor	1. Relatively narrow passageway permitting travel between two larger areas. 2. A fairly level and straight passage that

	links two or more rooms or chambers in a cave. 3. Intersecting linear depressions on the surface of the land, related to joints or dikes. See also bogaz; struga; zanjon.
corrosion	Erosion by solution or chemical action. See also corrosion.
coupole	(French.) hemispheric hill.
cove	(Southern Appalachians.) Narrow steep-sided karst valley flanking limestone plateaus.
covered karst	A terrane of karst features, usually subdued, resulting from the development of solution features in limestone covered by soil; contrasted with naked karst, which is soil free. See also buried karst.
crawl, crawlway.	A cave passage that must be negotiated on hands and knees.
crescentic wall niche	See meander niche.
crevice karst	An intricate irregular crevice system that has formed by solution widening of closely spaced joints. Crevices may be as much as 6 m across and 20 m deep. Especially well developed near rivers in lowland New Guinea.
crust stone	A fragile layer of flowstone covering portions of walls of caves; looks like a flaky crust. Found in some Kentucky caves.
cryokarst	European equivalent of thermokarst.
crystal cave	A cave in which much of the surface of the roof, walls, and floor is covered with well-formed mineral crystals.
crystal pool	In caves a pool, generally having little or no overflow, containing crystals.
cueva	(Spanish.) Cave, especially one that is horizontal or nearly so.
cul-de-sac	A subterranean passage having only one entry, a dead end.
cupola	A hemispheric hill of limestone. (French.) Cupole. (German.) Halbkugel.
current marking	Shallow asymmetrical hollows, caused by turbulent water flow, that are distributed in rather regular fashion over limestone surfaces. See also scallop.
curtain	A wavy or folded sheet of flowstone hanging from the roof or projecting from the wall of a cave; often translucent and resonant. See also bacon; blanket; drapery.
cutter	1. (Tennessee.) Solution crevice in limestone underlying residual phosphate deposits. 2. A Karren-like groove ices may be as much as 6 m across and 20 m deep. Especially well developed near rivers in lowland New Guinea.
crust stone	A fragile layer of flowstone covering portions of walls of caves; looks like a flaky crust. Found in some Kentucky caves.
cryokarst	European equivalent of thermokarst.

crystal cave	A cave in which much of the surface of the roof, walls, and floor is covered with well-formed mineral crystals.
daylight hole	A hole in the roof of a cave, reaching the surface.
dead cave	A dry cave in which all solution and precipitation has ceased.
dead end	See cul-de-sac.
decalcification	Removal of calcium carbonate from a rock, leaving a residuum of noncalcareous material.
Deckenkarren	(German.) Solutional pendant features in cave ceiling.
decoration	Cave features due to secondary precipitation of calcite, aragonite, gypsum, and other rarer minerals.
diffuse circulation	Circulation of ground water in karst aquifers under conditions in which all, or almost all, openings in the karstified rock intercommunicate and are full of water in the zone of saturation.
dissolution	See solution.
dog-tooth crystal/spar	A variety of calcite in the form of sharp pointed crystals.
doline	A basin or funnel-shaped hollow in limestone, ranging in diameter from a few meters to a kilometer and in depth from a few to several hundred meters. Some dolines are gentle grassy hollows; others are rocky cliff-bounded basins. A distinction may be made between those formed mainly by direct solution of the limestone surface zone, solution dolines, and those formed by collapse over a cave, collapse dolines, but it is generally not possible to establish the origin of individual examples. Closed depressions receiving a stream are known as swallow holes or stream sinks. In America most dolines are referred to as sinks or sinkholes.
doline lake	A small karst lake occupying a doline.
dolomite	1. A mineral composed of calcium magnesium carbonate, $CaMg(CO3)2$. 2. Rock chiefly composed of the mineral dolomite. Also called dolostone.
dolomitization	The process whereby limestone becomes dolomite by the substitution of magnesium carbonate for part of the original calcium carbonate.
dome pit	"Mammoth Cave" possesses several extraordinary vertical cavities of which the arched tops are called domes and the deep bottoms are called pits.
drapery	A thin sheet of dripstone, equivalent to curtain. See also bacon; blanket.
driphole	1. Hole in rock or clay produced by fast-dripping water. 2. Hollow space surrounded by precipitated material, such as the bottom of a stalactite.

dripstone	Calcium carbonate deposited from water dripping from the ceiling or wall of a cave or from the overhanging edge of a rock shelter; commonly refers to the rock in stalactites, stalagmites, and other similar speleothems; in some places composed of aragonite or gypsum. See also flowstone.
dry cave	A cave without a running stream. See also dead cave.
dry valley	A valley that lacks a surface water channel; common in the chalk of southern England.
duck-under	1. A place where water reaches the cave roof for a short distance and can be passed by quick submergence without swimming. 2. In cave diving, a longer stretch of passage where the water is so close to the roof that crawling or swimming beneath the water surface is needed to pass.
dune limestone	(Australia) see eolian calcarenite.
Durchgangshohle	(German.) See through cave.
ebb and flow spring/well	A spring or flowing well exhibiting. Periodic variation in volume of flow; the periodicity, which is often irregular, is attributed to siphonic action.
eccentric	Adjective or noun implying abnormal shape in speleothems, such as helictites.
effluent cave	See outflow cave.
emergence	Point at which an underground stream comes to the surface. See also exsurgence; resurgence; rise.
eolian calacernite	A terrestrial limestone formed by the cementation by carbonates of calcareous coastal dune sand. Often shortened to eolianite. Synonym, dune-limestone. Compare beachrock.
epiphreas, epiphreatic zone.	The zone in a cave system, immediately above the phreatic zone, affected morphologically and hydrologically by floods too large for the cave to absorb at once.
estavelle	(French.) An intermittent resurgence or exsurgence, active only in wet seasons. May act alternatively as a swallow hole and as a rising according to ground-water conditions.
etched pothole	See solution pan.
evorsion	Mechanical erosion by whirling water that may carry sand and gravel; pothole erosion.
exhumed karst	Karst features reexposed by erosion from beneath former covering strata.
exsurgence	Point at which an underground stream reaches the surface if stream has no knovn surface headwaters. See also emergence; resurgence.
facet	See scallop.
fault cave	A cave developed along a fault or fault zone.
fissure cave	A narrow vertical cave or cave passage along a fissure. Fissures widen out to become wells or shaft caves.

flattener	A cave passage, which though wide, is so low that movement is only possible in a prone position. See also crawl.
floe calcite	Very thin film of pure calcium carbonate floating on the surface of a subterranean pool of very calm water.
floor pocket	See pocket.
flowstone	Deposits of calcium carbonate, gypsum, and other mineral matter which have accumulated on the walls or floors of caves at places where water trickles or flows over the rock. See also dripstone.
fluorescein	A reddish-yellow crystalline compound that imparts a brilliant green fluorescent color to water in very dilute solutions; used to label underground water for identification of an emergence.
flute	See scallop.
fluviokarst	A predominantly karst landscape in which there is much evidence of past or present fluvial activity.
foiba	(Italian.) 1. A deep wide vertical cavity or the swallow point of a river at the beginning of its underground course. 2. A natural vertical shaft in soluble rock, tending toward cylindrical shape; it may or may not reach the surface. A dome pit.
formation	1. The fundamental unit in rock-stratigraphic classification, consisting of a distinctive mappable body of rock. 2. See cave formation; speleothem.
fossil karst	See paleokarst.
free-surface stream	In a cave, a stream that does not completely fill its passage.
fresco	A half-section of a stalactite on the wall of a cave.
fungling	(Chinese.) Isolated limestone hill in alluvial plain, probably similar to mogote.
gallery	A rather large, nearly horizontal passage in a cave.
geode	Hollow globular bodies varying in size from a few centimeters to several decimeters, coated on the interior with crystals.
geological organ	A group of solution pipes best seen in the walls of quarries.
glacier cave	Cave in ice formed within or at the base of a glacier.
glaciokarst	A glaciated limestone region possessing both glacial and karst characteristics.
glade	1. (Jamaica.) An elongate depression, having steep sides, in which a generally flat floor is divided into small basins separated by low divides. 2. (Tennessee.) Limestone pavement having extensive growth of cedar trees.
globularite	Small crystals of calcite tipped with spheres composed of radiating fibers.
gour	See rimstone barrage, rimstone barrier, rimstone dam.

grape formation	See botryoid.
grike, gryke	(England.) A vertical or subvertical fissure in a limestone pavement developed by solution along a joint. Synonym. (German.) Kluftkarren.
grotto	1. A small cave, natural or artificial. 2. A room, in a cave system, of moderate dimensions but richly decorated.
ground air	See soil air.
ground water	1. The part of the subsurface water that is in the zone of saturation. See also phreas, phreatic water. 2. Used loosely by some to refer to any water beneath the surface.
gryke	See grike.
guano	See cave guano.
gulf	Steep-walled closed depression having a flat alluviated bottom; in some gulfs a stream flows across the bottom.
gushing spring	See vauclusian spring.
gypsum	A mineral composed of hydrous calcium sulfate, CaSO4-H-O.
gypsum flower	See cave flower.
Halbhohle	(German.) See rock shelter.
Halbkugelkarst	(German.) Tropical karst topography containing dome shaped residual hills surrounding depressions, a kind of Kegelkarst. Also called Kugelkarst.
half-blind valley	Blind valley in which the stream overflows in flood time when the swallow hole cannot accept all the water.
half tube	Trace of a tube remaining in the roof or wall of a cave. See also tube.
hall	In a cave, a lofty chamber which is much longer than it is wide. See also gallery.
hanging blade	A blade projecting down from the ceiling. See also blade.
hardness	Property of water that prevents lathering because of the presence of cations (ions), mainly calcium and magnesium, which form insoluble soaps.
haystack hill	(Puerto Rico.) In the tropics, rounded conical hill of limestone developed as a result of solution. Term being replaced by mogote.
helictite	A curved or angular twiglike lateral projection of calcium carbonate having a tiny central canal, found in caves. Synonym, eccentric stalactite.
heligmite	An eccentric growing upward from a cave floor or from a shelf in a cave. A curved or angular thin stalagmite.
holokarst	A term for a karst area like that of the Dinaric Karst of Slovenia. Such areas have bare surfaces on thick deposits of limestone that extend below sea level, well developed Karren, dolines, uvalas, poljes, deep ponors, and extensive cave systems; they have little or no surface drainage. Contrast merokarst.

hoya, hoyo	(Spanish.) A very large closed depression. Used in Puerto Rico for doline, in Cuba for polje.
hum	Karst inselberg. Residual hill of limestone on a fairly level floor, such as the isolated hills of limestone in poljes. In some tropical areas, used loosely as synonym for mogote.
hydraulic conductivity	The ability of a rock unit to conduct water under specified conditions.
ice cave	1. A cave, generally in lava or limestone, in which the average temperature is below 0°C., and which ordinarily contains perennial ice. Ice may have the form of stalactites, stalagmites, or flowstone. 2. See glacier cave.
inflow cave	Cave into which a stream flows or formerly entered.
intake, recharge area	The surface area in which water is absorbed into an aquifer eventually to reach the zone of saturation. An inverted siphon. See water trap.
jama	1. (Slavic.) Verbal or steeply inclined shaft in limestone, known as abime or aven in France and as pothole in England. 2. Any crve.
joint-plane cave	A cavity high in relation to width developed along steeply dipping joint planes.
Kamenitza	(German, possibly of Slavic origin; plural, Kamer'ce.) See solution pan.
kankar, kunkar	(Australia.) See caliche.
Karren	(German.) Channels or furrows, caused by solution on massive bare limestone surfaces; they vary in depth from, a few millimeters to more than a meter and are separated by ridges. In modern usage, the terms are general, describing the total complex of superficial solution forms found on compact pure limestone.
Karrenfeld	(German.) An area of limestone dominated by Karren karst.
karst breccia	See also collapse breccia; solution breccia.
karst bridge	A natural bridge or arch in limestone.
karst fens	Marshes developed in sinkhole terrain; swampy solution fens.
karst fenster	See karst window.
karst hydrology	The drainage phenomena of karstified limestones, dolomites, and other slowly soluble rocks.
karst lake	A large area of standing water in extensive closed depression in limestone.
karst margin plain	A plain generally on limestone between higher country of limestone on one side and of less pervious rocks on the other, but having a cover of impervious detritus, which allows surface drainage. Synonym, Karstrandebene.

karst plain	A plain on which closed depressions, subterranean drainage, and other karst features may be developed. Also called karst plateau.
karst pond	Closed depression in a karst area containing standing water.
karst river	A river that originates from a karst spring.
karst spring	A spring emerging from karstified limestone. See also emergence; exsurgence; resurgent; rise.
karst topography	Topography dominated by features of solutional origin.
karst valley	1. Elongate solution valley. 2. Valley produced by collapse of a cavern roof.
karst window	1. Depression revealing a part of a subterranean river flowing across its floor, or an unroofed part of a cave. 2. A small natural bridge or arch which can be seen through.
karstic	Occasionally used as the adjective form of karst.
karstification	Action by water, mainly chemical but also mechanical, that produces features of a karst topography, including such surface features as dolines, Karren, and mogotes and such subsurface features as caves and shafts.
karstland	A region characterized by karst topography.
Karstrandebene	(German.) See karst margin plain.
katavothron	(Greek.) A closed depression or swallow hole.
Kegelkarst	(German.) A general term used to describe several types of tropical humid karst characterized by numerous, closely spaced cone hemispherical or tower-shaped hills having intervening closed depressions and narrow steep-walled karst valleys or passageways. See also cockpit karst; cone karst; Halbkugelkarst; tower karst.
keyhole	A small passage or opening in a cave; in cross section, rounded at the top, constricted in the middle, and rectangular or flared out below.
knobstone	Speleothem, larger, more pronounced and more widely separated than cave coral.
kras, krs	A Slavic word meaning bleak, waterless place, from which the term karst is derived. See also kars which is surrounded by higher country but is bordered on one side by impervious rock. Compare blind valley.
marl	Unconsolidated sedimentary rock consisting largely of calcium carbonate and clay; usage varies from calcareous clay to earthy limestone, and in some parts of the United States, the term has been used for any unconsolidated sedimentary rock containing fossil shells.
maze cave	See network.
meander	In a cave, an arcuate curve in a channel formed by lateral shifting of a cave stream. See ceiling meander; meander niche.

meander niche	A conical or crescent shaped opening in the wall of a cave formed by the downward and lateral erosion of a stream on the floor of a passage.
merokarst	Defined to indicate imperfect karst topography as found on thin, impure, or chalky limestone where surface drainage and dry valleys are present in addition to some karstic features. Contrast perfectly formed holokarst.
Mexican onyx	See onyx marble.
microkarst	Karst topography in which all superficial features are small; an area dominated by minor karst features.
minor karst features	See Karren; rill; solution pan.
Mischungskorrosion	(German.) Dissolving of limestone by water derived from the mixing of two saturated waters that differ in carbon dioxide partial pressure. Such a mixture is undersaturated because a nonlinear relation exists between calcite solubility and the partial pressure of carbon dioxide.
mogote	A steep-sided hill of limestone, generally surrounded by nearly flat alleviated plains; karst inselberg. Originally used in Cuba in referring to residual hills of folded limestone in the Sierra de los Organos but now used internationally for karst residual hills in the Tropics.
moonmilk	A white plastic calcareous cave deposit composed of calcite, untite, or magnesite. From Swiss dialect moonmilch, elf's milk.
mud stalagmite	Stalagmite composed principally of clay or sandy clay and commonly less than 30 per cent calcium carbonate.
naked, bare karst	Karst topography having much exposed bedrock.
natural arch	A rock arch or very short natural tunnel; contrasted with natural bridge, which spans a ravine or valley.
natural bridge	A rock bridge spanning a ravine and not yet eroded away.
natural tunnel	A nearly horizontal cave open at both ends, generally fairly straight in direction and fairly uniform in cross section.
natural well	(Jamaica.) A vertical shaft in limestone, open to the surface and having water at the bottom; similar to a cenote.
nested sinkholes	(American.) See uvala.
network	A complex pattern of repeatedly connected passages in a cave system. Synonym, labyrinth.
nip	An undercutting notch in rock, particularly limestone, along a seacoast between high and low tide levels, produced by erosion or solution.
ojo, ojo agua	(Spanish.) An artesian spring in limestone regions, especially one forming a small pond; a vauclusian spring.
onyx marble	Translucent layers of calcium carbonate from cave deposits, often called Mexican onyx or cave onyx; used as an ornamental stone.

Opferkessel	See solution pan.
oulopholite	See cave flower.
outflow cave	Cave from which stream flows out or formerly did so. Synonym, effluent cave.
ouvala	(French.) See uvala.
paleokarst	A karstified rock or area that has been buried by later sediments; in some places, ancient caves lave been completely filled by the later sediments.
palette	In a cave, a more or less flat protruding sheet of crystalline calcium carbonate spared during solution of the rock on each side of it. See also blade; shield.
panhole	See solution pan.
park	(Arizona.) Shallow broad solution depression.
partition	1. A nearly vertical residual rock mass in a cave. 2. A continuous rock span across a cave.
passage	In a cave, the opening between rooms or chambers.
pendant	One of a group of isolated similarly proportioned projections surrounded by a complex of connected cavities in the bedrock ceiling of a cave.
pepino hill	(Puerto Rico.) 1. Rounded or conical-shaped hill resulting from tropical humid karst action. Term generally replaced in Puerto Rico by mogote. 2. Elongate hill or ridge capped by mogotes.
permeability	The ability of a rock or soil to permit water or other fluids to pass through it. See also hydraulic conductivity.
phreas, phreatic water	(From the Greek word meaning well.) Water in the zone of saturation; water below the water table.
phreatic surface.	See water table.
phreatic zone	The region below, the water table, in which rock is saturated with water.
piezometric surface	The imaginary surface to which water from a given aquifer will rise under its full static head. Also called potentiometric surface.
pillar	1. A column of rock remaining after solution of the surrounding rock. See also rock pillar. 2. A stalactite stalagmite that reaches from roof to floor in a cave; more properly termed a column. 3. A tall thin stalagmite that does not reach the roof of a cave.
pipe	Small cylindrical hole in unconsolidated sediments, caused by removal of fine material by water.
piping	Formation of a passage by water under pressure in the form of conduits through permeable materials when the hydraulic head exceeds a certain critical value.
pisolite, pisolith	See cave pearl.

pit	A deep hole, generally circular in outline, having vertical or nearly vertical walls. See also pothole (definition 2); shaft.
piton	(French.) Limestone hill having sharply pointed peak.
pitted plain	Plain having numerous small closely spaced closed depressions.
pocket	Solution cavity in ceiling, floor, or walls of a cave, shaped like the interior of a round-bottomed kettle; unrelated to joints or bedding.
polje	(Slavic word for field.) In areas of karst topography, a very large closed depression, in some places several kilometers long and wide, having a flat floor either of bare limestone or covered by alluvium, and surrounded by generally steep walls of limestone. Synonym, interior valley.
polygonal karst	A karst area where the surface is completely pitted with closed depressions, the divides of which form a crudely polygonal network. Especially common in humid tropical cone-karst terrain, but also found in well-formed temperate doline-karst terrain.
ponor	(Slavic.) Hole in the bottom or side of a closed depression through which water passes to or from an underground channel. Compare swallet, swallow hole.
pool deposit	Crystalline material deposited in an isolated pool in a cave.
pore deposit	Mineral matter deposited on the interior of a cave from water entering the cave so slowly through pores and cracks that it does not form drops.
porosity	The ratio of the aggregate volume of interstices in a rock or soil to its total volume; generally stated as a percentage.
porthole	A nearly circular natural opening in a thin rock wall in a cave. See also window.
pothole	1. A small rounded hole worn into rock in a streambed, at a waterfall, or near sea level by sand, gravel, and stones being spun around by force of the currents; a natural 1 mill. 2. Term used in England for vertical or steeply inclined shaft in limestone. See also pit; shaft.
potholer	(British.)/See caver.
potholing	1. The process of scouring holes in rock in stream beds or near the strand line by rapid rotation of trapped pebbles or cobbles; evorsion. 2. (British.) See caving.
pozo	(Spanish.) See sima
pseudokarst	Karstlike terrain produced by a process other than the dissolving of rock, such as the rough surface above a lava field, where the ceilings of lava tubes have collapsed Features of pseudokarst include lava tunnels, lava tubes, lava stalactites, and lava stalagmites.

Randpolje	An enclosed plain at the edge of a karst area receiving surface water from the nonkarstic area. The water drains out through underground passages in the karst area. The plain is thus completely enclosed by higher ground. Compare blind valley; karst margin plain.
recharge area	See intake area.
residual clay	Clay or sandy clay remaining on a rock surface after removal of calcium carbonate by solution. Compare terra rossa.
resurgence	Point at which an underground stream reaches the surface and becomes a surface stream. In European literature, the term is reserved for the reemergence of a stream that has earlier sunk upstream; the term exsurgence is applied to a stream without known surface headwaters.
rice paddy	In a cave, a terraced rimstone pool.
rift	A long narrow high cave passage controlled by joints or faults.
rill, rille	Small solution groove on surface exposures of limestone; most common in arid or semiarid areas.
Rillenstein	(German.) Microsolution grooves and pitting on rock surface.
rimstone	Calcareous deposits formed around the rims of overflowing basins, especially in caves. rimstone barrage, rimstone barrier, rimstone rimstone dam. A wall-shaped deposit that impounds pools of water in caves, around springs, and in cascades of streams saturated with calcium bicarbonate. Synonym, gour.
rimstone pool	A pool kept in place by a rimstone dam.
rise	(Jamaica.) Spring rising from fractures in limestone. Point at which an underground stream comes to the surface.
rising	See emergence; exsurgence; resurgence.
rock fall	See cave breakdown.
rock pendant	See pendant.
rock pillar	A residual isolated mass of bedrock linking the roof or overhanging wall and floor of a cave, in contrast with a column, which is composed of dripstone or flowstone. See column; pillar.
rock pinnacle	A tall sharp projection of bedrock rising from the floor of a cave.
rock shelter	Shallow cave under an overhanging rock ledge. Many sea caves are rock shelters. Also found in limestone and other rock types where streams have undercut their banks at bends, or where there has been abrasion by blowing sand. Common in tropical areas at places where a secondarily hardened layer of limestone forms a ledge that projects over unindurated limestone.

roof crust	Flowstone deposited on ceilings of caves from thin films of water, which have crept over the rock from pore or crack sources.
roof slab	See ceiling slab.
room	A part of a cave system that is wider than a passage. Synonym (England.), chamber.
sand pipe	See solution pipe.
sand stalagmite	A stalagmite formed on sand and made of calcite cemented sandstone.
saturation, zone of	See zone of saturation.
scaling chip	A thin small rather irregular piece of limestone, commonly crumbly, which has fallen from the ceiling or wall of a cave. A form of cove breakdown.
scaling plate	A small flat piece of rock of rectangular or polygonal shape, that has fallen to the floor of a cave. A form of cave breakdown in thin- bedded impure limestone cut by closely spaced joints.
scallop	Oval hollow having an asymmetric cross section along its main axis. Scallops form patterns on the walls of caves and in streambeds and were used to determine direction of flow of turbulent water, since they are steeper on the upstream side. Commonly called flutes in America.
scar	(Northern England.) steep rock cliff in limestone country.
sea cave	A cave or cleft in a sea cliff eroded by waves or currents or dissolved by water.
shaft	1. A cylindrical tube generally steep sided, that forms by solution and (or) collapse. 2. A vertical passage in a cave. See also pit; pothole (definition 2).
shake, shakehole	(England; sometimes spelled shackhole.) 1. Term used mainly by cavers to indicate a doline, especially one formed by subsidence. 2. Hole formed by solution, subsidence, and compaction in loose drift or alluvium overlying beds of limestone.
shawl	Simple triangular-shaped curtain.
sheet	A thin coating of calcium carbonate formed on walls, shelves, benches, and terraces by trickling water.
shield	A disk-shaped speleothem standing edgewise at a high angle.
sima	(Spanish.) Natural well that has vertical sides.
sink, sinkhole	(American.) General terms for closed depressions. They may be basin, funnel, or cylindrical shaped. See also closed depression; doline; swallet, swallow hole.
sinkhole plain	(American.) Plain on which most of the local relief is due to closed depressions and nearly all drainage is subterranean.

sinkhole pond	(American.) Small lake in closed depression in limestone, due to an impervious clay floor or to intersection of depression with the water table.
sinking creek	A small stream that disappears underground. See also lost river; swallet, swallow hole.
sinter	A mineral precipitate deposited by a mineral spring, either hot or cold. Siliceous sinter, consisting of silica, is also called geyserite and fiorite; calcareous sinter, consisting of calcium carbonate, is also called tufa, travertine, and onyx marble.
siphon	In speleology, a cave passage in which the ceiling dips below a water surface. See also water trap.
slump pit	A hollow in the clay fill of a cave floor caused by erosion beneath the fill.
soil air	The air that fills soil and rock interstices above the zone of saturation.
solution	1. The change of matter from a solid or gaseous state to a liquid state by combination with a liquid. 2.The result of such change; a liquid combination of a liquid and a nonliquid substance.
solution breccia	A mass of rock composed of angular to rounded fragments of rock that have accumulated by solution of surrounding or underlying carbonate. See also collapse breccia.
solution lake	A lake whose origin is attributed largely to solution of underlying rock.
solution pan	Shallow solution basin formed on bare limestone, generally characterized by flat bottom and overhanging sides. Synonyms, Kamenitza; Opferkessel; panhole; tinajita.
solution pipe	A vertical cylindrical hole attributable to solution, often without surface expression, filled with debris, such as sand, clay, rock chips, and bones. Synonym, sand pipe. See also geological organ.
solution scarp	Escarpment formed by more active solution of lower area or by corrosional undercutting of the base of the escarpment.
solution subsidence	Any subsidence due to solution of underlying rock but particularly the subsidence of parts of a formation into hollows or pockets of an immediately underlying soluble formation.
sotano	(Spanish for cellar or basement.) Term used in Mexico for deep vertical shafts in limestone, which may or may not lead to a cave.
spelean	Of, pertaining to, or related to caves.
speleogen	A secondary cave structure formed by dissolving, such as a dome pit or a scallop.
speleologist	A scientist engaged in the study and exploration of caves, their environment, and their biota.

speleology	The scientific study, exploration, and description of caves and related features.
speleothem	A secondary mineral deposit formed in caves, such as stalactite or stalagmite. Synonym, cave formation.
spelunker	See caver.
spelunking	See caving.
Spitzkegelkarst	(German.) Tropical karst topography containing sharply pointed residual limestone hills.
splash cup	The shallow concavity in the top of a stalagmite.
spongework	A complex of irregular interconnecting cavities with intricate perforation of the rock. The cavities may be large or small.
spring	Any natural discharge of water from rock or soil onto the surface of the land or into a body of surface water.
squeeze	A narrow passage or opening just passable with effort. Differs from flattener in that there is little spare space in any direction.
stagmalite	A general term including stalactite and stalagmite. Superseded by dripstone.
stalactite	A cylindrical or conical deposit of minerals, generally calcite, formed by dripping water, hanging from the roof of a cave, generally having a hollow tube at its center. From the Greek word meaning exude drops.
stalagmite	A deposit of calcium carbonate rising from the floor of a limestone cave, formed by precipitation from a bicarbonate solution through loss of CO_2. The water drops on the stalagmite from above. From Greek word meaning drip. See also dripstone.
stalagmite, capillary	See capillary stalagmite.
steephead	A deeply cut valley, generally short, terminating at its upslope end in an amphitheater, at the foot of which a stream may emerge.
straw stalactite	Thin tubular stalactite, generally less than a centimeter in diameter and of very great length (examples as long as 4 m); also called soda straw.
stream sink	Point at which a surface stream sinks into the ground. See also ponor; swallet, swallow hole.
streamtube	A cave passage completely filled, or formerly filled, with fast-moving water and whose ceiling and walls normally possess scallops.
struga	(Slavic.) A corridor formed along a bedding plane in karst country.
subjacent karst	Karst landscape in noncarbonated rocks due to presence of karstified rocks beneath the surface formation.

submarine spring	Large offshore emergence, generally from cavernous limestone but in some areas from beds of lava.
Torricellian chamber	A submerged airfilled chamber of a cave at a pressure below atmospheric pressure, sealed by water, having an air-water surface above that of adjacent free air-water surfaces.
tourelle	(French.) A little tower; applied to small flat-topped buttes of limestone in karst areas. Contrasted with pitons, which have pointed tops, and with coupoles, which have rounded tops.
tower karst, Turmkarst	Karst topography characterized by isolated limestone hills separated by areas of alluvium or other detrital sand; towers are generally steep-sided forest-covered hills, and many have flat tops.
tracers	Materials, such as chemicals, dyes, radioactive salts, and light insoluble solids introduced into underground waters to determine points of egress of the water and its velocity.
Trap	See siphon; water trap.
travertine	Calcium carbonate, $CaCO_3$, light in color and generally concretionary and compact, deposited from solution in ground and surface waters. Extremely porous or cellular varieties are known as calcareous tufa, calcareous sinter, or spring deposit. Compact banded varieties, capable of taking a polish, are called onyx marble or cave onyx.
troglobite	An animal living permanently underground in the dark zone of caves and only accidentally leaving it.
troglodyte	A human cavedweller.
troglophile	An animal habitually entering the dark zone of a cave but necessarily spending part of its existence outside such as some species of bats.
troglophobe	An animal or person unable physically or psychologically to enter the dark zone of a cave or other underground area.
trogloxene	An animal entering a cave for various reasons but not living there permanently.
tube	A smooth-surfaced cave passage of elliptical or nearly circular cross section.
tufa	See sinter; travertine.
tunnel	See natural tunnel.
turlough	(Ireland.) A depression in limestone or in glacial drift over limestone that is liable to flood either from excess surface runoff or from rising ground water. From two Irish words tuar loch, meaning dry-lake.
Turmkarst	(German.) See tower karst.
underground river	See subterranean river.
upside-down channel	See ceiling channel.

uvala	Large closed depression formed by the coalescence of several dolines; compound doline. See also canyon; karst-valley; valley sink.
vadose water	Water in the zone of aeration; water above the zone of saturation.
valley sink	(American.) An elongated closed depression or series of interconnecting depressions forming a valley like depression. Compare karst valley; uvala.
vauclusian spring	A large spring or exsurgence of an underground river Generally from limestone, that varies greatly in output and is impenetrable except with diving apparatus. Synonym, gushing spring.
vug	A small cavity in rock usually lined with crystals. Adjective, vuggy. See also geode.
wall block	A roughly cubical joint-controlled large block of limestone or dolomite, which has rotated outward from a cave wall. See also cave breakdown.
wall slab	A thin but large block of rock, which has fallen outward from the wall of a cave in limestone in which the dip is nearly vertical. See also cave breakdown.
water table	The upper boundary of an unconfined zone of saturation, along which the hydrostatic pressure is equal to the atmospheric pressure.
water trap	A place where the roof of a chamber or passage of a cave dips under water but lifts again farther on. Synonym, trap.
well	1. A shaft or hole sunk into the ground to obtain water, oil, gas, or minerals. 2. A deep vertical rounded hole or shaft in the floor of a cave or at the bottom of a closed depression.
window	1. In speleology, a natural opening above the floor of a passage or a room, giving access to an adjoining cavity or to the surface; larger and less symmetrical than a porthole. 2. The opening under the arch of a small natural bridge. 3. See karst window.
zanjon	(Spanish.) In Puerto Pico, solution trench in limestone. Zanjones range from a few centimeters to about 3 m in width and from about 1 to 4 m in depth. Apparently they form by the widening and deepening of joints by solution. See also corridor; struga.
zone of aeration	The zone in permeable soil or rock that is above the zone saturated with water; the zone of vadose water.
zone of saturation	The zone in permeable soil or rock that is saturated with water; the phreatic zone.
Zwischenhohle	(German.) Cave in which a river passage, or former river passage, is entered from above or laterally and which can be followed upstream and downstream some distance but not to daylight.

References, Internet Links and Videos

Books by Springer-Praxis Books

Expedition Mars. Springer-Praxis by Martin J.L. Turner, 2004

Marswalk One: First Steps on a New Planet. Springer-Praxis by David J. Shayler, Andrew Salman and Michael D. Shayler, 2005

Martian Outpost: The Challenge of Establishing a Human Settlement on Mars. Springer-Praxis by Erik Seedhouse, 2009

Use of Extraterrestrial Resources for Human Space Mission to Moon or Mars. Springer-Praxis by Donald Rapp, November, 2012

Human Mission to Mars: Enabling Technologies for Exploring the Red Planet. Springer-Praxis by Donald Rapp, November 2015

Next Stop Mars: The Why, How, and When of Human Missions: Springer-Praxis by Giancarlo Genta, February, 2017

Books by Others

Tensile Structures, Frei Otto, 1962

Expandable Structures for Space Applications, F.W. Forbes, 1964

Intelligent Life in the Universe, Carl Sagan and I.S. Shklovskii, January 1966

The Science of Speleology, edited by T.D. Ford and C.H.D. Cullingford, Academic Press, 1976

Mission To Mars: Plans and Concepts for the First Manned Landing, James E. Oberg, 1982

Speleology: Caves and the Cave Environment, by George W. Moore and G. Nicholas Sullivan, Cave Books, 1997

Acid House, by Stephanie Pain, New Scientist, June 6, 1998

Mission to Mars: Michael Collins, November, 1990

The Case For Mars: The Plan To Settle the Red Planet and Why We Must. Robert Zubrin, 1996

The Human Use of Caves, British Archeological Reports International Series, by March 1, 2001

Caves: Exploring Hidden Realms, by Michael Ray Taylor, Ronal C. Kerbo, 2001

© Springer Nature Switzerland AG 2019

M. von Ehrenfried, *From Cave Man to Cave Martian*,

Springer Praxis Books, https://doi.org/10.1007/978-3-030-05408-3

Encyclopedia of Cave and Karst Science, John Gunn, Editor. Fitzroy-Dearborn Publisher, Ltd, London, UK 2009. https://sudartomas.files.wordpress.com/2012/11/encyclopediaof_cavesandkarstscience.pdf

Colonizing Mars. The Human Mission to the Red Planet: Robert Zubrin and Joel Levine, 2012

The Planet Mars: A History of Observation and Discovery: William Sheehan

Mission to Mars-My Vision for Space Exploration, Buzz Aldrin, 2013

Mission Mars: Get Ready for the Future of Space Exploration: Pascal Lee, 2013

Shaping Humanity: How Science, Art and Imagination help us understand our Origins, by John Gurche, 2013

Welcome to Mars: Making a Home on the Red Planet: Buzz Aldrin and Marianne Dyson, 2015

Beyond Earth: Charles Wohlforth and Amanda R. Hendrix, 2016

Mars One: Humanity's Next Great Adventure: Nobert Kraft, James R. Kraft, Raye Kayss, Gerard t Hooft

From Habitability to Life on Mars, edited by Nathalie A. Cabrol and Edmond A. Grin for SETI; published by Elsevier, 2018

Reports by NASA

Expandable Structures Technology for Manned Space Applications, by NASA Langley Research Center engineers, Osborne, R. S..Tynan, C. I., Jr. Williams, J. G., AIAA Paper 71-399, 1971

NASA N93-17442 Inflatable Habitation for the Lunar Base, M. Roberts, Presented at The Second Conference on Lunar Bases and Space Activities of the 21st Century held in Houston, TX, April 5-7, 1988. Edited by W. W. Mendell

Inflatable Structures Technology Handbook, JSC 7/5/2000. Available at: https://ntrs.nasa.gov/archive/nasa/casi.ntrs.nasa.gov/20110000798.pdf

NIAC Caves of Mars Grant: *Human Utilization of Subsurface Extraterrestrial Environments:* P. Boston, G. Frederick, S. Welch, J. Werker, T.R. Meyer, B. Sprungman, V. Hildreth-Werker, D. Murphy, S.L. Thompson Complex Systems Research, Inc., Boulder, CO., Final Report 2004 http://www.niac.usra.edu/files/studies/final_report/710Boston.pdf

NASA/TP-2005-213164; *Managing Lunar and Mars Mission Radiation Risks, Part I: Cancer Risks, Uncertainties, and Shielding Effectiveness*, Francis A. Cucinotta NASA Lyndon B. Johnson Space Center, Myung-Hee Y. Kim, Wyle Laboratories Houston, Texas, Lei Ren U.S.R.A., Division of Space Life Science Division, Houston, Texas

NASA Press Kit, 2009: *Lunar Reconnaissance Orbiter (LRO):Leading NASA's Way Back to the Moon Lunar Crater Observation and Sensing Satellite (LCROSS): NASA's Mission to Search for Water on the Moon*

NASA-HDBK-6022, 8.17.2010: *Handbook for the Microbial Examination of Space Hardware*

NASA Space Technology Roadmaps and Priorities: Restoring NASA's Technological Edge and Paving the Way for a New Era in Space. 2012 and revisited in 2015 and 2016

Human Spaceflight Architecture Team (HAT) Mars Destination Operations Team (DOT) FY 2013 Final Report, January 2014, Larry Toups, NASA JSC and Dr. Marianne Bobskill, NASA LRC. https://www.lpi.usra.edu/lunar/strategies/HAT_Mars-DOT-2013-Final-Report.pdf

Pioneering Space: NASA's Next Steps on the Path to Mars, 2014. http://www.nasa.gov/sites/default/files/files/Pioneering-space-final-052914b.pdf

NASA Publication No. NPI 8020.7. *Policy on Planetary Protection Requirements for Human Extraterrestrial Missions, 2014* http://planetaryprotection.nasa.gov/documents/

NASA-SP-2009-566: *Human Exploration of Mars: Design Reference Architecture 5.0* B.G. Drake Editor: with two addendums, the last in 2014. http://www.nasa.gov/pdf/373665main_NASA-SP-2009-566.pdf

Sustaining Human Presence on Mars Using ISRU and a Reusable Lander, NASA Langley Research Center, 2015 https://ntrs.nasa.gov/archive/nasa/casi.ntrs.nasa.gov/20160006324.pdf

NASA Astrobiology Strategy, 2015 Editor, Lindsay Hays, Jet Propulsion Laboratory, California Institute of Technology https://nai.nasa.gov/media/medialibrary/2016/04/NASA_Astrobiology_Strategy_2015_FINAL_041216.pdf

NASA Policy Directive (NPD) 8020.7, *"Biological Contamination Control for Outbound and Inbound Planetary Spacecraft,"* as a parallel document to NPR 8020.12, *"Planetary Protection Provisions for Robotic Extraterrestrial Missions."*

A Path to Planetary Protection Requirements for Human Exploration: A Literature Review and Systems Engineering Approach, James E. Johnson, NASA JSC, 2015 https://ntrs.nasa.gov/search.jsp?R = 20170006174

NPR 8020.12D: *Planetary Protection Provisions for Robotic Extraterrestrial Missions*

NPD 8020.7G: *Biological Contamination Control for Outbound and Inbound Planetary Spacecraft*

NPD 7100.10E: *Curation of Extraterrestrial Materials*

Radiation Shielding Materials Containing Hydrogen, Boron, and Nitrogen: A Systematic Computational and Experimental Study-Phase I. NIAC FINAL Sheila A. Thibeault, et al, Sep 2012

Radiation and Human Exploration of Mars: Briefing to the NASA Advisory Council by Richard Williams, January, 2015

NASA HRP-47065: *Human Research Program Integrated Research Plan*, NASA JSC, 2015

NP-2015-08-2018: *NASA's Journey to Mars: Pioneering Next Steps in Space Exploration*, October, 2015

NASA's Journey to Mars: Plans for Human Missions to the Red Planet. Assessment of Candidate Landing Sites and Exploration Zones. October, 2015 *Exploration Systems Development*: NASA Advisory Council (NAC) Meeting, November, 2015

NASA/TM—2013-216541 *Kilowatt-Class Fission Power Systems for Science and Human Precursor Missions*, Lee Mason and Marc Gibson Glenn Research Center, Cleveland, Ohio, Dave Poston Los Alamos National Laboratory, Los Alamos, New Mexico

NASA/TM—2015-218460, *Development of NASA's Small Fission Power System for Science and Human Exploration*, Marc A. Gibson, Lee S. Mason, and Cheryl L. Bowman, Glenn Research Center, Cleveland, Ohio, David I. Poston and Patrick R. McClure, Los Alamos National Laboratory, John Creasy and Chris Robinson, National Security Complex, Oak Ridge, Tennessee

NASA's Kilopower Reactor Development and the Path to Higher Power Missions, NASA Glenn Research Center

Books & Documents Related to Planetary Protection

Lessons Learned from the Viking Planetary Quarantine and Contamination Control Experience, NASA Contractor Report NASW-4355, L. Daspit, J. Stern & J. Martin, 1988

The Planetary Quarantine Program: Origins and Achievements, 1956-1973 Where No Man Has Gone Before: A History of Apollo Lunar Exploration Missions, 25 Years of Curating Moon Rocks

Safe on Mars: Precursor Measurements Necessary to Support Human Operations on the Martian Surface, NRC, Space Studies Board, 2002

Planetary Protection Issues in the Human Exploration of Mars, Pingree Park Report, NASA, 2005

ESA/NASA Workshop on *Planetary Protection & Human System Research and Technology*, 2007

Planetary protection for humans in space: Mars and the Moon, C.A. Conley & J.D. Rummel Acta Astronautica 63 1025–1030, 2008

IAA Report on *Protecting the Environment of Celestial Bodies* (2010)

Planetary Protection and Human Missions-Principles and Operating Guidelines COSPAR, 2011

Preparing for the Human Exploration of Mars: Health Care and Planetary Protection Requirements and Practices, J.D. Rummel & C.A. Conley IAC A1.5-15689, 2012

When Biospheres Collide: A History of NASA's Planetary Protection Programs, Michael Meltzer, 2014

Expanding Options for Implementing Planetary Protection During Human Space Exploration and Robotic Precursor Missions : Interim Report, IAA Study, 2014

Workshop on Planetary Protection Knowledge Gaps for Human Extraterrestrial Missions, NASA Ames Research Center, 2015

Workshop on Refining Planetary Protection Requirement for Human Missions, Lunar Planetary Institute, 2016

2nd COSPAR Workshop on Refining Planetary Protection Requirements for Human Missions, Lunar Planetary Institute, 2018

Bibliographies

Bibliographies of Planetary Protection (Quarantine) Research (1945-1977)
Bibliography of Planetary Quarantine. Volume I: Policy (1958-1967)
Bibliography of Planetary Quarantine. Volume II: Environmental Microbiology (1945-1967)
NASA Planetary Quarantine References, assembled by George Washington University (1966-1976)
Bibliography of Scientific Publications and Presentations Relating to Planetary Quarantine (1966-1971)
Bibliography of Scientific Publications and Presentations Relating to Planetary Quarantine (1972-1976)

NASA Lunar and Planetary Institute

Lava Tubes: Potential Shelters for Habitats, Lunar Bases and Space Activities of the 21st Century, ed. Horz, Friedrich, Mendell, W.W.; Symposium on Lunar Bases and Space Activities of the 21st Century, Washington D.C. 1984

Lunar Machining, Lewis, William; Lunar Bases and Space Activities of the 21st Century, 1984

Mechanical Properties of Lunar Materials Under Anhydrous, Hard Vacuum Conditions: Applications of Lunar Glass Structural Components, Lunar Bases and Space Activities of the 21st Century, ed. Blacic, James D., Mendell, W.W.; Washington D.C. 1984

Apollo 15 Lunar Base Site: Steep Slopes as an Energy Resource, Workshop on the Geology and Petrology of the Apollo 15 Landing Site, ed. Burke, J.D., Spudis, Paul D. and Graham Ryder; 1986) pp. 38-43

Constant Temperature Vessels for Lunar Base Applications, Papers Presented to the Second Symposium on Lunar Bases and Space Activities of the 21st Century, ed. Bergeron, Denis E., Mendell, W.W 1988

Regolith and Local Resources to Generate Lunar Structures and Shielding, Khalili, E. Nader, Papers Presented to the Second Symposium on Lunar Bases and Space Activities of the 21st Century, ed. Mendell, W.W, 1988

NASA-JPL

Robotic Vehicles for Planetary Exploration, Brian Wilcox, Journal of Applied Intelligence, 2nd Qtr 1992, Kluwer Academic Publishers, Boston MA., 1992, pp. 181-193

A Mars Rover For the 1990's, Brian H. Wilcox and Donald B. Gennery, Journal of the British Interplanetary Society, Vol. 40, No. 10, October 1987, pp. 483-488

Moon Diver: A Discovery Mission Concept for Understanding the History of the Mare Basalts through the Exploration of a Lunar Mare Pit. JPL, L. Kerber, et al. Astrogeology Science Center, Flagstaff, AZ, Brown Univ. Providence, RI, Johns Hopkins Applied Physics Laboratory, Laurel MD, Smithsonian Institution, University of Oxford, UK, NASA MSFC

Books by Associations and Other Organizations

Explore Mars, Inc.

The Humans to Mars Reports, 2015, 2016, 2017, 2018 https://h2m.exploremars.org/

National Research Council (NRC)

Safe on Mars: Precursor Measurements Necessary to Support Human Operations on the Martian Surface. Washington, DC: The National Academies Press. 2002
Vision and Voyages for Planetary Science in the Decade 2013-2022: National Research Council of the National Academies, 2011

U.S. Geological Survey

Candidate Cave Entrances on Mars, Glen E. Cushing, USGS Astrogeology Science Center, 2010
Possible Cave Skylights On Mars, by G.E. Cushing1,2, T.N. Titus1, J.J. Wynne1,2, P.R. Christensen. USGS/Arizona State University
Cave Detection Using Oblique Thermal Imaging, T. N. Titus, J. J. Wynne, M. D. Jhabvala, G. E. Cushing, P. Shu, N. A. Cabro; U.S. Geological Survey, Astrogeology Science Center, Flagstaff, AZ 86001, 2011
Complex Bedding Geometry in the Upper Portion of Aeolis Mons, Gale Crater, Mars, 2018 Anderson, R.B., Edgar, L.A., Rubin, D.M., Lewis, K.W., Newman, C., 2018
A Chronology of Activities from Conception through the End of Project Apollo (1960-1973), Gerald G. Schaber, The U.S. Geological Survey, Branch of Astrogeology
Mars Global Digital Dune Data Base: MC2-MC29 https://pubs.usgs.gov/of/2007/1158/
Atypical pit craters on Mars: New insights from THEMIS, CTX, and HiRISE observations, Glen E. Cushing, Timothy N. Titus, Chris Okubo, USGS. https://www.researchgate.net/publication/276520605_Atypical_pit_craters_on_Mars_New_insights_from_THEMIS_CTX_and_HiRISE_observations

The Planetary Society

The Mars Link Teacher's Guide: The Planetary Society, September, 1993
Humans Orbiting Mars: The Planetary Society (with input from JPL), 2015

The Mars Institute

Phobos-Deimos ASAP: A Case for the Human Exploration of the Moons of Mars: by
   Pascal Lee at the Workshop on the Exploration of Phobos and Deimos, 2007
Terrestrial Analogs for Lunar Science and Exploration: A Systematic Approach.
   Pascal Lee, Mars Institute, et al. SETI Institute, Ames Research Center, Johnson
   Space Center, National Space Biomedical Research Institute, 2008

Lunar and Planetary Institute

Concepts and Approaches for Mars Exploration, 2000

The SETI Institute

The Scientific Importance of Caves in Our Solar System, J. Judson Wynne Northern
   Arizona University and the SETI Institute, Highlights of the 2nd International
   Planetary Caves Conference, Flagstaff, AZ. 2015
*Planetary Protection Knowledge Gaps for Future Mars Human Missions: Stepwise
   Progress in Identifying and Integrating Science and Technology Needs*, 48th
   International Conference on Environmental Systems, 2018, J. Andy Spry and
   Margaret S. Race, SETI Institute. Gerhard Kminek ESA-ESTEC, Noordwijk,
   Netherlands and Bette Siegel and Cassie Conley, NASA Headquarters, Wash-
   ington D.C.

National Speleological Society/International Union of Speleology

*Mega-caves of Mars Revisited: Speleological Information Systems in Planetary
   Science and Technology*, William R. Halliday, Commission on Volcanic Caves of
   the International Union of Speleology, Nashville, TN

Association of Space Explorers (ASE)

To Mars Together: Proceeding of the 8th Planetary Congress of the ASE, August
   24-29, 1992

International Astronautical Federation (IAF)

An International Mars Exploration Program: Presented to the 40th Congress of the
   IAF in Malaga, Spain October 7-12, 1989

Other Reports/Papers

Expandable Structures for Space Applications, by F. W. Forbes, Technical Support
   Division, Air Force Aero Propulsion Laboratory, Research and Technology
   Division, 1964

On the Origin of Lunar Sinuous Rilles: Oberbeck, V.R., Quaide, W.L., and Greeley, R., 1969, Modern Geology

Comparing Structural Metals for Large Lunar Bases, Kelso, Hugh, et al., Engineering, Construction, and Operations in Space II: Proceedings of Space 90, ed. Johnson, Stewart W. and John P. Wetzel; New York: American Society of Civil Engineers 1990

Erosion by Flowing Lava: Field Evidence. Ronald Greeley, Sarah A. Fagents, Robert Scott Harris, Steven D. Kadel, and David A. Williams, 1998, Department of Geology, Arizona State University, Tempe Arizona

A Search for Intact Lava Tubes on the Moon: Possible Lunar Base Habitats, Coombs, Cassandra R. and B. Ray Hawke, Planetary Geosciences Division, Hawaii Institute of Geophysics, University of Hawaii, 1988

Remote Sensing of Lunar Lavatubes from Earth, Billings, Thomas L., Journal of the British Interplanetary Society (JBIS) 44, 1991

Lava tube Options, Billings, Thomas L., Cheryl Lynn York (Singer), and Bryce Walden, Lockheed Large Habitable Volumes Study, ed. (Lockheed Engineering & Sciences Corporation, under NASA Contract), 1989

A Preliminary Assessment of the Potential of Lava Tube-situated Lunar Base Architecture, Engineering, Construction, and Operations in Space II: Proceedings of Space 90, ed. Daga, Andrew W., Meryl A. Daga, and Wendel R. Wendel Johnson, Stewart W. and John P. Wetzel; American Society of Civil Engineers, 1990

Preliminary Geologic Characterization of Young's Cave Lava Tube System, Bend, Oregon, and Implications for Lunar Base Lava Tube Siting, Site Characterization and Phase One Development Plan for the Oregon Moonbase, ed. York (Singer), Gillett, Stephen L., Cheryl Lynn, Bryce Walden, and Thomas L. Billings; The Oregon L-5 Society, Inc., under NASA NASW-4460, 1990

Geology of Selected Lava Tubes in the Bend Area, Oregon, Greeley, Ronald, State of Oregon, Department of Geology and Mineral Industries, 1971

Planetary Landscapes, Second Edition, Greeley, Ronald (New York: Chapman & Hall, 1994)

Lunar Lava Tube Sensing, York, Cheryl Lynn, Bryce Walden, Thomas L. Billings, and P. Douglas Reeder; Poster Presentation at Joint Workshop on New Technologies for Lunar Resource Assessment, Santa Fe, New Mexico 1992

Lava Tube Exploration Robot and Payload Development, Presented at the International Planetary Caves Conference in 2015, H.S. Kelly, A. J. Parness and P. J. Boston

Reference Mission Architecture for Lunar Lava Tube Reconnaissance Missions, Presented at the First International Planetary Cave Research Workshop in 2011, Exploration Architecture Corp., JPL, Sasakawa International, the Aerospace Corporation

Ice Caves on Mars: A Good Place for Life !?: A. Pflitsch1, Ch. Grebe, D. Holmgren1and M. Steinrücke, 1Ruhr-University Bochum, Workgroup of Cave and Subway Climatology, Universitätsstr.150, 44801 Bochum, Germany, andreas.pflitsch@rub.de)

On The Development of Additive Construction Technologies for Application to Development of Lunar/Martian Surface Structures Using In-Situ Materials, Presented at the AIAA SPACE 2015 Conference and Exposition Pasadena, California, Mary J. Werkheiser, NASA Ames Research Center; Michael Fiske, Jacobs; Jennifer Edmunson, Jacobs; Behrokh Khoshnevis, University of Southern California. Read more at: https://arc.aiaa.org/doi/abs/10.2514/6.2015-4451

Cave Biosignature Suites: Microbes, Minerals, and Mars, a research papter published in ASTROBIOLOGY Volume 1, Number 1, 2001, by P.J. Boston, M.N. Spilde, D.E. Northup, et al

A Prototype Mass Spectrometer for In-Situ Analysis of Cave Atmospheres, Rev. Sci. Instrum. 83, 105116; E.L. Patrick, K.E. Mandt, E.J. Mitchell, J.N.Mitchell, K.N. Younkin, C.M.Seifert, and G.C. Williams. (2012). http://dx.doi.org/10.1063/1.4761927

In-Situ Measurements of Air Samples from Groundwater Wells with a Field Deployable Mass Spectrometer, Geological Society of America Abstracts with Programs. Vol. 44, No. 7, p.313, Mitchell, E. J., Patrick, E.L., Mandt, K.E., Mitchell, J.N., and Williams, G.C., (2012)

Ongoing Development of Hands-On Demonstrations for Environmental Science Outreach at the Elementary Level, Geological Society of America Abstracts with Programs. Vol. 44, No. 7, p.98, Sobery, A.C., Mitchell, E.J., and Turner, D.R., (2012)

Development of Earth Science Hands-on Exhibits for STEM Enrichment at the Elementary Level. Geological Society of America Vol. 44, No. 1., Sobery, A., D. R. Turner, E.J. Mitchell. (2012)

Comparison of Infrared and Quadrupole Mass Spectrometer Measurements of Carbon Dioxide in Cave Environments, Geological Society of America, Vol. 44, No. 1., Seifert, C., E.J. Mitchell, K.E. Mandt, E.L Patrick. (2012)

In-Situ Mass Spectrometer Measurements of Cave Atmospheres as an Analogue to Future Planetary Cave Missions. Proceedings of the 43rd Lunar and Planetary Science Conference. Abstract #1442., Mandt, K.E., E.L. Patrick, E.J. Mitchell, J. N. Mitchell and K.N. Younkin. (2012)

In-Situ Meteorology and Elemental Composition of Cave Atmospheres in Texas. First International Planetary Caves Workshop Program and Abstract Volume, p. 17., Mitchell, E. J., J.N. Mitchell, E.L. Patrick, K.E. Mandt and K.N. Younkin. (2011)

Meteorological Measurements of Three South Texas Caves in Differing Geological Formations. Geological Society of America Abstracts with Programs, Vol. 42, No. 2, p. 67., Duran, L. and E. J. Mitchell. (2010)

A Study of Meteorological Differences Between Caves in Different Geologies. St. Mary's University McNair Scholars Research Journal, Volume II, 2009-2010., Duran, L. E. J. Mitchell – Mentor. (2010)

Age of Caves in the Cordillera De La Sal (Atacam, Chile), Joe De Waelei, Vincenzo Picotti, Paolo Forti, George Brook, Cucchi Franco, Zini Luca, Italian Institute of Speleology, University of Bologna, Italy, Department of Geography, University of Georgia, Athens GA, Department of Geological, Environmental and Marine Sciences, University of Trieste, Italy. https://www.researchgate.net/publication/290889248_Age_of_caves_in_the_Cordillera_de_la_Sal_Atacama_Chile

Analysis of Cave Atmospheres by Comprehensive Two-Dimensional Gas Chromatography (GC×GC) With Flame Ionization Detection (FID), 2014. Ryan Blasé, Southwest Research Institute, Edward Patrick, Southwest Research Institute, Mark Libardoni, Southwest Research Institute and Joseph N. Mitchell

Aerosol forcing of the position of the intertropical convergence zone since AD 1550, Harriet E. Ridley, Yemane Asmerom, James U. L. Baldini, Sebastian F. M. Breitenbach, Valorie V. Aquino, Keith M. Prufer, Brendan J. Culleton, Victor Polyak, Franziska A. Lechleitner, Douglas J. Kennett, Minghua Zhang, Norbert Marwan, Colin G. Macpherson, Lisa M. Baldini, Tingyin Xiao, Joanne L. Peterkin, Jaime Awe & Gerald H. Haug, Nature Geoscience volume 8, pages 195-200 (2015)

NASA's New Robotic Lunar Program...Seeking Water on the Moon, David M. Harland, *Space Exploration 2008*, Springer-Praxis, pages 116-127, (2008)

Japan Space Agency (JAXA)

Detection of Intact Lava Tubes at Marius Hills on the Moon by SELENE (Kaguya) Lunar Radar Sounder. T. Kaku J. Haruyama W. Miyake A. Kumamoto K. Ishiyama T. Nishibori K. Yamamoto Sarah T. Crites T. Michikami Y. Yokota R. Sood H. J. Melosh L. Chappaz K. C. Howell, First published: 17 October 2017 https://doi.org/10.1002/2017GL074998

University Reports

Technologies for Exploring Skylights, Lava Tubes and Caves, The Robotics Institute, William Whittaker, 2012 https://www.nasa.gov/pdf/637136main_Whittaker_Presentation.pdf

Space weathering effects on lunar cold traps, D. H. Crider and R. R. Vondrak. *Proceedings of the Lunar and Planetary Science Conference*, page 1922, held March 12-16, 2001

Modeling the stability of volatile deposits in lunar cold traps, D. H. Crider and R. R. Vondrak. *Workshop on Moon beyond 2002: Next steps in lunar science and exploration*, Taos, New Mexico, 3006, September 12-14, 2002

Space weathering of ice layers in lunar cold traps, D. H. Crider and R. R. Vondrak. *Advances in Space Research*, volume 31, pages 2293-2298, 2003

Internet Links

NASA

https://www.nasa.gov/ (enter the Center you want then search for the subject)

Erosion by Flowing Lava: Field Evidence. Arizona State University. file:///C:/Users/dvonehrenfried/Downloads/Greeley_et_al-1998-Journal_of_Geophysical_Research__Solid_Earth.pdf

About Mars and THEMIS http://themis.asu.edu/topic

NASA Technical Reports Server (NTRS) https://ntrs.nasa.gov/search.jsp

NASA Technology Roadmaps https://www.nasa.gov/offices/oct/home/roadmaps/index.html

NASA Scientific & Technical Information Program https://www.sti.nasa.gov/

NASA Planetary Data System

NASA-JPL Planetary Data System Image Atlas https://pds-imaging.jpl.nasa.gov/search/?q = *%3A*

Lunar and Planetary Institute https://www.lpi.usra.edu/lunar/

Access to the LPI data base https://www.lpi.usra.edu/search/?cx = 002803415602668413512%3Acu4craz862y&cof = FORID%3A11&q = cave + research&sa = Search&siteurl = https%3A%2F%2Fwww.lpi.usra.edu%2Fpublications%2Fabsearch%2F

Lunar Orbiter Image Recovery Project (LOIRP) The Lunar Orbiter Image Recovery Project Online Data Volumes were published online for public access by NASA at the Planetary Data System Cartography and Imaging Sciences Node on 31 January 2018. You can access all of the imagery recovered by LOIRP here: https://pds-imaging.jpl.nasa.gov/volumes/loirp.html

National Innovative Advanced Concepts (NIAC) https://www.nasa.gov/directorates/spacetech/niac/index.html

Arizona State University

Lava Tube Data Base https://rpif.asu.edu/index.php/ltdb/

JMARS Data Base https://jmars.asu.edu/

Speleology Organizations

National Speleological Society https://caves.org/

World's Longest Caves http://www.caverbob.com/wlong.htm

World's Deepest Caves http://www.caverbob.com/wdeep.htm

A 50-year-old astronaut training film of the USGS Geologic Rover/Lunar Rover. https://astrogeology.usgs.gov/rpif/videos/grover

Robotics History: Narratives and Networks Oral Histories: Brian Wilcox https://ieeetv.ieee.org/history/robotics-history-narratives-and-networks-oral-histories-brian-wilcox

Videos

Making Common Sensors for Mars, Sarah Fagents University of Hawaii. https://www.youtube.com/watch?v = AHU37Ts3TSg

Mars Insight Scientists https://mars.nasa.gov/news/8359/meet-the-people-behind-nasas-insight-mars-lander/

A Passage to Mars (Trailer-Full Movie on YouTube as well) https://www.youtube.com/watch?v=dumOXFpD6Jk

Houghton Mars Project | Project Mars On Earth: https://www.youtube.com/watch?v=4D3Yozt-A70&t=2s

Manned Mars Rover concept car https://www.youtube.com/watch?v=fZddqJ7y5TI

Caves and Lava Tubes

Planetary scientist Pascal Lee on candidate impact-melt lava tube skylights near the north pole of the Moon: https://www.youtube.com/watch?v=doMXNGc_N-I

TED talk: Inside the world's deepest caves with Bill Stone https://www.youtube.com/watch?v=-Bn6Gel7yEs

Dr. Penelope Boston https://www.youtube.com/watch?v=NWvXyobH9fg

Dr. Diana Northrup https://www.youtube.com/watch?v=XuIYzZcvW7w&t=20s

Dr. Jut Wynne Colonizing the Caves of Mars https://www.youtube.com/watch?v=BgbNzqKYcnQ

Life Beyond Earth Dr. Linda Spilker https://www.youtube.com/watch?v=VUrVPjrdobw

Mike Dunn on Mars Lava Tube Habitat Simulations https://www.youtube.com/watch?v=CJI32sfCzuw

Caves of Mars https://www.youtube.com/watch?v=_30csyj44zc

ESA Cave Video http://www.esa.int/spaceinvideos/Videos/2013/05/All_about_CAVES

ESA CAVE Kit https://esamultimedia.esa.int/HSO/caves/CAVES-Information-Kit-2012.pdf

Ape To Man: Evolution Documentary History Channel https://www.youtube.com/watch?v=5sMqFivWTmk&index=3&list=PL4hRRF5TQrT9IN_C7q9EbDwQWu69ir0tE

Lunar Reconnaissance Orbiter Video https://svs.gsfc.nasa.gov/11155

The Moon to Debussey's Clair de Lune (using LRO images) https://www.youtube.com/watch?v=zNpsy6lBPBw

University of Hawaii, Geology and Geophysics Department. https://www.youtube.com/embed/TGzwFmmCeCY

The creation of lava tubes in Hawaii, go to: https://www.youtube.com/watch?v=vGZ5KNe94bI

Robots

Dr. "Red" Whittaker lecture on Extreme Robots https://www.youtube.com/watch?v=7X4-jozFVFo

NASA's Space Robotics Challenge https://www.youtube.com/watch?v=aTpDj5hDO6s

Robonaut https://www.youtube.com/watch?annotation_id = annotation_106268106
    5&feature = iv&src_vid = g3u48T4Vx7k&v = ePWjFlSdB4U
X-1 https://www.youtube.com/watch?v = ldl-V1n7Efw
TALISMAN https://www.youtube.com/watch?v = fLKqvAsHfss
Atlas https://www.bostondynamics.com/atlas
Cassie https://www.youtube.com/watch?v = Is4JZqhAy-M
Several types of robots https://www.youtube.com/watch?v = 8vIT2da6N_o
20 minute video of many of the above mentioned robots https://www.youtube.com/
    watch?v = kbaDdg4LA9k
Japanese Robots https://www.youtube.com/watch?v = r3GMGkFZFzI

Radiation

Dr. Sheila Thibeault on Radiation Shielding https://www.youtube.com/watch?v =
    ADA-FtQ_Vno
Dr. Catharine Fay on TED Talk https://www.youtube.com/watch?v = CoHSNiZ
    qwEY
Dr. Catharine Fay on Boron Nitride Nanotubes (BNNT) https://www.youtube.
    com/watch?v = r25RMceegKM
Marco Durante TED talk https://www.youtube.com/watch?v = q-9Avd__dQ4
Dr. Jingnan Guo TEDxKielUniversity https://www.youtube.com/watch?v = QBi8B
    TK71Mk

Inflatables/Expandables

Astronauts entering the ISS BEAM https://www.youtube.com/watch?v = 5kZZ
    dp727ek
Expanding BEAM operations https://www.youtube.com/watch?v = gARj5wmlFKg
ILC's landing bags for rovers on Mars https://www.youtube.com/watch?v = Kyktv
    C7w7Js

Power

Kilowatt https://www.youtube.com/watch?v = 5WEiMk8eeAs
KRUSTY https://www.youtube.com/watch?v = 6K8SEkr9I3o
The Problems of Power in Space https://www.youtube.com/watch?v = m2IiI4UVZ
    P8

Instruments

Mass Spectrometers https://www.youtube.com/watch?v = _L4U6ImYSj0
Dextre using the IRELL https://www.youtube.com/watch?v = nNcRDBK8zxY

Glossary

AAES	Aeroassist, Aerocapture, and Entry Systems
AC	Alternating Current
ACS	Attitude Control System (similar to/same as RCS)
AD	anno domini (after the year of Jesus' birth)
AD #2	Addendum #2 to the DRA 5
AES	Advanced Exploration Systems
AHP	Analytical Hierarchy Process
AI	Artificial Intelligence
ALARA	As Low as Reasonably Achievable
ALSEP	Apollo Lunar Science Experiments Package
AMADEE	Flagship research program of the ÖWF
AMO	Autonomous Mission Operations
Aouda.X	Austrian space suit simulator
AR	Atmospheric Revitalization
ARC	Ames Research Center
AR&D	Automated Rendezvous and Docking
ARM	Asteroid Redirect Mission
ASE	Airborne Support Equipment
ATV	Automated Transfer Vehicle
AU	Astronomical Unit
AUV	Autonomous Underwater Vehicle
BAA	Broad Agency Announcement
BAC	Broad Area Cooling
BCE	Before the Christian Era
BCM	Biological Countermeasures
BEAM	Bigelow Expandable Activity Module
BHP	Behavioral Health and Performance
BIM	Building Information Modeling
BNTEP	Bi-modal Nuclear Thermal Electric Propulsion
BPLF	Black Point Lava Flow
BP	Before Present

© Springer Nature Switzerland AG 2019
M. von Ehrenfried, *From Cave Man to Cave Martian*,
Springer Praxis Books, https://doi.org/10.1007/978-3-030-05408-3

BPP	Bubble Point Pressure
BRAILLE	Biologic and Resource Analog Investigations in Low Light Environments

CaSSIS	Colour and Stereo Surface Imaging System (ExoMars)
CATALYST	Cargo Transportation and Landing by Soft Touchdown program
CaveR	CaveR Cave Rover
C&DH	Command and Data Handling
C&T	Communications and Telemetry, Communications and Tracking
CAD/CAM	Computer Aided Design/Computer Aided Manufacturing
CAVES	Cooperative Adventure for Valuing and Exercising human behavior and performance Skills (ESA)
CBT	Computer Based Training
CCD	Charged Coupled Device
CDF	Capability Driven Framework
CDR	Critical Design Review
CES	Consumer Electronics Show
CESR	French Center for the Study of Radiation in Space
CFEET	Compact Fuel Element Environmental Test
CFM	Cryogenic Fluid Management
CFP	Conceptual Flight Profile
CG	Center of Gravity
CH4	Methane
CIAP	Centre International d'Art Pariétal/International Center of Cave Art
CIRAS	Commercial Infrastructure for Robotic Assembly and Services
CISAS	Center of Studies and Activities for Space - University of Padova
ClO4	perchlorate
CM	Crew Module
CME	Coronal Mass Ejections
CMU	Carnegie Mellon University
CNC	Computer Numerical Control
CNS	Central Nervous System
CNES/IRAP	Centre National d'Etudes Spatiales, Institut de Recherche en Astrophysique et Planetologie (Mars 2020)
ConOps	Concept of Operations
COPV	Composite Overwrapped Pressure Vessel
COSPAR	Committee on Space Research
COTS	Commercial Off-The-Shelf
CPS	Cryogenic Propulsion Stage
CRaTER	Cosmic Ray Telescope for the Effects of Radiation
CRISM	Compact Reconnaissance Imaging Spectrometer (MRO)
CS	Core Stage
CSA	Canadian Space Agency

CTB	Cargo Transfer Bag
CTV	Crew Transport Vehicle
CTX	The Context Camera Mars Climate Sounder (MRO)
CWR	Center for Water Research (UTSA)
CxP	Constellation Program
DAC	Design Analysis Cycle
DAF	Device Assembly Facility (NASA Langley)
DARPA	Defense Advanced Research Projects Agency
DAV	Descent/Ascent Vehicle
DC	Direct Current
DDT&E	Design, Development, Test and Evaluation
DDU	Direct Drive Unit
DHL	Dalsey, Hillblom and Lynn (founders of DHL Worldwide)
DM	Descent Module
DMC	Destination Mission Concept
DOE	Department of Energy
DOF	Degree of Freedom
DOT	Destination Operations Team/Department of Transportation
DSPSE	Deep Space Program Science Experiment (Clementine)
DRA	Design Reference Architecture - Mars Design Reference Architecture 5.0 – Addendum #2
D-RATS	Desert Research and Technology Studies
DRL	Dexterous Robotics Laboratory (NASA JSC)
DRM	Design Reference Mission
DSITMS	Direct Sampling Ion Trap Mass Spectrometry
DSH	Deep Space Habitat
DSN	Deep Space Network
DSOC	Deep Space Optical Communications
DSV	Deep Space Vehicle
DU	Depleted Uranium
ΔV	Delta Velocity (m/s or km/s)
ECLS	Environmental Control and Life Support
EDAC	Earth Data Analysis Center
EDL	Entry, Descent, and Landing
EDL-SA	Entry, Descent, and Landing- Systems Analysis
EDM	Entry, Descent and Landing Demonstrator Module (Schiaparelli)
EI	Entry Interface
ELI	Electrical Load Interface
ELV	Expendable Launch Vehicle
EMAT	Exploration Maintainability Analysis Tool
EMU	Extravehicular Mobility Unit
EMVE	Earth-Mars-Venus-Earth

EP	Electric Propulsion
EPSCoR	Established Program to Stimulate Competitive Research
EPO	Earth Parking Orbit/Education and Public Outreach
ERWG	Exploration Roadmap Working Group
ESA	European Space Agency
ESF	European Science Foundation
ETDP	Exploration Technology Development Program
ETO	Earth-To-Orbit
EVA	Extravehicular Activity
EZ	Exploration Zone
FAP	Flight Analogs Project
FARU	Flight Analogs Research Unit
FCR	Flight Control Room
FDL	Frontier Development Lab (at SETI)
FDIR	Fault Detection, Fault Isolation, and Recovery
FELDSPAR	Field Exploration and Life Detection Sampling for Planetary Analog Research
FLT	Flight Laser Transceiver
FM&T	Future Missions and Technology
FPGA	Field Programmable Gate Array
FPS	Fission Power Source
FRC	Field Robotics Center
FSP	Fission Surface Power
FSPS	Fission Surface Power System
FSPU	Fission Surface Power Unit
ft	foot or feet
FTIR	Fourier-transform infrared spectroscopy
GC	gas chromatography
GC/MS	gas chromatography-mass spectrometer
GCR	Galactic Cosmic Rays
GCR&A	Ground Rules, Constraints, and Assumptions
DLR	German Aerospace Center
GEO	Geosynchronous Orbit/Geospatial Environment Online
GER	Global Exploration Roadmap
GIFT	Goddard Instrument Field Team
GIS	Geographic Information Science
GLOW	Gross Liftoff Weight
GMT	Greenwich Mean Time
GN&C	Guidance, Navigation and Control
GOX	Gaseous Oxygen
GPR	Ground Penetrating Radar
GPS	Global Positioning System

GRAIL	Gravity Recovery and Interior Laboratory (Ebb & Flow)
GRC	Glenn Research Center
GRS	Gamma Ray Spectrometer (Mars Odyssey Orbiter)
GSC	Grab Sample Container
GSDO	Ground Systems Development and Operations
ha	Apoapsis or Apogee Altitude
hp	Periapsis or Perigee Altitude
HAT	Human Spaceflight Architecture Team
HEM-SAG	Human Exploration of Mars-Science Analysis Group
HEFT	Human Exploration Framework Team
HEND	High Energy Neutron Detector (Mars Odyssey Orbiter)
HERA	Human Exploration Research Analog
HEO	High-Earth Orbit
HEOMD	NASA's Human Exploration and Operations Mission Directorate
HEU	High Enriched Uranium
HIAD	Hypersonic Aerodynamic Inflatable Decelerator
HIGP	Hawaii Institute of Geophysics and Planetology
HIP	Hot Isostatic Press
HiRISE	High Resolution Imaging Science Experiment (MRO)
HI-SEAS	Hawaii Space Exploration Analog and Simulation
HMM	Human Mars Mission
HMO	High Mars Orbit
HMP	Haughton-Mars Project
HMPRS	HMP Research Station
HP3	The Heat Flow and Physical Properties Package (InSight Lander)
HRP	Human Research Program
HRSC	Hi Resolution Stereo Camera
HSI	Human Systems Integration
HSIR	Human Systems Integration Requirements
IBMP	Institute for Biomedical Problems
IAA	International Academy of Astronautics
IAD	Inflatable Aerodynamic Decelerator
IAWG	International Architecture Working Group
ICSU	International Council for Science
IDA	Instrument Deployment Arm (InSight)
IDC	Instrument Deployment Camera (InSight)
IGEO	Inst. of Geociencias - CSIC - Universidad Complutense de Madrid
IHMC	Florida Institute for Human and Machine Cognition
IHX	Intermediate Heat Exchanger
IMLEO	Initial Mass in Low Earth Orbit
ILH	Inflatable Lunar Habitat
ILMAH	Inflatable Lunar/Mars Analog Habitat
IMM	Integrated Medical Model

IMU	Inertial Measurement Unit
INL	Idaho National Laboratory
IoT	The Internet of Things
IP	International Partner
IRELL	ISS Robotic External Leak Locator
IRG	Intelligent Robotics Group (NASA Ames)
IRU	Inertial Reference Unit
ISECG	International Space Exploration Coordination Group
ISRO	Indian Space Research Organization
IMG	Cartography and Imaging Sciences Node
IMS	ion mobility spectrometer
IRVE-3	Inflatable Reentry Vehicle Experiment
ISRU	In-Situ Resource Utilization
Isp	Specific Impulse (seconds)
ISS	International Space Station
ISTAR	International Space Station Test-bed for Analog Research
UIS	International Union of Speleology
IUVS	Imaging Ultraviolet Spectrometer (MAVEN)
IVA	Intravehicular Activity
JAXA	Japan Aerospace Exploration Agency
JPL	Jet Propulsion Laboratory
JSA	Johnson Space Center
JSC	Johnson Space Center
KBO	Kuiper Belt Object
kg	kilogram(s)
KG	Knowledge Gaps
km	kilometer(s)
KRUSTY	Kilopower Reactor Using Stirling Technology
KSC	Kennedy Space Center
KSU	Kansas State University
kWe	kilowatt electric
L1-5	Libration Points 1-5
LABE	Lava Beds National Monument
LADEE	Lunar Atmosphere and Dust Environment Explorer
LAMP	Lyman-Alpha Mapping Project
LANL	Los Alamos National Laboratory
LaRC	Langley Research Center
LaRRI	Laser Retro Reflector (InSight)
LAT	Lunar Architecture Team
lbf	Pound-force
LCC	Launch Control Center

LCCR	Lunar Capability Concept Review
LCH4	Liquid Methane
LCG	Liquid Cooling Garment
LCROSS	Lunar Crater Observation and Sensing Satellite
L/D	Lift to Drag
LDAC	Lander Design Analysis Cycle
LEE	Latching End Effectors
LEM	Lunar Excursion Module
LEND	Lunar Exploration Neutron Detector
LEO	Low Earth Orbit
LH2	Liquid Hydrogen
LIBS	Laser-Induced Breakdown Spectrometer
LIDAR	Light Detection and Ranging
LIDS	Low Impact Docking System
LLO	Low Lunar Orbit
LMO	Low Mars Orbit
LO2	Liquid Oxygen
LOLA	Lunar Orbiter Laser Altimeter
LOR	Lunar Orbit Rendezvous
LOS	Line-Of-Sight
LOx	Liquid Oxygen
LPC	Local Power Controller
LPI	Lunar and Planetary Institute
LPW	Langmuir Probe and Waves (MAVEN)
LRE	Liquid Rocket Engines
LROC	Lunar Reconnaissance Orbiter Camera
LRS	Lunar Radar Sounder (SELENE)
LRV	Lunar Rover Vehicle
LSRG	Large-Scale Stirling Radioisotope Generator
LSS	Lunar Surface Systems
LPS	The Lunar and Planetary Science Conference
LV	Launch Vehicle(s)
m	meter
M3	Moon Mineralogy Mapper
MAG	Magnetometer (MAVEN)
MALTO	Mission Analysis Low Thrust Optimization
MARCI	Mars Color Imager (MRO)
MARIE	Mars Radiation Environment Experiment (Mars Odyssey Orbiter)
MARSH	Multi-mission Artificial Gravity Reusable Habitat
MAT	Mars Architecture Team
MatiSSE	The Maturation of Instruments for Solar System Exploration
MAV	Mars Ascent Vehicle
MAVEN	The Mars Atmosphere and Volatile EvolutioN
MBSE	Model-Based Systems Engineering

MBSU	Main Bus Switching Unit
MCC	Midcourse Correction/Mission Control Center
MCS	Mars Climate Sounder (MRO)
MDM	Mars Descent Module
MDPI	Molecular Diversity Preservation International
MEDA	Mars Environmental Dynamics Analyzer
MEIT	Multi-Element Integrated Testing
MEL	Master Equipment List
MEP	Mars Exploration Program
MEPAG	Mars Exploration Program Analysis Group
MER	Mass Estimating Relationship
MFPF	Mobile Fission Power Center
MHH	Marius Hills Hole
ML	Mobile Launcher
MLI	Multi-Layer Insulation
MLTPP	The Mars Lava Tube Pressurization Project
MMH	Monomethyl Hydrazine
MMOD	Micrometeoroid and Orbital Debris
MMRTG	Multi-Mission Radioisotope Thermoelectric Generator
MMSEV	Multi-Mission Space Exploration Vehicle
MOI	Mars Orbit Insertion
MOU	Memorandum of Understanding
MPCV	Multi-Purpose Crew Vehicle
MPD	Magnetoplasmadynamic
MPD	Mars-Phobos-Deimos
MOMA	Mars Organic Molecule Analyzer (ExoMars Rover)
MOXIE	Mars Oxygen ISRU Experiment (Mars 2020)
MPO	Mars Parking Orbit
MPPG	Mars Program Planning Group
MPS	Main Propulsion Subsystem
MRO	Mars Reconnaissance Orbiter
MS	Mass Spectrometer
MSA	Multi-Purpose Crew Vehicle Stage Adapter
MSL	Mars Science Laboratory (Curiosity)
MSFC	Marshall Space Flight Center
MSR	Mars Sample Return
mt	metric tons
MTH	Mars Transit Habitat
MTBF	Mean Time Between Failures
MTO	Mars-to-Orbit
MTV	Mars Transfer Vehicle
N	Newtons
NAC	Narrow-Angle Cameras
NaK	Sodium-Potassium

NAS	National Academy of Sciences
NASA	National Aeronautics and Space Administration
NAU	Northern Arizona University
NCPS	Nuclear Cryogenic Propulsion Stage
NCRP	National Council on Radiation Protection and Measurements
NDX-1, 2	North Dakota Experimental-1 and -2 (space suit)
NDS	NASA Docking System
NE	Nuclear Electric
NEA	Near-Earth Asteroid
NEEMO	NASA Extreme Environment Missions Operations
NEO	Near-Earth Object
NEP	Nuclear Electric Propulsion
NextSTEP	Next Space Technologies for Exploration Partnerships
NGIMS	Neutral Gas and Ion Mass Spectrometer (MAVEN)
NIA	National Institute of Aerospace
NIAC	NASA Innovative Advance Concepts program
NINJAR	NASA Intelligent Jigging and Assembly Robot
nmi	nautical miles
NPD	NASA Policy Directive
NPI	NASA Procedural Instruction
NPRs	NASA Procedural Requirements
NRC	National Research Council
NREC	National Robotics Engineering Center
NSF	National Science Foundation
NSRL	NASA Space Radiation Laboratory
NSS	National Speleological Society
NTE	Nuclear Thermal Engine
NTEC	NASA's Technology Executive Council
NTO	Nitrogen Tetroxide (N2O4)
NTP	Nuclear Thermal Propulsion
NTR	Nuclear Thermal Rocket
OCT	NASA's Office of the Chief Technologist
ODMSP	Orbital Debris Mitigation Standard Practices
OF	Oxidizer to Fuel Ratio
OMS	Orbital Maneuvering System
OpNav	Optical Navigation
OPP	Office of Planetary Protection
ORD	Operations Requirements Document
ORNL	Oak Ridge National Laboratory
ORSC	Oxygen-Rich Staged Combustion
ORU	Orbital Replacement Unit
OSIRIS-REx	Origins-Spectral Interpretation-Resource Identification-Security
OSMA	Office of Safety and Mission Assurance (NASA Headquarters)
OST	Outer Space Treaty (United Nations)

OTES	Thermal Emission Spectrometer (OSIRIS-Rex)
OTF	Operations Technology Facility
OTS	Orbital Transfer Stage
ÖWF	The Austrian Space Forum
PANGAEA	Planetary ANalogue Geological and Astrobiological Exercise for Astronauts (ESA)
P&O	Production and Operations
PAD	Physical Architecture Diagram
PDR	Preliminary Design Review
PDS	Planetary Data System (USGS)
PDU	Power Distribution Unit
PEC	Pulsed Electric Current
PEL	Power Equipment List
PEM	Proton Exchange Membrane
PEL	Permissible Exposure Limits
Perilune	The lowest altitude in an orbit around the Moon
PERISCOPE	PERIapsis Subsurface Cave OPtical Explorer
PICASSO	The Planetary Instrument Concepts for the Advancement of Solar System Observations
PISCES	Pacific International Space Center for Exploration System
PIT	Pulsed Inductive Thruster
PIXL	Planetary Instrument for X-ray Lithochemistry (Mars 2020)
PLA	Payload Launch Adapter
pLOC	Probability of Loss of Crew
PLR	Parasitic Load Radiator
PLRP	Pavilion Lake Research Project
PLSS	Portable Life Support System
PM	Powder Metallurgy
PMAD	Power Management and Distribution
PP	Planetary Protection
PRD	Program Requirements Document
POL	Permissible Outcome Limits
Pol Ares	An early ÖWF program
Portrillo	A NASA analog site in the Rio Grande Rift
POST	Program to Optimize Simulated Trajectories
PPU	Power Processing Unit
PRA	Probabilistic Risk Assessment
PROM	Programmable Read-Only Memory
P-SAG	Precursor Science Analysis Group
PSR	Perennially Shadowed Regions
PSTAR	Planetary Science and Technology Through Analog Research
PTC	Parametric Technology Corporation
PV	Photovoltaic
PVA	Photovoltaic Array

RA	Robotics Automation
RAAN	Right Ascension of the Ascending Node
RAAI	Robotics Automation Artificial Intelligence
RAC	Requirements Analysis Cycle
RAD	Radical Assessment Detector
RATS	Research And Technology Studies
RBO	Reduced Boil Off
RCS	Reaction Control System
RF	Radio Frequency
RFC	Regenerative Fuel Cell
RGIS	Resource Geographic Information System
RI	The Robotics Institute
RIMFAX	Radar Imager for Mars' Subsurface Exploration (Mars 2020)
RISE	Rotation and Interior Structure Experiment (InSight Lander)
RLS	Raman Laser Spectrometer (InSight Lander)
RMC	Robotic Mining Competition
RMI	Remote Micro-Imager
ROV	Remotely Operated Vehicle
RP	Rocket Propellant
RRT	Rapid Response Tool
RSA	Russian Space Agency
Rx	Reactor
SA	Spacecraft Adapter
SAFE	Subsurface Active Filtering of Exhaust
SAIC	Science Applications International Corporation
SAM	Sample Analysis at Mars (Curiosity)
SAMURAI	Strut Assembly, Manufacturing, Utility & Robotic Aid
SARJ	Solar Alpha Rotary Joint
SBAG	Small Bodies Assessment Group
SBIR	Small Business Innovation Research
SDR	System Definition Review
SEIS	Seismic Experiment for Interior Structure (InSight Lander)
SELENE	The SELenological and ENgineering Explorer Spacecraft
Serenity	Next generation ÖWF space suit
SESE	School of Earth and Space Exploration (ASU)
SETI	Search for Extraterrestrial Intelligence
SEP	Solar Electric Propulsion
SEP	Solar Energetic Particle
SEV	Space Exploration Vehicle
SFHSS	Space Flight Human Systems Standards
SHAB	Surface Habitat
SHARAD	Shallow Subsurface Radar (MRO)
SHERLOC	Scanning Habitable Environments with Raman & Luminescence for Organics and Chemicals (Mars 2020)

SIAD	Supersonic Inflatable Aerodynamic Decelerator
SKG	Strategic Knowledge Gap
SKUs	Stock Keeping Unit
SLAM	Simultaneous Localization and Mapping
SLIME	Subsurface Life In Mineral Environments
SLOC	Source Lines of Code
SLS	Space Launch System
SM	Service Module
SMD	Science Mission Directorate
SMEs	Subject Matter Experts
SNAP-10A	Systems for Nuclear Auxiliary Power (A 1965 spacecraft)
SNC	Sierra Nevada Corporation
SNL	Sandia National Laboratory
SOA	State of the Art
SOEST	School of Ocean and Earth Science and Technology (U of Hawaii)
SOI	Sphere of Influence
SORT	Simulation to Optimize Rocket Trajectories
SPE	Solar Particle Events
SPEL	Space Permissible Exposure Limits
SPR	Small Pressurized Rover
SPTO	Single Phase to Orbit
SRB	Solid Rocket Boosters
SRC	Short-Radius Centrifuge
SRM	Solid Rocket Motor
SRP	Supersonic Retro Propulsion
SRR	Strategic Readiness Review/Systems Requirements Review
SSA	Space Situational Awareness
SSAS	Solid Sorbent Air Sampler
S&T	Science and Technology
STATIC	SupraThermal And Thermal Ion Composition (MAVEN)
STEM	Science, Technology, Engineering and Mathematics
STM	Space Traffic Management
STMD	Space Technology Mission Directorate
STS	Space Transportation System
STIP	Strategic Technology Investment Plan
STTR	Small Business Technology Transfer
SuperCam	Mars 2020 upgrade to the Curiosity ChemCam
SWEA	Solar Wind Electron Analyzer (MAVEN)
SWIA	Solar Wind Ion Analyzer (MAVEN)
TA	Technology Area
TABS	Technology Area Breakdown Structure
TALISMAN	The Tension Actuated In Space MANipulator
TBD	To be Determined
TBR	To be Reviewed

TCA	Thrust Chamber Assembly
TCM	Trajectory Correction Maneuver
TDU	Technology Demonstration Unit
TEI	Trans-Earth Injection
TES	Thermal Emission System (Mars Global Surveyor)
TGO	Trace Gas Orbiter (ExoMars)
THEMIS	Thermal Emission Imaging System (Mars Odyssey Orbiter)
TIM	Technical Interchange Meeting
TMI	Trans-Mars Injection
TNSC	Tanegashima Space Center
TPS	Thermal Protection System
TPTO	Two Phase to Orbit
TRL	Technology Readiness Level
T/W	Thrust-to-Weight
TWINS	Temperature and Winds for InSight
UA	University of Arizona
UAV	Unmanned Aerial Vehicle (AKA drone)
UML	Unified Modeling Language
UNESCO	United Nations Educational, Scientific and Cultural Organization
UO2	Uranium Oxide
UND	University of North Dakota
UPR	Unpressurized Rover
USDA	United States Department of Agriculture
USGS	United States Geological Survey
USRA	Universities Space Research Association
UTSA	University of Texas at San Antonio
Va	Velocity at Apoapsis
Vp	Velocity at Periapsis
VHP	Hyperbolic Excess Velocity
VAB	Vehicle Assembly Building
VAB HB	Vehicle Assembly Building High Bay
VASIMR	VAriable Specific Impulse Magnetoplasma Rocket
VCS	Vapor Cooled-Shield System
Vdc	Direct Current Voltage
VIS	Visible imager subsystem of THEMIS
VOA	Volatile Organic Analysis
VOCs	Volatile Organic Compounds
VRAD	Very Long Baseline Interferometer
VSAPR	Vienna Statement for Planetary Analog Research
WAC	wide-angle camera
WBS	Work Breakdown Structure
WVRTC	West Virginia Robotic Technology Center

WR   Water Recovery

XBASE   Expandable Bigelow Advanced Station Enhancement
X-Hab   eXploration Systems and Habitation

Z   Atomic number, e.g. Hydrogen = Z-1
ZBO   Zero Boil Off

Image Links

(where appropriate)

Frontispiece   https://spaceart.photoshelter.com/image/I0000aaDVK_b0z6U

Chapter 2

Fig. 2.1   http://cdn.sci-news.com/images/enlarge3/image_4452e-Sima-del-Elefante-Hominins.jpg

Fig. 2.2   https://sipofstarrshine.files.wordpress.com/2017/02/dsc02605.jpg?w = 1024

Fig. 2.3   http://humanorigins.si.edu/sites/default/files/styles/full_width/public/images/square/neanderthalensis_JG_Recon_Head_CC_3qtr_lt_sq.jpg?itok = Kmky0dPB

Fig. 2.4   https://imgur.com/gallery/ohfJQjd

Fig. 2.5   https://4sport.ua/_upl/2/1490/0001foto-1_c.jpg

Fig. 2.6   https://upload.wikimedia.org/wikipedia/commons/a/a7/Mammoth_Cave_Historic_Entrance.jpg

Fig. 2.7   https://upload.wikimedia.org/wikipedia/commons/8/8e/Mammoth_Cave_travertine_formation.jpg

Fig. 2.8   http://speleologija.eu/velebita/foto/2004-Velebita-DivkaGromovnica-Foto_D_Baksic.JPG

Fig. 2.9   https://1.bp.blogspot.com/-KP73f42T_ZY/U1Tv-sl4GdI/AAAAAAAAePI/d0bw4fWLl5w/s1600/10.jpg

Fig. 2.10   https://www.nps.gov/npgallery/GetAsset/FCFF7F5D-155D-451F-678B0E5CBAD91FBB/proxy/hires

Fig. 2.11   https://www.urban75.net/forums/proxy.php?image = http%3A%2F%2Fupload.wikimedia.org%2Fwikipedia%2Fcommons%2F0%2F02%2FLechuguilla_Cave_Pearlsian_Gulf.jpg&hash = 36156b515fbcf089b86d7cecc2f0311c

Fig. 2.12   https://cdn.onlyinyourstate.com/wp-content/uploads/2018/06/TearsoftheTurtleCave-700x933.jpg

Fig. 2.13   https://cdn.onlyinyourstate.com/wp-content/uploads/2018/06/Tears2-700x467.jpg

Fig. 2.14   https://www.zenja.si/wp-content/uploads/2017/12/jamski-svet.jpg

Fig. 2.15   https://www.zenja.si/wp-content/uploads/2017/12/jamski-svet.jpg

© Springer Nature Switzerland AG 2019
M. von Ehrenfried, *From Cave Man to Cave Martian*,
Springer Praxis Books, https://doi.org/10.1007/978-3-030-05408-3

Fig. 2.16	https://upload.wikimedia.org/wikipedia/commons/f/fa/Son_Doong_Cave_DB_%283%29.jpg
Fig. 2.17	https://i.pinimg.com/originals/f4/96/68/f49668d403e1408a91173873c67af494.jpg
Fig. 2.18	https://i.pinimg.com/originals/1c/13/b7/1c13b770c120cdf8e0c69d9859e03e93.jpg
Fig. 2.19	http://www.leisureopportunities.co.uk/images/HIGH15864_109470.jpg
Fig. 2.20	https://images.newscientist.com/wp-content/uploads/2017/06/07170031/figure-3.jpg
Fig. 2.21	https://upload.wikimedia.org/wikipedia/commons/5/5f/Zhoukoudian_Peking_Man_Site.jpg
Fig. 2.22	https://upload.wikimedia.org/wikipedia/commons/thumb/f/ff/Zhoukoudian_Entrance.JPG/800px-Zhoukoudian_Entrance.JPG
Fig. 2.23	https://upload.wikimedia.org/wikipedia/commons/thumb/2/21/Cave_houses_shanxi_1.jpg/1200px-Cave_houses_shanxi_1.jpg
Fig. 2.24	https://www.familiscope.fr/assets/fiches/38000/38055-rochemenier-village-troglodytique.jpg
Fig. 2.25	https://www.hotelamericagranada.com/wp-content/uploads/2018/06/40b7db65-703e-465f-a490-c4d990cbfc89-sacromonte-1024x726.jpg
Fig. 2.26	https://www.irantravelingcenter.com/wp-content/uploads/2018/01/kandovan_village.iran_.jpg
Fig. 2.27	http://1.bp.blogspot.com/-lAGquRNmPhQ/US2RmDa9u3I/AAAAAAAASA/rMR1XgP0oFE/s1600/094026_bawah5.jpg
	https://imgcy.trivago.com/c_limit,d_dummy.jpeg,f_auto,h_650,q_auto,w_1000/itemimages/44/99/4499598_v2.jpeg
	http://www.venushill.com.au/images/bedbreafast_page/bed_and_breafast_coober_pedy2.jpg
	https://images.helionews.ru/1483232760000/1485972311000/kuber-pedi-gorodok-pod-zemlei-cogqnu.jpg
Fig. 2.28	https://www.smh.com.au/content/dam/images/g/p/o/d/l/3/image.imgtype.articleLeadwide.620x349.png/1466487642850.jpg
Fig. 2.29	https://lh3.googleusercontent.com/3tQMXmy0sj1CVZfOtxyJOeAjxMSpqydy9sRVL_kgy1ya7OlgZDlorlPpoO_3vcg5mOJy=s170

Chapter 3

Fig. 3.1	http://upload.wikimedia.org/wikipedia/commons/thumb/3/36/Moon_names.jpg/600px-Moon_names.jpg
Fig. 3.2	https://3c1703fe8d.site.internapcdn.net/newman/gfx/news/hires/1-downthelunar.jpg
Fig. 3.3	https://assets.atlasobscura.com/article_images/26305/image.jpg
Fig. 3.4	http://1.bp.blogspot.com/-3daBSfNMd2Y/UDIVzPKdWiI/AAAAAAAAFSU/5YmoqqwFlI8/s1600/Post+-+August+2012+(6C)+Marius+Hills+Pit.jpg
Fig. 3.5	http://lroc.sese.asu.edu/ckeditor_assets/pictures/112/content_fresh_but_not_clean_context.png
Fig. 3.6	https://clipground.com/images/mare-tranquillitatis-clipart-6.jpg
Fig. 3.7	https://www.nasa.gov/images/content/463367main_thomsonM.jpg

Fig. 3.8   https://www.btlv.fr/wp-content/uploads/2017/10/grotte-sur-la-lune-btlv.fr_-1.
              jpg
Fig. 3.9   https://www.treking.cz/astronomie/ascraeus-mons.jpg
Fig. 3.10   https://upload.wikimedia.org/wikipedia/commons/thumb/b/b7/Tharsis_
              Montes_MOLA_zoom_64.jpg/390px-Tharsis_Montes_MOLA_zoom_64.jpg
Fig. 3.11   https://www.jpl.nasa.gov/spaceimages/images/wallpaper/PIA09929-
              1920x1200.jpg
Fig. 3.12   https://upload.wikimedia.org/wikipedia/commons/e/ec/Mars%3B_Arsia_
              Mons_cave_entrance_-MRO.jpg
Fig. 3.13   https://mars.jpl.nasa.gov/odyssey/gallery/martianterrain/images/20070921_
              Cave_04.jpg
Fig. 3.14   https://upload.wikimedia.org/wikipedia/commons/8/84/Hadriaca_Patera_
              based_on_THEMIS-DAY-IR.png
Fig. 3.15   https://upload.wikimedia.org/wikipedia/commons/thumb/6/63/Hadriacus_
              Mons_THEMIS_day_IR_100m_v11.5_0.5.jpg/600px-Hadriacus_Mons_
              THEMIS_day_IR_100m_v11.5_0.5.jpg

Chapter 4
Fig. 4.1   https://si.wsj.net/public/resources/images/RV-AD414_CREATI_G_
              20110628233754.jpg
Fig. 4.2   https://www.astrobio.net/wp-content/uploads/2016/03/naica_cave_gypsum_
              crystals_image_courtesy_of_dr._tom_kieft_nmt-519x600.jpg
Fig. 4.3   https://asu.pure.elsevier.com/files-asset/82285562/pchrist
Fig. 4.4   https://i.ytimg.com/vi/ZbWh75XJXsk/hqdefault.jpg
Fig. 4.5   https://cpb-us-e1.wpmucdn.com/sites.northwestern.edu/dist/b/306/files/2014/
              10/lavatuberover-297fwf6.png
Fig. 4.6   https://4.bp.blogspot.com/-FAhWN7lVJms/WfdGLWbwfcI/AAAAAAAAJ_
              8/u9BOtrY1KkQR226qq4-FUVYeEDQhB8mZQCEwYBhgL/s1600/lb13.png
Fig. 4.7   http://4.bp.blogspot.com/-k_rqRhrb-F0/UmyiXA14otI/AAAAAAAAAcg/
              XybvZP9kmrg/s1600/100_8005.JPG
Fig. 4.8   http://www.hawaiicaves.org/2008/img6diana.jpg
Fig. 4.12   https://4thplanetlogistics.files.wordpress.com/2016/12/chalmer-dunn_2016.jpg

Chapter 5
Fig. 5.1   https://steemitimages.com/0x0/https://www.nasa.gov/sites/default/files/
              thumbnails/image/niac_2011_thibeault.jpg
Fig. 5.2   https://ir0.mobify.com/980/https://www.bnnt.com/images/bnnt/gallery/
              BNNT-closeup980.jpg
Fig. 5.3   https://forum.nasaspaceflight.com/index.php?action = dlattach;topic = 34018.
              0;attach = 571810;image
Fig. 5.4   https://upload.wikimedia.org/wikipedia/commons/0/09/BEAM_mockup.jpg
Fig. 5.5   https://img.purch.com/h/1400/aHR0cDovL3d3dy5zcGFjZS5jb20vaW1hZ2
              VzL2kvMDAwLzAwMi8zNDEEvb3JpZ2luYWwv
              MDcwMzI4X2luZmxhdGGVfaGFiXzAyLmpwZ w= =
Fig. 5.6   https://upload.wikimedia.org/wikipedia/commons/9/9f/Senator_John_Heinz_
              History_Center_-_IMG_7802.JPG
Fig. 5.7   https://i.pinimg.com/236x/00/d9/58/00d95817587365aa28e98e6eca4a1375–
              alternate-history-classic-movies.jpg

Fig. 5.8	https://img.purch.com/h/1400/aHR0cDovL3d3dy5zcGFjZS5jb20vaW1 hZ2V zL2kvMDAwLzA2Ni8yNzcvb3JpZ2luYW wvbmFzYS1tYXJzLTIwMjAtcm 92ZXItY29uY2 VwdC1hcnQuanBnPzE0OTU2MzIyMDY =
Fig. 5.9	https://mars.nasa.gov/system/resources/detail_files/21365_mars_2020_cameras_labeled_web.jpg
Fig. 5.10	https://www.nasa.gov/sites/default/files/thumbnails/image/ksc-20170605-ph_kls01_0135a.jpg
Fig. 5.11	https://upload.wikimedia.org/wikipedia/commons/f/fd/Robonaut_2_and_Centaur_2.jpg
Fig. 5.12	https://www.nasa.gov/sites/default/files/thumbnails/image/valkyrie-robot-3.jpg
Fig. 5.13	https://www.sciencemag.org/sites/default/files/styles/article_main_image_-_1280w__no_aspect_/public/msl-rover-wheel-damage-pia21486-full-copy_16x9.jpg?itok = ev7KsFfr
Fig. 5.14	https://readtiger.com/img/wkp/en/LATMOS_LHASTEC_-_005.jpg
Fig. 5.15	https://i2.wp.com/www.yourlourdes.com/wp-content/uploads/2015/11/DaVinci-SI-Lourdes.jpg?ssl = 1
Fig. 5.16	https://upload.wikimedia.org/wikipedia/commons/2/23/Fueling_of_the_MSL_MMRTG_001.jpg
Fig. 5.17	https://upload.wikimedia.org/wikipedia/commons/8/88/MMRTG_after_fit_check_with_Curiosity_from_angular_above.jpg
Fig. 5.18	https://upload.wikimedia.org/wikipedia/commons/thumb/6/6a/Kilopower_experiment.jpg/1200px-Kilopower_experiment.jpg
Fig. 5.19	https://www.nasa.gov/sites/default/files/styles/full_width_feature/public/thumbnails/image/101_0202-dip.jpg
Fig. 5.20	https://www.nasa.gov/sites/default/files/styles/ubernode_alt_horiz/public/thumbnails/image/stmd_kilopower.jpg
Fig. 5.21	https://www.jpl.nasa.gov/spaceimages/images/largesize/PIA22562_hires.jpg
Fig. 5.22	http://earth-chronicles.com/wp-content/uploads/2018/08/pic_e0e6853a271bae34f2fe8d39e3153acc.jpg
Fig. 5.23	https://forum.nasaspaceflight.com/index.php?action = dlattach;topic = 41688.0;attach = 1393656;image

Chapter 6

Fig. 6.1	http://www.patrawlings.com/detail.cfm?id = 1416
Fig. 6.2	https://spaceflight.nasa.gov/gallery/images/mars/lunaractivities/lores/s88_33646.jpg
Fig. 6.3	http://www.patrawlings.com/detail.cfm?id = 1355
Fig. 6.4	http://www.patrawlings.com/detail.cfm?id = 1415
Fig. 6.5	https://www.ilcdover.com/wp-content/uploads/2018/03/top-pic2.jpg
Fig. 6.6	http://patrawlings.com/detail.cfm?id = 1071
Fig. 6.7	https://upload.wikimedia.org/wikipedia/commons/9/9d/The_image_shows_the_distribution_of_surface_ice_at_the_Moon%27s_south_pole_%28left%29_and_north_pole_%28right%29.webp
Fig. 6.8	https://img.purch.com/h/1400/aHR0cDovL3d3dy5zcGFjZS5jb20vaW1h Z2VzL2kvMDAwLzA3My8yOTYvb3JpZ2luYW wvbWFycy1pY2Utc2hlZXQtbXJvLmpwZw = =
Fig. 6.9	https://www.jpl.nasa.gov/missions/web/mars_sample_return.jpg

Fig. A.2.5	https://mir-cdn.behance.net/v1/rendition/project_modules/1400/4926e070724867.5bacc6d7c1f0d.jpg
Fig. A.2.6	https://lh3.googleusercontent.com/f946lvNW_xB3Hq3AhsD6P2Tp9jCoRcsxYy65w21J_VXu6caSffPZcOwhKkUvYkk39ZkjKA = s128
Fig. A.3.1	https://www.nasa.gov/sites/default/files/thumbnails/image/penny_boston_cover_photo_0.jpg
Fig. A.3.2	https://mars.nasa.gov/images/mep/PhilipChristensen.jpg
Fig. A.3.3	http://www.digitalspace.com/presentations/leag-ssr-2005/moon-suit/DSC09680.JPG
Fig. A.3.4	https://lh3.googleusercontent.com/BEdnWxmIR73iT_vPbGE6qNH4QhOuj3kxdP8GuMe0OEpauIoRcxvNA2OGYO3Bw7yVe2_8xQ = s85
Fig. A.3.5	http://www.k-state.edu/geology/images/Saugata2.jpg
Fig. A.3.6	https://www.aate.org/DeLeonPablo.jpg
Fig. A.3.7	https://www.soest.hawaii.edu/soest_web/images/Ruapehu/ClimbingRuapehu.jpg
Fig. A.3.8	https://upload.wikimedia.org/wikipedia/commons/thumb/f/f2/Ronald_Greeley.tif/lossy-page1-1920px-Ronald_Greeley.tif.jpg
Fig. A.3.9	https://prd-wret.s3-us-west-2.amazonaws.com/assets/palladium/production/s3fs-public/styles/full_width/public/thumbnails/image/db0be74dbeb07edd6dab6c21e1aeee5c_Laz_Kestay_Aug_2017.jpg?itok = DKnrnRS_
Fig. A.3.10	https://s18874.pcdn.co/wp-content/uploads/2016/06/Evelynn-Mitchell-2017.jpg
Fig. A.3.12	https://lh3.googleusercontent.com/IWmGZHEDZmUj11CORpX0odrooYoTq5lqE4nFuAt9VdLlOvpdXORXja7pWujgvfZ8a9O-Og = s151
Fig. A.3.13	https://firstyear.unm.edu/faculty/faculty-images/northup.jpg
Fig. A.3.14	https://www-robotics.jpl.nasa.gov/images/people-1046.jpg
Fig. A.3.15	https://www.ri.cmu.edu/drives-red-whittaker/
Fig. A.3.16	https://lh3.googleusercontent.com/G5tNnu–pKEASpgGcnpiJxMt5TFhQTEdKCFcNRyO9UxOoI7BdnNvHUVKA_a6-ZD3O974 = s85
Fig. A.3.17	https://media.licdn.com/dms/image/C5603AQFCJdZPmvE_MA/profile-displayphoto-shrink_200_200/0?e = 1552521600&v = beta&t = txalRdD_Tp55zeLKeocQCZXKFYz6WVdnFZUXxJVC8so

About the Author

Manfred "Dutch" von Ehrenfried had the very good fortune to have interviewed with the NASA Space Task Group the day before Alan Shepard was launched on MR-3 in 1961. At the time, he had very little knowledge of Project Mercury and thought that since his degree was in physics he would be working in that area. As fate would have it, he was assigned to the Flight Control Operations Section. His supervisor and mentor there was the redoubtable Eugene F. Kranz.

Most of Dutch's work for Project Mercury was in the areas of mission rules, countdowns, operational procedures, and coordination with the remote tracking station flight controllers. During his first six months, he was in training to be a flight controller and spent MA-4 and MA-5 at the Goddard Space Flight Center learning communications between the Mercury Control Center and the Manned Space Flight Network. His first mission as a flight controller was in the Mercury Control Center for John Glenn's flight on MA-6, learning the Procedures flight control position under Kranz. He then supported the remaining orbital flights of Carpenter, Schirra, and Cooper from his console in the Mercury Control Center.

After the transfer of the Space Task Group from Langley to Houston, Dutch supported Gemini missions 1–3 as a Procedures Officer and was then Assistant Flight Director for missions 4–7 (which included the first EVA by Ed White on GT-4 and the first rendezvous in space on GT-6/7).

In 1966, Dutch became a Guidance Officer for Apollo 1, which was unable to fly owing to a launch pad accident in which the crew died. After the stand-down, he was the Mission Staff Engineer for Apollo 7 and backup for Apollo 8. During that same period, he was an Apollo Pressure Suit Test Subject. This afforded him the opportunity to test pressure suits in the vacuum chamber to altitudes of over 400,000 ft, including one test with Neil Armstrong's suit. He also experienced nine g's in the centrifuge and flew in the zero-g aircraft. Later, he was also fitted for his own Apollo A7-LB Skylab suit. These activities led to him joining the Earth Resources Aircraft Program, becoming the first sensor equipment operator and Mission Manager for the high altitude RB-57F. In his role as the sensor operator, he worked with scientists to operationally achieve their objectives. He wore a full-pressure suit for these flights because they generally flew at altitudes in the range 65,000–67,000 ft (one was to 70,000 ft).

© Springer Nature Switzerland AG 2019
M. von Ehrenfried, *From Cave Man to Cave Martian*,
Springer Praxis Books, https://doi.org/10.1007/978-3-030-05408-3

The author in late 1961 as a young STG flight controller. Center left: At the console to the left of Gene Kranz and George Low. Center right: Testing Neil Armstrong's suit to an equivalent altitude of 400,000 ft in the vacuum chamber at the Manned Spacecraft Center. Bottom: The author wearing the A/P22S-6 full-pressure suit required for the RB-57F. All photographs courtesy of NASA.

The author in the Inner Space Cave in Georgetown, Texas, gaining inspiration for this book. Photograph courtesy of Ken Kamka.

During 1970 and 1971, Dutch was the Chief of the Science Requirements and Operations Branch at NASA-JSC. This Branch was responsible for the definition, coordination, and documentation of science experiments assigned to Apollo and Skylab missions. Its competence included the Apollo Lunar Science Experiment Packages (ALSEP) left on the Moon and the experiments in lunar and Earth orbit. The ALSEP packages included seismic sensors, magnetometers, spectrometers, ion detectors, heat flow sensors, charged particle and cosmic ray sensors, gravity sensors, and more. The lunar orbit experiments included the Scientific Instrument Module (SIM) Bay cameras and sensors, and the Particle and Fields subsatellites that were released prior to the Apollo spacecraft departing lunar orbit. The work also defined the procedures for deploying packages and conducting experiments on the Moon and in lunar orbit. Later, Dutch spent one year with a contractor at NASA-GSFC on the Earth Resources Technology Satellite (ERTS) later known as Landsat 1.

Dutch also worked in the nuclear industry for seven years and wrote *Nuclear Terrorism—A Primer*. As a contractor, he also worked with the original NASA Headquarters Space Station Task Force for ten years. He has written several books about his experiences (see www.dutch-von-ehrenfried.com) including several for Springer-Praxis. For the past 20 years, he has worked in the finance and insurance fields.

Index

© Springer Nature Switzerland AG 2019
M. von Ehrenfried, *From Cave Man to Cave Martian*,
Springer Praxis Books, https://doi.org/10.1007/978-3-030-05408-3